Introduction to Plant-Based Latex

Introduction to Plant-Based Latex

Phytochemistry, Ethnopharmacology, Biological and Pharmacological Activities

edited by

Laleh Khodaie | Ajay Sharma

Saurabh Bhatia | Ahmed Al-Harrasi

JENNY STANFORD
PUBLISHING

Published by

Jenny Stanford Publishing Pte. Ltd.
101 Thomson Road
#06-01, United Square
Singapore 307591

Email: editorial@jennystanford.com
Web: www.jennystanford.com

British Library Cataloguing-in-Publication Data
A catalogue record for this book is available from the British Library.

Introduction to Plant-Based Latex: Phytochemistry, Ethnopharmacology, Biological, and Pharmacological Activities

ISBN 978-981-5129-12-0 (Hardcover)
ISBN 978-1-003-57071-4 (eBook)
DOI 10.1201/9781003570714

Contents

Acknowledgments

We would like to dedicate this book to all students, researchers, academicians, and scholars who have sincerely contributed to the area of natural products, depicted in the book, at national as well as international levels. We would also like to acknowledge the University of Nizwa, Sultanate of Oman, for extending its support for accomplishing this book project successfully. We are grateful to the Natural and Medical Sciences Research Center (NMSRC) housed at the University of Nizwa for providing central resources of advanced analytical instruments for our research work and for promoting interdisciplinary research studies. Moreover, we would like to thank the School of Health Sciences, University of Petroleum and Energy Studies, Dehradun, Uttarakhand, India, for offering their support in the completion of the book project.

Natural and Medical Sciences Research Center, University of Nizwa

Last but not least, we express our sincere gratitude to the whole team of Jenny Stanford Publishing for their active cooperation and support.

Preface

Exudates as a diverse group of phytochemicals are known to possess diverse pharmacological effects. In this regard, natural latex, a sticky emulsion that oozes from damaged tissues, has captured the attention of researchers, as it has physiological and pharmacological effects and also alleviates various diseases in the folk and herbal medicine. These effects are attributed to the complex and highly concentrated mixture of latex containing some phytoconstituents, namely, proteins, enzymes, terpenoids, cardenolides, alkaloids, and phenolics.

Owing to the growing demand for and surging medicinal use of herbals, research and credible information in the field of secondary metabolites, in particular, latex, are vital. Significantly, the literature regarding herbal latex is scattered and unorganized. Thus, this book attempts to fill the gap by means of authentic material.

This textbook meets the requirements of scholars, researchers, instructors, and students interested in herbal latex. It contains a total of six chapters that cover all aspects of plant latex, including occurrence, habitat, distribution, and ethnopharmacology of lactiferous plants. Also, it elaborates the production and function of latex in latex-bearing plants. Furthermore, it covers phytochemicals and enzymes of latexes, their structures, and their quantitative and qualitative analyses. This book also concentrates on the processing of latexes, for instance, collection, and various drying approaches as well as their effects on physical and chemical properties. Furthermore, storage and preservation conditions of latex and relevant tests to qualify its durability are discussed.

The book also presents a thorough investigation on the pharmacological activities of latex by means of in vitro and in vivo assessments. In vitro assessments evaluate wound healing, anti-HIV, estrogenic and anti-estrogenic, anti-leishmanial, antibacterial, antifungal, and antiviral activities. In vivo assessments analyze anti-diarrhea, gastro-protective and anti-gastric ulcer, sedative, anticonvulsant, and anti-diabetic effects.

Since latex and its isolated compounds, with remarkable biological and pharmacological properties, have the potential to be good candidates for formulating and manufacturing new medicines, this book has devoted a chapter to drug discovery, commercial applications, regulations, safety and toxicity of latexes, and the relevant guidelines.

It is hoped that the book will provide readers with exciting new knowledge about natural latex and familiarize them with the recent developments. It will serve as a useful reference that provides substantial background information.

Chapter 1

Occurrence, Habitat, Distribution, and Traditional Uses of Laticiferous Plants

Laleh Khodaie,[a,b] Saurabh Bhatia,[c,d] Ahmed Al-Harrasi,[c] and Ajay Sharma[e]

[a]*Department of Traditional Pharmacy,*
Faculty of Traditional Medicine, Tabriz University of Medical Sciences, Tabriz, Iran
[b]*Department of Pharmacognosy, Faculty of Pharmacy,*
Tabriz University of Medical Sciences, Tabriz, Iran
[c]*Natural & Medical Sciences Research Center,*
University of Nizwa, Birkat Al Mauz, Nizwa, Oman
[d]*School of Health Sciences, University of Petroleum and Energy Studies,*
Dehradun, Uttarakhand, India
[e]*Department of Pharmacognosy & Phytochemistry,*
School of Pharmaceutical Sciences,
Delhi Pharmaceutical Sciences and Research University,
Mehrauli-Badarpur Road, New Delhi, India

khodaiel@gmail.com

Laticiferous species produce latex which is white and milky exudate or a complex colloidal suspension. Latex is the source of isolated compounds with various pharmacological activities. Latex-yielding plants occur in approximately 10% of angiosperms. They comprise 12500 species belonging to nearly 40 families. These families are mainly affiliated with dicotyledons. The mentioned flora contain secretary tissues concerned with the secretion of

Introduction to Plant-Based Latex: Phytochemistry, Ethnopharmacology, Biological, and Pharmacological Activities
Edited by Laleh Khodaie, Ajay Sharma, Saurabh Bhatia, and Ahmed Al-Harrasi

ISBN 978-981-5129-12-0 (Hardcover), 978-1-003-57071-4 (eBook)
www.jennystanford.com

latex. These tissues are highly specialized and heterogeneous cells regarding their function, structure, and morphology. These cells both synthesize and store latex. Based on their cellular organization, they fall into two fundamental categories: non-articulated and articulated laticifers. Near 70% of latex-bearing families belong to tropical regions due to the pressure of herbivores and microorganisms there. Latex production is a physiological adaptation as a first line of plant defense to restrict herbivores. Latex as a defensive agent could deter phytophagous insects living predominantly in tropical and subtropical regions. Latex-producing taxa are scarce in colder climates owing to the decreased pressure of herbivores and microorganisms. Herbal latex is used by indigenous people living in different regions. They are applied for various ailments from head to toe. This inherited knowledge acquired from indigenous people could be the base of modern pharmacological studies and the development of novel natural medicines.

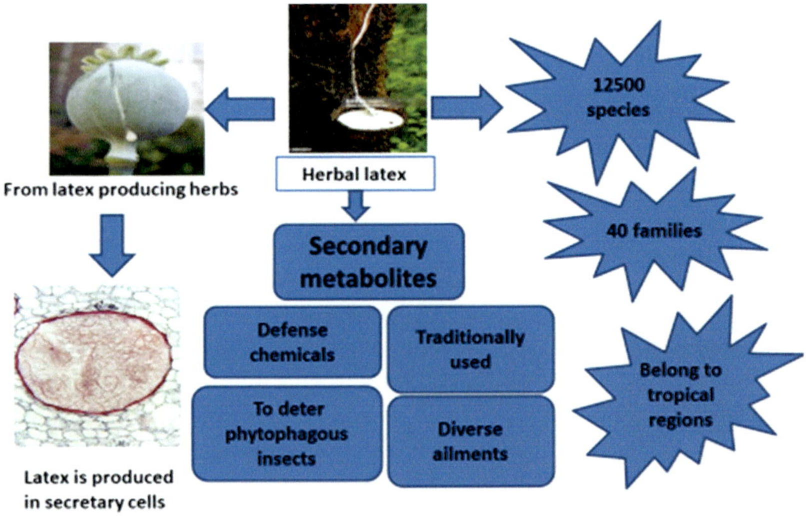

1.1 General Introduction

In the plant's kingdom, products of metabolic processes are either stored in special cells or transferred to typical tissues.

Secretion is defined as the transfer of intermediates or end products of plant metabolism from one plant body to another. Some plant secretions are induced in response to injuries on plant tissues, for instance, attacks of herbivores resulting in wounds or environmental stresses. Exudates of plants are categorized as gums, mucilage, resins, wax, and latex [1]. Latex is a typical type of plant exudate. It is a viscous liquid harvested from various trees and shrubs, especially from their points of damage. Latex has a wide array of applications, ranging from daily items to specialized commodities. Historically, there are some ancient documents indicating the use of latex to produce artifacts. In several sites in Mexico, some rubber artifacts, such as play balls were found belonging to 1600BC. Further investigations showed that these artifacts were made of latex from *Castilla elastica* tree [2]. This exploitable plant exudate could be obtained from laticiferous plants. These plants contain specialized cells, named laticifers, assembled in a tubular structure producing and storing latex. This sap occurs in different plants distributed in wide taxonomic groups. Latex is produced in 40 families of angiosperms and approximately 21500 species [3]. From an ecological perspective, latex-bearing taxa are more highly distributed in tropical than temperate regions. As an ecosystem service highly distributed in various regions, indigenous people and people of many communities used latex to treat ailments. As a natural remedy, it is important as a part of traditional knowledge. This chapter addresses the occurrence of laticiferous plants, their natural habitat, ecological setting, and biogeography or geographic distribution patterns, as well as traditional and regional uses of latex-bearing plants.

1.2 General Features

Latex means juice in Latin. Natural latex could be either milky or clear fluid secreting from plants. In other words, it is a type of suspension that consists of a liquid dispersion medium encompassing diverse dispersed particles. Refractive indexes of latexes obtained from species are highly diverse. Latex-bearing families producing milky latex are Asclepias, Euphorbia, Ficus, and Lactuca. Families of plants oozing clear or translucent latex include Morus, Nerium, and oleander [4]. The white and milky appearance of the latex reveals that it is a complex colloidal suspension composed of a matrix that encompasses characteristic phytochemicals. Similarly, these phytochemicals dispersed in a matrix are highly diversified due to the family and genus of plants. The latex obtained from different plants has varied refractive indexes. In some cases, its color could be yellow (Cannabis), orange (Papaver), red, green, or colorless [4]. Oxidation of white latex or its exposure to air causes it to be darkened. This plant secretion is solidified when exposed to air [5]. According to previous studies, in some taxa, latex oozed from different parts of a plant may have different colors [4]. Regarding latex particle size, they differ in diameter in various taxa. Furthermore, latex particles vary in shape ranging from spherical to pear or rod-shaped. Latex particles' size and shape are mainly dependent on age and the plant's stage of growth. That is to say, the older the plants are, the larger the particle size of latex will be [4].

1.3 Overview of the Latex-Bearing Plants

If we want to place laticiferous plants in a hierarchical structure and consider their taxonomic order, the following taxonomic units should be considered for each plant. Latex-yielding plants occur in approximately 10% of angiosperms (flowering plants bearing seeds in fruits). They comprise 12500 species belonging to nearly 40 families. These families are mainly affiliated with dicotyledons; however, families belonging to monocotyledons are also included. As shown in Table 1.1, monocotyledon subclasses are Alismatidae, Arecidae, Liliidae, and Zingiberidae. Some of the herbs in the monocot group, such as the ones in Alismatale, are aquatic and herbaceous [6].

Table 1.1 Biological classification of laticiferous plants in the taxonomic hierarchy

No.	Division	Class	Order	Family
1	Angiosperms	Dicotyledon	Boraginales	Boraginaceae
2	Angiosperms	Dicotyledon	Cucurbitales	Cucurbitaceae
3	Angiosperms	Dicotyledon	Asterale	Compositae, Campanulaceae
4	Angiosperms	Dicotyledon	Gentianale	Apocynaceae, Asclepediaceae
5	Angiosperms	Dicotyledon	Solanale	Convolvulaceae
6	Angiosperms	Dicotyledon	Caryophyllale	Cactaceae
7	Angiosperms	Dicotyledon	Ericale	Sapotaceae
8	Angiosperms	Dicotyledon	Guttiferale	Guttiferae
9	Angiosperms	Dicotyledon	Malvale	Bixaceae
10	Angiosperms	Dicotyledon	Brassicale	Caricaceae
11	Angiosperms	Dicotyledon	Rosale	Cecropiaceae, Moraceae Urticaceae, Cannabidaceae
12	Angiosperms	Dicotyledon	Aquifoliales	Aquifoliaceae, Cardiopteridaceae
13	Angiosperms	Dicotyledon	Magnolidae	Nelumbonaceae, Nymphaeaceae, Papaveraceae, Lardizabalaceae
14	Angiosperms	Dicotyledon	Malpighiale	Euphorbiaceae
15	Angiosperms	Dicotyledon	Fabales	Leguminosae
16	Angiosperms	Dicotyledon	Myrtales	Myrtaceae
17	Angiosperms	Dicotyledon	Celastrales	Celastraceae
18	Angiosperms	Dicotyledon	Santalales	Loranthaceae, Olacaceae
19	Angiosperms	Dicotyledon	Sapindales	Anacardiaceae, Burseraceae
20	Angiosperms	Monocotyledon	Asparagales	Amaryllidaceae
21	Angiosperms	Monocotyledon	Alismatale	Alismataceae, Limnocharitaceae, Aponogetonaceae, Butomaceae, Araceae
22	Angiosperms	Monocotyledon	Liliale	Liliaceae
23	Angiosperms	Monocotyledon	Zingiberale	Musaceae
24	Angiosperms	Monocots	Pandanales	Cyclanthaceae

1.4 Accumulation of Latex within Plant Body (Plant Secretary Tissue)

Secretory tissues in plants are concerned with the secretion of secondary metabolites, for example, gums, resins, volatile oils, and latex. They vary from single cell to accumulation of cells constructing complex structures such as secretary cavities and ducts. Technically, they fall into two categories: laticiferous and glandular tissues. Laticifers could be in both forms of single cells scattered among other types of cells in plant tissues and secretary ducts. Laticifers, the organization of specialized cells, are highly differentiated in terms of metabolism, structure, and development. Laticifers or vessels are types of conducting elements in laticiferous plants. They are a group of highly specialized and heterogeneous cells in terms of their function, structure, and morphology. These cells both synthesize and store latex. Based on their cellular organization, they fall into two fundamental categories: first, non-articulated, and second, articulated laticifers. Non-articulated laticifers are multinucleate and derive from a single coenocytic cell (a multinucleate cell originated from several nuclear divisions). In other words, similar single lactiferous cells are not fused. While plant growth, extensive elongation of these laticifers forms either network of branched vessels or unbranched ones. In the second type of laticifers, articulated ones, several simple cells are interconnected. In crude terms, intervening walls of adjacent cells are dissolved and a series of cells become joined and united. Articulated laticifers are subdivided into two categories: anastomosing and non-anastomosing ones. Articulated non-anastomosing laticifers are comprised of a chain of long cells or tubes that are not laterally bridged, while articulated anastomosing laticifers incorporate laterally connected tubes presenting a reticulum or network. According to previous reports, in several cases, two types of articulated and non-articulated laticifers occur in plants of a family, for instance in Asclepiadaceae and Euphorbiaceous. In rare cases, both mentioned types of laticifers occur in plant species [4, 7]. Table 1.2 presented some orders containing each type of laticifer.

Table 1.2 Classification of plant order and family according to the structure of latex-producing cells

No.	Type of laticifer	Plant order	Families
1	Non-articulated unbranched	Urticales, Cornales, Gentianales, Rosales, Pandanales,	Moraceae, Cyclanthaceae, Urticaceae, Apocynaceae, Cannabaceae, Asclepiadaceae
2	Non-articulated branched	Urticales, Euphorbiales, Gentianales, Cornales, Rosales	Apocynaceae, Asclepiadaceae, Moraceae, Apocynaceae, Euphorbiaceae
3	Articulated non-anastomosing	Zingiberales, Liliales, Nymphaelals, Papaverales, Euphorbiales, Ebenales, Asterales, Cornales, Solanales, Ericales, Solanales	Sapotaceae, Liliaceae, Convolvulaceae, Musaceae, Papaveraceae, Sapotaceae, Liliaceae, Musaceae
4	Articulated anastomosing	Arales, Papaverales, Euphorbiales, Campanulales, Ranunculales, Asterales, Brassicales, Malpighiales	Papaveraceae, Caricaceae, Compositae, Euphorbiaceae, Campanulaceae, Asteraceae

1.5 Development of Laticifers

There is a developmental distinction between articulated and non-articulated laticifers, with the latter being elaborately studied. Branched, non-articulated latex-producing cells are derived from primordium. Primordial cells are initial and detectable foundations in undeveloped embryos which lead to organ development [8]. Elongation and division of primary laticifer cells occur during the growth of the seedling. The growth of these cells proceeds in the meristemic tissue of an immature plant. The network of branched and non-articulated latex-producing cells develops from a specified number of initials. The number of initials differs in various plants. For instance, in embryos of *Nerium oleander* 28 initials were detected [9], while in the embryo of *Euphorbia engelmanni*, 4 initials were reported to develop to the network of laticiferous cells. Regarding unbranched, non-articulated laticifers, primordia or primary cells appear and elongate beneath apical meristemic tissue. The development of latex-bearing cells initiates in the stems and leaves of some plants, yet the initiation of unbranched ones is not elaborately recognized [10]. As far as articulated development is concerned, in some species initials could be observed in embryos. Articulated cells are originally engaged in the procambium, a set of meristematic cells providing original vascular tissue, later they are recruited in the vascular cambium. Differentiation of laticifers concurrently takes place with adjacent tissues of the phloem, forming distinct rows of longitudinal laticiferous cells. Mature articulated laticifers form a group of superimposed cells with perforate lateral walls that bring about adjoining vessels. In addition, perforated end walls generate anastomosing articulate laticifers. In crude terms, this intricate structure leads to a network of cytoplasm of latex-producing cells. Following the development process, the middle lamella undergoes both lateral and transverse perforations. For the next stage, simultaneous degeneration of cytoplasm and emergence of reshaped plastids occur. Fully developed articulated and non-articulated cells morphologically resemble, due to the perforations on the cell walls they alter into one cell.

Even though numerous pieces of research have been carried out to elucidate laticifers penetration in meristemic tissue, is yet to understand the process elaborately. During development, intrusive growth occurs; laticiferous cells elongate across lateral walls, extending between joining meristemic cells. The branches of cells intrude into newly constituted organs. The mentioned process of growth requires the disintegration of cell wall constituents. Presumably, the decomposition of the cell wall and its synthesis occur simultaneously to interpret the balance between penetration to meristematic cells and the generation of new connections between laticifer and neighboring cells [11]. It has been indicated that once differentiation of articulated laticifers occurs, some vesicles are formed in plasmalemma (plant cytoplasmic membrane). Being implicated in the process of endocytosis, wall fragments are degraded consequently [12].

Regarding cell wall alterations, cell wall-related enzymes such as pectinase have been identified in the latex of *Asclepias* [13]. Since the main ingredient of plant cell walls and middle lamella are pectins, pectinase, and its associates play a key role in their alterations. Polygalacturonase was detected in *Hevea brasiliensis* latex [14]. Similarly, in the latex of opium poppy transcripts of enzymes altering pectins, namely pectin methylesterases were indicated [15]. Further pectin remodeling enzymes have been detected in *Populous spp.* wood cells and pollen tubes of *Arabidopsis thaliana* [16]. Thus, modification of pectin by mentioned enzymes is the most probable mechanism to disintegrate neighboring latex-producing cells. In addition to pectin-related enzymes, cellulase has been detected in species containing articulated and non-articulated laticifers. Detection of cellulase by electron microscopy elucidates that the wall of laticiferous cells is enzymatically perforated. In addition, it has been suggested that auxin (hormone of growth regulation) is associated with the elevation of cellulase activity for diminishing cell walls [17]. According to previous manuscripts, cellulase was enormously identified in articulated laticiferous plants compared to non-articulated ones. For instance, in *Hevea brasiliensis*, the amount of cellulase is 50–150 times higher than in other taxa. Typically, in immature trees of *Hevea brasiliensis*, cellulase content is much higher compared to mature ones, indicating that the cells are actively differentiating [17]. According to previous findings, in *Lactuca sativa*, by isolation of latex and measurement of glycolytic activity, a large amount of β-D-fucosidase was identified. It is highly likely that due to the specificity of the enzyme and the existence of β-fucose connections in the plants' cell wall, the mentioned enzyme fulfill a crucial function in the wall perforation of *L. sativa* [12].

1.6 Geographical Distribution and Ecological Adaptation

Latex-bearing plants, similar to other taxa, are adapted to particular geographic regions. This is following biogeographic variation. In terms of biogeographic variation, species with special features are distributed in typical habitats with particular geographic variables for instance, climate or other ecologic settings. In crude terms, genetically based characteristics of taxa contribute to the natural selection of species favoring some geographic areas. As a result, ecological characteristics could affect the occurrence of phytochemicals in plants.

Near 70% of latex-bearing families belong to tropical regions. However, there is an exception to the fact that latex-bearing plants belonging to Campanulaceae, Cannabidaceae, Papaveraceae, and Lardizabalaceae are found in temperate regions. There are widely distributed latex-producing plants belonging to Compositae, Convolvulaceae, Nelumbonaceae, Nymphaeaceae, Euphorbiaceae, Leguminosae, Alismataceae, and Liliaceae, which are adapted to various climates. Likely, the main reason for a high proportion of llaticiferous plants in warm and humid regions is considered advantageous for those taxa. In tropical regions, there is a pressure of herbivores and microorganisms. Latex production is a physiological adaptation as a first line of plant defense to restrict herbivores. Latex as a defensive agent could deter phytophagous insects living predominantly in tropical and subtropical regions. Thus, latex-producing taxa are scarce in colder climates due to the decreased pressure of herbivores and microorganisms.

1.7 Ethno Pharmacological Applications

Throughout history, ethnomedicine and ethnobotany have provided mutual assistance to survive mankind from several diseases. In the modern age, people rely on various herbal remedies as primary healthcare to both prevent and cure several pathologies and ailments. Herbs and herbal products popular in communities or used by local healers revealed their medicinal effects. This folklore knowledge contributes to many modern and pharmacological studies bringing about clinically used medicine and market products. Moreover, ethnomedical knowledge plays a vital role in providing lead compounds and materials for the synthesis of new compounds. Thus, to prevent the gradual extinction of indigenous information they should be collected and documented systematically. Systematically collected folklore information, as a first and significant phase, could captivate the attention of scientists to conduct modern research on them, for instance phytochemical, pharmacological, and pharmacognosy screenings [18].

Documentation of traditional knowledge used by local folks offers a good opportunity to recognize secondary metabolites of plants, such as latex to treat ailments of certain body organs. The ethnomedical information of latex-bearing taxa was summarized in Table 1.3. Table 1.3 documents the botanical name, country of origin, plant part utilized for treating ailments, and the type of disease or biological activities.

Regarding the preparation and route of administration, there were several routes in previous manuscripts; according to previous reports, the most frequent preparation modes were pastes, decoctions, and sometimes unprepared latex of herbs in both forms fresh and dried. Some herbal latex was mixed with water to make juices [19]. Preparations were used both topically and internally. Ethiopian indigenous healers administered taking a coffee cup of fresh leaf latex of *Aloe monticola* each morning for 30 days or longer to reduce symptoms of diabetes and its complications. In Yemen, healers prescribed *Euphorbia balsamifera* Aiton to the patients to chew dry latex without swallowing to reduce infection in the mouth. Latex droplets of *Euphorbia cactus* were applied to the tooth or oral cavity to control the infection of the region. Or droplets of Euphorbia granulate latex were utilized to control eye infection. Dry latex of *Gnidia somalensis* was orally ingested in the form of a mixture with water to alleviate gastrointestinal disorders such as chronic constipation. The latex obtained from *Ficus vasta* is applied to the chest to alleviate painful coughs and chest [19].

Table 1.3 Ethnomedical data of latex-producing medicinal plants

Family	Botanical name	Habitat	Part used	Category of disease	Ref.
Anacardiaceae	*Mangifera indica*	Togo	Fruits	**FM**, **IA**, **GID**: jaundice, **V**	[20]
Asclepiadaceae	*Leptadenia hastata*	Togo	NM	**IA:** microbial infection, **RD:** cough	
	Pergularia daemia	Togo	NM	**BD:** Anemia, **GD**: Agalactia, female infertility, abortion risk, dysmenorrhea, **IA**: vaginal candidiasis, microbial infection, **DD**: dermatitis, **FM**, **GID:** stomachache, abdominal pain, chronic hiccough, And diarrhea, **D**, **O**: ocular pains. **RD:** cough, tuberculosis.	[20]
	Secamone afzelii	Togo	NM	**GD:** Agalactia, dystocia, **RD:** cough, **V**, **IA:** microbial infection, amebiasis	[20]
Asteraceae	*Lactuca taraxacifolia*	West Africa	NM	**BD:** Anemia, **D**, **GID:** abdominal bloating, **USD:** urinary retention, **GD:** dystocia,	[20]
Convolvulaceae	*Ipomoea batatas*	Togo		**GD:** abortion risk, **H:** bleeding	[20]
Liliaceae	*Aloe citrina*	Ethiopia	Leaf	**WH**, **O**: eye infections, **FM**, **GID**: abdominal pain.	[21]
	Aloe harlana	Ethiopia	Leaf	**ID**, **IA**, **WH**	[22]
	Aloe macrocarpa	Ethiopia	Leaf	**FM**, **GID**: jaundice, **DD**	[23]
	Aloe megalacantha	Ethiopia	Leaf	**D**, **ID**, **WH**	[24–27]
	Aloe monticola	Ethiopia	Leaf	**D**, **IA**	[25–27]
	Aloe pulcherrima	Ethiopia	Leaf	**IA**, **FM**	[28]
	Aloe trigonantha	Ethiopia	Leaf	**IA, ID**	[29]

Family	Botanical name	Habitat	Part used	Category of disease	Ref.
	Aloe vera	Togo	Leaf	**IA, ID, GID**: abdominal pain, jaundice	[20]
	Aloe weloensis	Ethiopia	Leaf	**WH**, **DD**, **IA**: ear infection, **ID**: rheumatism, **PK**: for headache	[30]
	Aloe yavellana	Ethiopia	Leaf	**LA**, control of ticks on livestock	[31]
Moraceae	*Ficus benjamina*	Brazil	Apex and branches	**V**	[32]
	Ficus capensis	Togo	NM	**GD:** Infertility, **FM**	[20]
	Ficus carica	Iran, Italy, Turkey, Brazil, China	Leaves and fruits	**DD:** warts, papillomatosis, **V, D, IA**	[33–36]
	Ficus exasperata	Togo	NM	**FM, GID:** icterus, intestinal functional troubles	[20]
	Ficus insipida	Amazon	Stem	**LA**	[35, 37, 38]
	Ficus platyphylla	Togo	NM	**BD:** Anemia, **ID:** microbial infection, **FM**, **GD:** ovarian cyst	[20]
	Ficus polita	Togo	NM	**ID:** microbial infection, **FM**	[20]
	Ficus racemosa	India	Stem	**WH, ID:** hemorrhoids, **GID:** diarrhea and dysentery in children, **A**	[39, 40]
	Ficus religiosa	India	Branch	**GD:** abortifacient, treating hemorrhage, **DD**, **PK:** neuralgia	[41]

(*Continued*)

Table 1.3 (*Continued*)

Family	Botanical name	Habitat	Part used	Category of disease	Ref.
	Ficus thonningii	Tanzania Mali, Togo	leaf Stem Root	**FM**, **DOD**: gingivitis, **V**, **O:** cataract treatment, **GD:** agalactia, **GID:** diarrhea, stomach conditions, jaundice, **IA:** gonorrhea, **D**, **WH**, **RD:** bronchitis, **UTI**, **ID:** arthritis, rheumatism, **GD:** induction of menstruation.	[42]
	Ficus umbellata	Togo	NM	**IA:** abscess, **BD:** anemia, **FM**, **GID**: diarrhea, amoebic dysentery, hemorrhoid, **GD**: polymenorrhea, **RD:** cough, **A:** oligospermia	[20]
	Milicia excelsa	Togo	NM	**PK:** Chronic headache	[20]
Euphorbiaceae	*Euphorbia antiquorum*	Thailand	NM	**IA:** cutaneous infections, **RD, PK, NSD, DOD:** toothache	[43, 44]
	Euphorbia argillicolq		NM	**WH:** Vulnerary wounds	[44]
	Euphorbia armena	Turkey	NM	**GID:** constipation, **WH**	[45]
	Euphorbia atoto	S. Central Pacific	NM	**PK**	[44]
	Euphorbia coniosperma	Turkey	NM	**DD:** warts, **I:** scorpion and snake bite	[46]
	Euphorbia corollata	USA, Georgia	Root latex	**GD:** Hemorrhages	[47]
	Euphorbia dracunculoides	Spain North Africa	NM	**GID:** constipation, **USD:** diuretic	[44]
	Euphorbia duceimata		NM	**NSD:** Mental illness	[44]

Family	Botanical name	Habitat	Part used	Category of disease	Ref.
	Euphorbia esula	China	NM	**DD:** wart removal	[48]
	Euphorbia falcata	Turkey	NM	**DD:** eczema, **IA:** fungal infection	[45]
	Euphorbia grandifolia	Togo	NM	**RD:** Cough, **IA:** microbial infection, whitlow, dermatosis, **GID**: stomach ache, intestinal functional troubles, hemorrhoid, **GD:** infertility,early menopause, **NSD**: epilepsy	[20]
	Euphorbia heteradena	Turkey, Iran	NM	**GID:** constipation	[45]
	Euphorbia heterophylla	Togo		**IA:** Microbial infection, **GD:** early menopause	[20]
	Euphorbia hirta	India and Malaysia	Apical part	**SD, FM, IA:** gonorrhea, amebiasis, and candidiasis, **DOD:** tooth decay, **V, RD:** bronchitis, asthma, cough, **GD:** agalactia, dysmenorrhea, ovarian cyst, **GID:** diarrhea, dysentery, stomach ache, stomach ulcer	[20, 49, 50]
	Euphorbia humifusa	Eastern Asia, South, and Central Europe	NM	**CD**, **RD:** influenza	[44]
	Euphorbia ingens	Southern Africa	NM	**CD:** Cancerous lesion, **O:** eye tumor, **I:** snake bites, **NSD:** mental disorders	[47]
	Euphorbia lactea	Dominican Republic	Leave latex	**ID**	[51]
	Euphorbia lagascae	Turkey, Spain	NM	**ID:** Rheumatism, **USD:** bladder Stones.	[44]

(Continued)

Table 1.3 (*Continued*)

Family	Botanical name	Habitat	Part used	Category of disease	Ref.
	Euphorbia lathyris	Turkey, Pakistan	NM	**GID:** purgative, emetic, diarrhea, constipation, **BD:** blood purifier	[47]
	Euphorbia macroclada	Turkey	NM	**GID:** constipation, **DD:** Inflamed wounds, **I:** scorpion and bee bites, **DD:** eczema, **IA:** fungal infection, **DOD:** dental disease	[45]
	Euphorbia maculata	Canada, USA	NM	**WH, DD, I**	[47]
	Euphorbia milli	Africa, Brazil, China, India	NM	**DD:** wart remover, **GID:** hepatitis and abdominal edema, **IA:** schistosomiasis in sheep, cattle, and humans	[52–54]
	Euphorbia myrsinites	Turkey, Italy	NM	**V:** stomach worms	[55]
	Euphorbia neriifolia	Iran, India, Malaysia,	Stem	**BD:** anemia, **GID:** carminative, laxative, abdominal pain, **RD:** expectorant, asthma, **CD**, **IA:** gonorrhea, leprosy, **DD:** wart removal, leukoderma, **USD:** removal of kidney and bladder stones, **FM**	[56, 57]
	Euphorbia nicaensis	Turkey	NM	**GID:** constipation, **V**	[45]
	Euphorbia nicaeensis	Turkey	NM	**WH, DD**	[45]
	Euphorbia nivulia	India	Stem and trunk	**WH, H, FM, GID**: dysentery, enteritis **NSD**, **I:** poisonous snake bites	[56]
	Euphorbia obtusifolia	Spain	NM	**DD:** wart, skin infections, **CD**	[58]

Family	Botanical name	Habitat	Part used	Category of disease	Ref.
	Euphorbia oncoclada	Togo		**RD:** Bronchitis	[20]
	Euphorbia parviflora	Nepal	NM	**RD:** asthma, coryza, **I:** snake bites	[47]
	Euphorbia peplus	North America	NM	**CD, GID:** purgative, **DD:** removal of warts, **RD:** asthma, bronchitis	[47]
	Euphorbia prostrata	South America	NM	**USD, D:** diabetes	[44]
	Euphorbia royleana	India	NM	**V, IA, PK, DOD:** to fill tooth cavities	[59]
	Euphorbia resinifera	Thailand, Morroco	Stem	**DD, O, WH**	[60, 61]
	Euphorbia rigida	Southern Europe and Southwest Asia	NM	**RD:** asthma	[44]
	Euphorbia seguieriana	Turkey	NM	**GID:** constipation, **DD:** Inflamed wounds, **FM**	[62]
	Euphorbia stricta	Turkey	NM	**DOD, GID:** constipation, **USD:** diuretic	[44]
	Euphorbia tirucalii	Brazil, African countries	Branches and trunk	**IA:** syphilis, **ID, RD, CD, V, GID:** jaundice, **ID:** rheumatism, **PK:** neuralgia, **DOD:** toothache, **I**	[63, 64]
	Euphorbia umbellata	Brazil	Trunk	**CD, ID**	[65]
	Euphorbia wallichii	Turkey	NM	**GID:** constipation	[44]

(*Continued*)

Table 1.3 (*Continued*)

Family	Botanical name	Habitat	Part used	Category of disease	Ref.
	Hevea brasiliensis	Southeast Asia, Brazil, Thailand	Trunk	**WH**	[66]
	Jatropha curcas	Togo	NM	**IA:** amebiasis, whitlow, **ID:** rheumatism, **GD:** abortion, dystocia, **DOD:** tooth decay, buccal wounds, buccal mycosis, **WH, PK:** lumbago, **GID:** abdominal pain, icterus, **USD:** urinary infection, **FM**	[20]
	Jatropha gossypifolia	Togo	NM	**BD:** anemia, **NSD:** anorexia, **GID:** hepatitis, icterus, **WH, IA:** microbial infection, **FM**	[20]
	Manihot esculenta	Togo	NM	**BD:** Anemia, **PK:** headache, **GID:** dysentery, stomach ache, **WH, O**: ocular pains, **FM, IA:** microbial infection, **I:** snake bite, **V**	[20]
Apocynaceae	*Alstonia boonei*	Togo	NM	**DD:** yaws, **GID:** hemorrhoids, constipation, **IA:** chickenpox, vaginitis, dermatitis, **V**	[20, 67]
Caricaceae	*Carica papaya*	Central and South America	Fruit latex	**DD**, **IA:** infected burns	[68]
	Calotropis procera	Brazil	aerial parts	**ID, PK, IA:** leprosy, **CD:** piles and tumors, **IA:** sinusitis, **I:** snake bites	[69]
	Calotropis gigantia	India Thailand China, and Southeast Asia	aerial parts	**H, GD:** easy delivery or abortion, **GID:** constipation	[70, 71]
	Cryptolepis sanguinolenta	Togo		**IA:** amoebic dysentery, microbial infection, **FM, GID:** stomach ache, intestinal functional troubles	[20]

Family	Botanical name	Habitat	Part used	Category of disease	Ref.
	Himatanthus drasticus	Brazil	trunk	**GID:** hemorrhoids, gastritis, **CD, PK, DD, D, WH**	[72, 73]
	Himatanthus sucuuba	Amazon regions of Brazil	NM	**V, CD, PK, WH, ID**	[74, 75]
	Holarrhena floribunda	Togo	NM	**V, IA:** candidiasis, **GID:** dystocia, diarrhea, **GD:** infertility, **PK:** lumbago, **USD:** hematuria	[20]
	Pulmeria rubra	Brazil	Arial parts	**FM, DD, IA, GID:** hepatitis, **USD:** ureterolithiasis,	[76, 77]
	Rauvolfia vomitoria	Togo	NM	**GD**: amenorrhea, infertility, **PK**: headache, **NSD**: convulsive attacks, mental diseases, **IA, GID**: stomach ulcer, stomach ache, **BD:** anemia	[20]
	Thevetia neriifolia	Togo	NM	**PK:** headache, **NSD:** mental diseases, madness	
Periplocaceae	Parquetina nigrescens	Togo	NM	**NSD:** Epilepsy, **USD:** sexual impotence, **GID:** hemorrhoid, **IA:** microbial infection, **NSD:** insomnia, **GD:** abortion risk, **PK:** heart pains	[20]
Sapotaceae	Vitellaria paradoxa	Togo	NM	**IA:** microbial infection, **GD:** female infertility	[20]
Vitaceae	Cissus populnea	Togo	NM	**A:** Impotence, male infertility, oligospermia	[20]

GID: Gastrointestinal Disease; **FM:** Fever and Malaria; **CD:** Cancer Diseases; **DOD:** Dental and Oral Diseases; **D:** Dermal Disease; **ID:** Inflammatory Disease; **IA:** Infectious Ailments; **D:** Diabetes; **PK:** Pain Killer; **LA:** Larvicidal Activity; **A:** Aphrodisiac; **GD:** Gynecological Disease; **WH:** Wound Healer; **V:** Vermifuge; **O:** Ophthalmology; **RD:** Respiratory Disease; **USD:** Urinary System Disorders; **H:** Hemostasis; **NSD:** Nervous System Diseases; **I:** Injuries, **BD:** Blood Disorders.

The ethnomedical information is categorized and summarized in Table 1.3. Laticiferous herbs were alphabetically listed by species names. As could be seen in Table 1.3, 12 families comprising 90 laticiferous plants were recorded. Among the families, Euphorbiaceae encompassed the largest number of latex-bearing plants (47 plants) documented as an applicable herb in ethnomedicine. Moreover, there is a dominance of ethnomedical uses attributed to its diversity and occurrence in many regions and countries. Euphorbiaceae plants The remaining substantial families were Moraceae (13 plants), Apocynaceae (10 plants), Liliaceae (9 plants), Asclepiadaceae (3 plants), Anacardiaceae (1 plant), Asteraceae (1 plant), Convolvulaceae (1 plant), Caricaceae (1 plant), Periplocaceae (1 plant), Sapotaceae (1 plant), and Vitaceae (1 plant). The plant parts used to exploit latex were leaves, roots, stems and branches, fruits, and whole plant. In some cases, the herbal body for the exploitation of latex was not mentioned. Among the mentioned plant bodies, the most commonly used part was leaves, due to easier collection compared to other plant bodies. It is highly likely that the indigenous ones predominantly used leaves due to the conservation of taxa and not to dig out the root and cause the death of herbs. Documented taxa are extremely versatile in terms of ethnomedical uses which are indicative of various phytochemical constituents which will be discussed in the following chapters. The therapeutic potential of some documented herbs has not been scientifically proven. Pharmacological studies of the mentioned taxa are required to be confirmed both experimentally and clinically [20]. The recorded families and species fall into two categories in terms of toxicity: first, taxa producing toxic latex applicable in topical form to cure dermal disorders, second, herbs producing non-toxic latex used both in topical and oral forms. Laticiferous plants belonging to Euphorbiaceae and Apocynaceae families due to containing toxic phytochemicals are utilized for dermal disorders, such as wart removal, leukoderma, eczema, and fungal infection. Due to having broad therapeutic activities, they were also used for gastrointestinal diseases such as jaundice, icterus stomachache, abdominal pain, chronic hiccough, diarrhea, purgative, emetic, constipation, abdominal bloating, intestinal functional troubles, amoebic dysentery, hemorrhoid, hepatitis, abdominal edema, and stomach ulcer. Cancer diseases: cancerous lesions, piles and tumors. Dental and oral diseases: toothache, dental disease, to fill tooth cavities, buccal wounds, buccal mycosis. Dermal disease: dermatitis, warts, papillomatosis, leukoderma, yaws. Inflammatory Disease: rheumatism, arthritis, and inflamed wounds. Infectious ailments: microbial infection, vaginal candidiasis, amebiasis, ear infection, gonorrhea, cutaneous infections, fungal infection, whitlow, dermatosis, schistosomiasis in sheep, cattle, and humans, leprosy, amebiasis, syphilis, and candidiasis, leprosy, chickenpox, vaginitis, dermatitis, infected burns. Pain killer: earache, neuralgia, headache, lumbago, heart pains. Gynecological disease: agalactia, female infertility, abortion risk, dysmenorrhea, dystocia, ovarian cyst, polymenorrhea, abortifacient, treating hemorrhage, early menopause, easy delivery, or abortion. Ophthalmology: ocular pains, eye infections, cataract treatment, eye tumor. Respiratory disease: cough, tuberculosis. influenza, asthma, coryza, and bronchitis. Ureteral system disease: urinary retention, removal of bladder and kidney stones, diuretic, urinary infection, ureterolithiasis, hematuria, sexual impotence. Nervous System Diseases: epilepsy, mental disorders, anorexia, convulsive attacks, madness, insomnia. Injuries: scorpion and snake bite, blood disorders: anemia, blood purifier. aphrodisiac: Impotence, male infertility, oligospermia. Besides the mentioned ethnomedical uses, exudate of some of the laticiferous plants was used to manage diabetes, malaria, and hemostasis. The dominant use for latex belonged to dermal and gastrointestinal disorders. This inherited knowledge used by indigenous people of a region is still incomplete and requires more investigation. However, this documentation could be a valuable lead to further investigate laticiferous plants and their exudates for biological and pharmacological effects.

1.8 Conclusion

Latex-producing herbs benefitted mankind belonging to different cultures and geographical places. These herbs produce latex with a complex composition comprising diverse phytochemicals with various pharmacological effects which will be discussed in ensuing chapters. Phytochemicals are comprised of complex macromolecules and low-molecular ingredients. The latex obtained from the aforementioned plants is a rich supply of phyto-constituents used in folk medicine. Thus, they are worth to be explored to verify their traditional uses through modern science.

References

1. Licá ICL, Soares A, de Mesquita LSS, Malik S. Biological properties and pharmacological potential of plant exudates. *Food Research International (Ottawa, Ont)*. 2018; 105: 1039–53.
2. Ownby DR. A history of latex allergy. *The Journal of Allergy and Clinical Immunology*. 2002; 110(2 Suppl): S27–32.
3. Lewinsohn TM. The geographical distribution of plant latex. *Chemoecology*. 1991; 2(1): 64–8.
4. Abd Razak SBB. Development and distribution of laticifers in plants. Doctoral thesis, Durham University; 2000.
5. Flores JS, Ricalde RV. The secretions and exudates of plants used in Mayan traditional medicine. *Journal of Herbs, Spices & Medicinal Plants*. 1996; 4(1): 53–9.
6. Naidoo CM, Munsamy A, Naidoo Y, Dewir YH. Chemistry, biological activities, and uses of latex from selected species of apocynaceae. In *Gums, Resins and Latexes of Plant Origin: Chemistry, Biological Activities and Uses*, Springer. 2022; pp. 845–68.
7. Hagel JM, Yeung EC, Facchini PJ. Got milk? The secret life of laticifers. *Trends in Plant Science*. 2008; 13(12): 631–9.
8. Curtis J, Lersten N. Internal secretory structures in Hypericum (Clusiaceae): H. perforatum L., H. balearicum L. *The New Phytologist*. 1990; 114(4): 571–80.
9. Mahlberg PG. Embryogeny and histogenesis in Nerium oleander II. Origin and development of the non-articulated laticifer. *American Journal of Botany*. 1961; 48(1): 90–9.
10. Lee KB, Mahlberg PG. Ultrastructure and development of non-articulated laticifers in seedlings of Euphorbia maculata L. *Journal of Plant Biology*. 1999; 42(1): 57–62.
11. Serpe MD, Muir AJ, Driouich A. Immunolocalization of β-D-glucans, pectins, and arabinogalactan-proteins during intrusive growth

and elongation of nonarticulated laticifers in Asclepias speciosa Torr. *Planta*. 2002; 215(3): 357–70.

12. Giordani R, Noat G. Isolation, molecular properties and kinetic studies of a strict β-fucosidase from Lactuca sativa latex: Its possible role in the cell-wall degradation of articulated laticifers. *European Journal of Biochemistry*. 1988; 175(3): 619–25.
13. Wilson KJ, Nessler CL, Mahlberg PG. Pectinase in Asclepias latex and its possible role in laticifer growth and development. *American Journal of Botany*. 1976; 63(8): 1140–4.
14. Ko J-H, Chow K-S, Han K-H. Transcriptome analysis reveals novel features of the molecular events occurring in the laticifers of Hevea brasiliensis (para rubber tree). *Plant Molecular Biology*. 2003; 53(4): 479–92.
15. Pilatzke-Wunderlich I, Nessler CL. Expression and activity of cell-wall-degrading enzymes in the latex of opium poppy, Papaver somniferum L. *Plant Molecular Biology*. 2001; 45(5): 567–76.
16. Siedlecka A, Wiklund S, Péronne M-A, Micheli F, Leśniewska J, Sethson I, et al. Pectin methyl esterase inhibits intrusive and symplastic cell growth in developing wood cells of Populus. *Plant Physiology*. 2008; 146(2): 554–65.
17. Sheldrake A, Moir G. A cellulase in Hevea latex. *Physiologia Plantarum*. 1970; 23(2): 267–77.
18. Maregesi SM, Ngassapa OD, Pieters L, Vlietinck AJ. Ethnopharmacological survey of the Bunda district, Tanzania: Plants used to treat infectious diseases. *Journal of Ethnopharmacology*. 2007; 113(3): 457–70.
19. Al-Fatimi M. Ethnobotanical survey of medicinal plants in central Abyan governorate, Yemen. *Journal of Ethnopharmacology*. 2019; 241: 111973.
20. Hoekou YP, Tchacondo T, Karou SD, Koudouvo K, Atakpama W, Pissang P, et al. Ethnobotanical study of latex plants in the maritime Region of Togo. *Pharmacognosy Research*. 2016; 8(2): 128.
21. Girma B, Bisrat D, Asres K. Antimalarial evaluation of the leaf latex of Aloe citrina and its major constituent. *Ancient Science of Life*. 2015; 34(3): 142–6.
22. Asamenew G, Bisrat D, Mazumder A, Asres K. In vitro antimicrobial and antioxidant activities of anthrone and chromone from the latex of Aloe harlana Reynolds. *Phytotherapy Research*. 2011; 25(12): 1756–60.

23. Tewabe Y, Kefarge B, Belay H, Bisrat D. Antileishmanial evaluation of the leaf latex of Aloe macrocarpa, Aloin A/B, and its semisynthetic derivatives against two leishmania species. *Evidence-Based Complementary and Alternative Medicine.* 2019; 2019: 4736181.
24. Hammeso WW, Emiru YK. Antidiabetic and antihyperlipidemic activities of the leaf latex extract of Aloe megalacantha Baker (Aloaceae) in Streptozotocin-Induced diabetic model. *Evidence-Based Complementary and Alternative Medicine.* 2019; 2019: 8263786.
25. Tekulu GH, Araya EM, Mengesha HG. In vitro α-amylase inhibitory effect of TLC isolates of Aloe megalacantha baker and Aloe monticola Reynolds. *BMC Complementary and Alternative Medicine.* 2019; 19(1): 206.
26. Hintsa G, Sibhat GG, Karim A. Evaluation of antimalarial activity of the leaf latex and TLC isolates from aloe megalacantha baker in plasmodium berghei infected mice. *Evidence-Based Complementary and Alternative Medicine.* 2019; 2019: 6459498.
27. Gebremeskel L, Bhoumik D, Sibhat GG, Tuem KB. In vivo wound healing and anti-inflammatory activities of leaf latex of Aloe megalacantha Baker (Xanthorrhoeaceae). *Evidence-Based Complementary and Alternative Medicine.* 2018; 2018: 5037912.
28. Teka T, Bisrat D, Yeshak MY, Asres K. Antimalarial activity of the chemical constituents of the leaf latex of Aloe pulcherrima Gilbert and Sebsebe. *Molecules.* 2016; 21(11), 1415.
29. Megeressa M, Bisrat D, Mazumder A, Asres K. Structural elucidation of some antimicrobial constituents from the leaf latex of Aloe trigonantha L.C. Leach. *BMC Complementary and Alternative Medicine.* 2015; 15: 270.
30. Emiru YK, Siraj EA, Teklehaimanot TT. Antibacterial potential of Aloe weloensis (Aloeacea) leaf latex against gram-positive and gram-negative bacteria strains. *International Journal of Microbiology.* 2019; 2019: 5328238.
31. Hailesillassie T, Bisrat D, Asres K. Larvicidal effect of the leaf latex of Aloe yavellana Reynolds and its major compounds against Amblyomma variegatum (Ixodidae). *Veterinary Parasitology.* 2018; 263: 23–6.
32. Wanderley LF, Soares A, Silva CRE, Figueiredo IM, Ferreira A, Perales J, et al. A cysteine protease from the latex of Ficus benjamina has in vitro anthelmintic activity against Haemonchus contortus. Revista brasileira de parasitologia veterinaria = Brazilian

journal of veterinary parasitology: Orgao Oficial do Colegio Brasileiro de Parasitologia Veterinaria. *Revista Brasileira de Parasitologia Veterinaria*. 2018; 27(4): 473–80.

33. Ghandehari F, Fatemi M. The effect of Ficus carica latex on 7, 12-dimethylbenz (a) anthracene-induced breast cancer in rats. *Avicenna Journal of Phytomedicine*. 2018; 8(4): 286–95.
34. Lazreg-Aref H, Mars M, Fekih A, Aouni M, Said K. Chemical composition and antibacterial activity of a hexane extract of Tunisian caprifig latex from the unripe fruit of Ficus carica. *Pharmaceutical Biology*. 2012; 50(4): 407–12.
35. de Amorin A, Borba HR, Carauta JP, Lopes D, Kaplan MA. Anthelmintic activity of the latex of Ficus species. *Journal of Ethnopharmacology*. 1999; 64(3): 255–8.
36. Hemmatzadeh F, Fatemi A, Amini F. Therapeutic effects of fig tree latex on bovine papillomatosis. *Journal of Veterinary Medicine B, Infectious Diseases and Veterinary Public Health*. 2003; 50(10): 473–6.
37. Hansson A, Zelada JC, Noriega HP. Reevaluation of risks with the use of Ficus insipida latex as a traditional anthelmintic remedy in the Amazon. *Journal of Ethnopharmacology*. 2005; 98(3): 251–7.
38. Gonzales A, Santos GG, Tavares-Dias M. Anthelminthic potential of the Ficus insipida latex on monogeneans of Colossoma macropomum (Serrasalmidae), a medicinal plant from the Amazon. *Acta Parasitologica*. 2019; 64(4): 927–31.
39. Ahmed F, Urooj A. Traditional uses, medicinal properties, and phytopharmacology of Ficus racemosa: A review. *Pharmaceutical Biology*. 2010; 48(6): 672–81.
40. Yadav RK, Nandy BC, Maity S, Sarkar S, Saha S. Phytochemistry, pharmacology, toxicology, and clinical trial of Ficus racemosa. *Pharmacognosy Reviews*. 2015; 9(17): 73–80.
41. Singh D, Singh B, Goel RK. Traditional uses, phytochemistry and pharmacology of Ficus religiosa: A review. *Journal of Ethnopharmacology*. 2011; 134(3): 565–83.
42. Dangarembizi R, Erlwanger KH, Moyo D, Chivandi E. Phytochemistry, pharmacology and ethnomedicinal uses of Ficus thonningii (Blume Moraceae): A review. *African Journal of Traditional, Complementary and Alternative Medicines*. 2012; 10(2): 203–12.
43. Siritapetawee J, Limphirat W, Wongviriya W, Maneesan J, Samosornsuk W. Isolation and characterization of a galactose-specific lectin (EantH) with antimicrobial activity from

Euphorbia antiquorum L. latex. *International Journal of Biological Macromolecules.* 2018; 120(Pt B): 1846–54.

44. Kemboi D, Peter X, Langat M, Tembu J. A Review of the ethno-medicinal uses, biological activities, and triterpenoids of Euphorbia species. *Molecules.* 2020; 25(17): 4019.
45. Altundag E, Ozturk M. Ethnomedicinal studies on the plant resources of east Anatolia, Turkey. *Procedia-Social and Behavioral Sciences.* 2011; 19: 756–77.
46. Sezik E, Yeşilada E, Honda G, Takaishi Y, Takeda Y, Tanaka T. Traditional medicine in Turkey X. Folk medicine in central Anatolia. *Journal of Ethnopharmacology.* 2001; 75(2–3): 95–115.
47. Ernst M, Grace OM, Saslis-Lagoudakis CH, Nilsson N, Simonsen HT, Rønsted N. Global medicinal uses of Euphorbia L. (Euphorbiaceae). *Journal of Ethnopharmacology.* 2015; 176: 90–101.
48. Fu ZY, Han XD, Wang AH, Liu XB. Apoptosis of human gastric carcinoma cells induced by Euphorbia esula latex. *World Journal of Gastroenterology.* 2016; 22(13): 3564–72.
49. Salehi B, Iriti M, Vitalini S, Antolak H, Pawlikowska E, Kręgiel D, et al. Euphorbia-derived natural products with potential for use in health maintenance. *Biomolecules.* 2019; 9(8), 337.
50. Patel GK, Kawale AA, Sharma AK. Purification and physicochemical characterization of a serine protease with fibrinolytic activity from latex of a medicinal herb Euphorbia hirta. *Plant Physiology and Biochemistry.* 2012; 52: 104–11.
51. Fernandez-Arche A, Saenz MT, Arroyo M, de la Puerta R, Garcia MD. Topical anti-inflammatory effect of tirucallol, a triterpene isolated from Euphorbia lactea latex. *Phytomedicine.* 2010; 17(2): 146–8.
52. Dias-Baruffi M, Sakamoto M, Rossetto S, Vozári-Hampe MM, Roque-Barreira MC. Neutrophil migration and aggregation induced by euphorbin, a lectin from the latex of Euphorbia milii, var. milii. *Inflammation Research: Official Journal of the European Histamine Research Society.* 2000; 49(12): 732–6.
53. Fonseca KC, Morais NC, Queiroz MR, Silva MC, Gomes MS, Costa JO, et al. Purification and biochemical characterization of Eumiliin from Euphorbia milii var. hislopii latex. *Phytochemistry.* 2010; 71(7): 708–15.
54. Yadav SC, Pande M, Jagannadham MV. Highly stable glycosylated serine protease from the medicinal plant Euphorbia milii. *Phytochemistry.* 2006; 67(14): 1414–26.

55. Polat R, Satıl F. An ethnobotanical survey of medicinal plants in Edremit Gulf (Balıkesir–Turkey). *Journal of Ethnopharmacology.* 2012; 139(2): 626–41.

56. Badgujar SB. Evaluation of hemostatic activity of latex from three Euphorbiaceae species. *Journal of Ethnopharmacology.* 2014; 151(1): 733–9.

57. Mali PY, Panchal SS. Euphorbia neriifolia L.: Review on botany, ethnomedicinal uses, phytochemistry and biological activities. *Asian Pacific Journal of Tropical Medicine.* 2017; 10(5): 430–8.

58. Betancur-Galvis L, Palomares E, Marco JA, Estornell E. Tigliane diterpenes from the latex of Euphorbia obtusifolia with inhibitory activity on the mammalian mitochondrial respiratory chain. *Journal of Ethnopharmacology.* 2003; 85(2–3): 279–82.

59. Bani S, Kaul A, Jaggi BS, Suri KA, Suri OP, Sharma OP. Anti-inflammatory activity of the hydrosoluble fraction of Euphorbia royleana latex. *Fitoterapia.* 2000; 71(6): 655–62.

60. Wang S, Liang H, Zhao Y, Wang G, Yao H, Kasimu R, et al. New triterpenoids from the latex of Euphorbia resinifera Berg. *Fitoterapia.* 2016; 108: 33–40.

61. Daoubi M, Marquez N, Mazoir N, Benharref A, Hernández-Galán R, Muñoz E, et al. Isolation of new phenylacetylingol derivatives that reactivate HIV-1 latency and a novel spirotriterpenoid from Euphorbia officinarum latex. *Bioorganic & Medicinal Chemistry.* 2007; 15(13): 4577–84.

62. Yang D-S, Peng W-B, Li Z-L, Wang X, Wei J-G, He Q-X, et al. Chemical constituents from Euphorbia stracheyi and their biological activities. *Fitoterapia.* 2014; 97: 211–8.

63. Silva AC, de Faria DE, Borges NB, de Souza IA, Peters VM, Guerra Mde O. Toxicological screening of Euphorbia tirucalli L.: Developmental toxicity studies in rats. *Journal of Ethnopharmacology.* 2007; 110(1): 154–9.

64. de Oliveira LF, Fuentefria AM, Klein Fda S, Machado MM. Antifungal activity against Cryptococcus neoformans strains and genotoxicity assessment in human leukocyte cells of Euphorbia tirucalli L. *Brazilian Journal of Microbiology.* 2014; 45(4): 1349–55.

65. Cruz LS, de Oliveira TL, Kanunfre CC, Paludo KS, Minozzo BR, Prestes AP, et al. Pharmacokinetics and cytotoxic study of euphol from Euphorbia umbellata (Bruyns) Pax latex. *Phytomedicine.* 2018; 47: 105–12.

66. Mendonça RJ, Maurício VB, Teixeira Lde B, Lachat JJ, Coutinho-Netto J. Increased vascular permeability, angiogenesis and wound healing induced by the serum of natural latex of the rubber tree Hevea brasiliensis. *Phytotherapy Research*. 2010; 24(5): 764–8.

67. Adotey JPK, Adukpo GE, Opoku Boahen Y, Armah FA. A review of the ethnobotany and pharmacological importance of Alstonia boonei De Wild (Apocynaceae). *International Scholarly Research Notices*. 2012; 2012, 587160.

68. Gurung S, Skalko-Basnet N. Wound healing properties of Carica papaya latex: In vivo evaluation in mice burn model. *Journal of Ethnopharmacology*. 2009; 121(2): 338–41.

69. Kumar S, Dewan S, Sangraula H, Kumar VL. Anti-diarrhoeal activity of the latex of Calotropis procera. *Journal of Ethnopharmacology*. 2001; 76(1): 115–8.

70. Rajesh R, Raghavendra Gowda CD, Nataraju A, Dhananjaya BL, Kemparaju K, Vishwanath BS. Procoagulant activity of Calotropis gigantea latex associated with fibrin(ogen)olytic activity. *Toxicon*. 2005; 46(1): 84–92.

71. Parhira S, Yang ZF, Zhu GY, Chen QL, Zhou BX, Wang YT, et al. In vitro anti-influenza virus activities of a new lignan glycoside from the latex of Calotropis gigantea. *PLoS One*. 2014; 9(8): e104544.

72. Lucetti DL, Lucetti EC, Bandeira MA, Veras HN, Silva AH, Leal LK, et al. Anti-inflammatory effects and possible mechanism of action of lupeol acetate isolated from Himatanthus drasticus (Mart.) Plumel. *Journal of Inflammation*. 2010; 7: 60.

73. de Moura DF, Rocha TA, Barros DM, da Silva MM, de Lira M Dos Santos Souza TG, et al. Evaluation of the cytotoxicity, oral toxicity, genotoxicity, and mutagenicity of the latex extracted from Himatanthus drasticus (Mart.) Plumel (Apocynaceae). *Journal of Ethnopharmacology*. 2020; 253: 112567.

74. Soares DC, Andrade AL, Delorenzi JC, Silva JR, Freire-de-Lima L, Falcão CA, et al. Leishmanicidal activity of Himatanthus sucuuba latex against Leishmania amazonensis. *Parasitology International*. 2010; 59(2): 173–7.

75. de Miranda AL, Silva JR, Rezende CM, Neves JS, Parrini SC, Pinheiro ML, et al. Anti-inflammatory and analgesic activities of the latex containing triterpenes from Himatanthus sucuuba. *Planta Medica*. 2000; 66(3): 284–6.

76. Viana CA, Oliveira JS, Freitas CD, Alencar NM, Carvalho CP, Nishi BC, et al. Thrombin and plasmin-like activities in the latices of Cryptostegia grandiflora and Plumeria rubra. *Blood Coagulation & Fibrinolysis.* 2013; 24(4): 386–92.

77. de Alencar NM, Pinheiro RS, de Figueiredo IS, Luz PB, Freitas LB, de Souza Tde F, et al. The preventive effect on ethanol-induced gastric lesions of the medicinal plant Plumeria rubra: Involvement of the latex proteins in the NO/cGMP/K ATP signaling pathway. *Evidence-Based Complementary and Alternative Medicine.* 2015; 2015: 706782.

Chapter 2

Production and Function of Latex within Latex-Bearing Plants

Arya Lakshmi Marisetti,[a] Ajay Sharma,[a] Vijender Kumar,[a] Abhishek Tiwari,[b] Varsha Tiwari,[b] Vineet Mittal,[c] Saurabh Bhatia,[d,e] Ahmed Al-Harrasi,[d] and Laleh Khodaie[f,g]

[a]*Department of Pharmacognosy & Phytochemistry, School of Pharmaceutical Sciences, Delhi Pharmaceutical Sciences and Research University, Mehrauli-Badarpur Road, New Delhi, India*

[b]*Faculty of Pharmacy, Pharmacy Academy, IFTM University, Lodhipur Rajput, Moradabad, Uttar Pradesh, India*

[c]*Department of Pharmaceutical Sciences, Maharshi Dayanand University, Rohtak, Haryana, India*

[d]*Natural & Medical Sciences Research Center, University of Nizwa, Birkat Al Mauz, Nizwa, Oman*

[e]*School of Health Sciences, University of Petroleum and Energy Studies, Dehradun, Uttarakhand, India*

[f]*Department of Traditional Pharmacy, Faculty of Traditional Medicine, Tabriz University of Medical Sciences, Tabriz, Iran*

[g]*Department of Pharmacognosy, Faculty of Pharmacy, Tabriz University of Medical Sciences, Tabriz, Iran*

ajaysharmapharma1979@gmail.com

Introduction to Plant-Based Latex: Phytochemistry, Ethnopharmacology, Biological, and Pharmacological Activities
Edited by Laleh Khodaie, Ajay Sharma, Saurabh Bhatia, and Ahmed Al-Harrasi

ISBN 978-981-5129-12-0 (Hardcover), 978-1-003-57071-4 (eBook)
www.jennystanford.com

2.1 General Introduction

Latex is a milky fluid secreted by plants as a part of their defense mechanism. It is a suspension or emulsion comprising a biphasic system and is stored within the highly specialized latex-bearing cells called laticifers. The laticifers may vary with respect to their properties like distribution, appearance, origin as well as their anatomy. Laticifers are distributed extensively among different plant organs, including leaves, roots, fruits, stems and flowering tops. Chemically they are alkaloids, terpenoids, cardenolides, polyphenols, proteins, carbohydrates, tannins, resins and gums. The rubber obtained from *H. brasiliensis* is a polymer comprising isoprene units [1–4]. The color of latex in plant material is typically white, orange, cream or yellow. However, it is worth noting that the color and its intensity may undergo changes when exposed to air. The stability of latex may also be affected when exposed to air as latex is the dispersion comprising of polymeric microparticles [5]. Generally latex shows multiple colors that gets changed on exposure to air. Latex is stable in nature and coagulates when exposed to air. It comprises various catalytic enzymes such as cysteine proteases, profilins and chitin-related proteins [6] which provide defense against microbial infections [7, 8]. Plant latex serves as a defense mechanism, deterring herbivorous insects from feeding [9]. The plant latex may include potentially harmful chemicals that could cause hypersensitivity reactions in some people. Additionally, natural plant latex exhibits a remarkable variety of bioactive compounds, which contribute to its diverse biological activities. These activities include antiparasitic, antimicrobial, anti-carcinogenic, anti-inflammatory, antioxidant, insecticidal, and vasodilatory properties [10]. Plant latex has gained significant importance in the medical field and industrial applications. It is used in the production of pharmaceuticals, drugs, adhesives, films, nursing and other materials of diagnostic use. The versatile nature of plant latex makes it a valuable resource in the area of pharmaceuticals, agriculture as insecticides and pesticides as well as miscellaneous applications in the rubber and toy industries. With the advancement of technology, there is a rising opportunity to investigate plant latex as a key raw resource for the manufacture of home products and immunodiagnostic materials. This presents new opportunities for innovative applications across different sectors.

2.1.1 Geographical Distribution of Plant Latex

The production and distribution of latex among the different plant parts is a common defense mechanism found in plants to protect against natural enemies. A thorough examination of the existing literature reveals that latex-producing angiosperms have been documented in 40 different families, with an estimated number exceeding 20,000 species possessing some form of laticiferous structures. This count surpasses the commonly cited figure of 12,500 species. Tropical regions exhibit a greater abundance and proportion of latex-bearing plant families compared to temperate regions when it comes to distribution. This disparity is evident in both absolute numbers and relative proportions. The geographical distribution of latex is wide and it is distributed among the tropical and cosmopolitan families. This pattern of distribution may be attributed to a greater biodiversity of natural enemies and higher percentage of herbivores in tropical ecosystems. Geographical and climatic conditions affect latex production in latex bearing plants like *Hevea brasiliensis.*

The geographical distribution of latex-bearing plants among the plant families was compiled and revised by researchers [11, 12]. The study of evolutionary relationship of laticiferous along with resiniferous taxa has been reported previously [12–14]. Plant latex, consisting of various components, has wide-ranging applications in meeting human needs and desires. Over time, its uses have been discovered in diverse fields such as industry, healthcare, and production of vital items for human consumption. For example, natural rubber, the primary component derived from plant latex, is extensively utilized in the manufacturing of various everyday products. These products undergo processing within the rubber industry [15]. Additionally, plant latex has been employed for its medicinal properties by traditional tribal communities. It has been utilized for treating burns, facilitating wound healing, relieving joint pain, and addressing conditions such as worm infections and skin diseases [15]. In the field of medicine, dried latex from *Papaver somniferum* contains therapeutically important secondary metabolites like morphine, codeine, thebaine, papaverine and noscapine known for their analgesic, antitussive and vasodilator. It is worth noting, that plant latex can also be associated with the illegal production of substances like heroin [16]. Another derivative of latex, styrene, finds widespread application in immunoassays for the preparation of immunodiagnostic materials like ELISA plates [17]. Extracts of latex containing medicinal plants and phytoconstituents present in them exhibits activity against bacteria, virus, protozoa, fungus and used in cancer, inflammation and pain. Moreover, the integration of plant latex with nanoparticles, enhances their antimicrobial capabilities [18].

2.1.2 Production and Function of Latex in Latex-Bearing Plants

Latex is produced by more than twenty thousand plant species widely distributed among 43 families. The majority of these species are angiosperms, encompassing 41 families, while one family is found in ferns and one in gymnosperms. It is interesting to note that the occurrence of latex demonstrates instances of convergent evolution in phylogenetically unrelated plant orders. Laticifers, specialized cells responsible for latex production, have evolved in various clades including both monocotyledonous and dicotyledonous plants, spanning different evolutionary branches such as basal clades, magnoliids, monocots, basal eudicots, rosids and asterides.

Structure wise laticifers can be classified into two main categories: articulated and non-articulated laticifers. Both types of laticifers can elongate as the organ develops and form complex networks of tubes. Accurate classification of laticifer systems usually requires the examination of plant material containing embryos or meristems. The distinction between articulated and non-articulated morphotypes is typically based on ontogeny, involving the analysis of precursor cell numbers and the developmental stage of laticifers. The activity of proteolytic enzymes can also be helpful in distinguishing between the two types. Chitinases are specifically active in articulated laticifers. However, there have been instances where misclassification has occurred, as seen in mulberry, *Ficus montana* and *Maclura tinctoria L.* It is noted that different morphotypes may exist within different species of the same plant family making it difficult to rely solely on laticifer types as a definitive distinguishing character for specific taxa. It is now generally known that latex has a biological purpose in plants as a part of defense system against foreign invaders such as pathogens. It is important to remember that the function of latex was still unknown in 1989. Large vacuoles in latex contain a variety of metabolites that are produced when plant tissues are physically damaged. Some of these metabolites have poisonous and dissuasive properties. When plant tissues are mechanically disrupted, latex is rapidly released. The sticky and coagulative properties of latex create a protective barrier against pathogen invasion, while its quick coagulation and high viscosity can hinder the movement of herbivores and immobilize their mouthparts and sensory organs. This defensive mechanism gives latescent plants an edge, especially in situations with high levels of herbivory, such as tropical or subtropical forests [19].

Latex serves as a vital defense mechanism in plants, offering protection against herbivores. The latex of certain plants such as sandhill milkweed can effectively entrap and eliminate newly hatched caterpillars, particularly those of the monarch variety. Analyses of latex composition have revealed significantly higher concentrations of defense proteins and other substances compared to other plant tissues with levels ranging from 50 to 1000 times greater. While some latex chemicals may be poisonous to the parent plant. They are typically isolated within the plant structure and can even have negative effects on herbivores as well. Moreover, latex possesses remarkable coagulation properties. In species like *Cryptostegia grandiflora*, latex rapidly coagulates at the site of injury, minimizing sap loss and effectively trapping the mouthparts of insects due to its adhesive nature.

Furthermore, latex serves as a medium for the movement and storage of various essential nutrients within the plant including sugars, salts, alkaloids, tannins, enzymes and plant waste. It plays a role in regulating water concentration in different plant parts and facilitates the complex blending of substances like waxes, fats, resins and gums. The chemical constituents present in latex can circulate via plant circulatory system enabling the transportation of various substances between different regions. Additionally, latex functions as a reservoir for excreting plant waste products, serving as a solution for their elimination from the plant.

2.1.3 Articulated Laticifers

The laticiferous system, which is made up of cells known as laticifers, can evolve in two ways. The laticiferous system in many plant species develops from rows of cells generated in the stem or root meristem. The cell wall between these cells disintegrate, forming continuous tubes known as latex vessels. These vessels are known as articulated laticifers since they are made up of numerous cells. This manner of development is observed in several plant groups, including the poppy family, rubber trees (such as the Para rubber tree and certain members of the Euphorbiaceae, mulberry, and fig families), and some members of Asteraceae family. For example, some members of family Asteraceae like *Parthenium argentatum* and *Taraxacum kok-saghyz* are latex-producing species contains articulated laticifers.

2.1.4 Non-Articulated Laticifers

The milkweed and spurge families contain specialized types of laticifers cells. These cells differentiate during seedling development and expand throughout the entire plant. This results in the formation of the entire structure from a single cell in many euphorbs, commonly known as non-articulated laticifers. The laticiferous system in mature plants develops from unicellular or multicellular embryo. Non-articulated laticifers can be distributed throughout the plant parts and cortical tissues. The color of the latex produced by these cells are of different types like white liquid, transparent, yellow and crimson. This has been observed in the member of Cannabaceae, Euphorbiaceae and Papaveraceae family [20].

2.2 Biosynthetic Pathway

Biosynthesis is a complicated and detailed process in which living organisms convert basic molecules enzymatically into more complex products. This metabolic pathway involves the modification, conversion or joining of substrates to form macromolecules. Biosynthesis might take place within a single cellular organelle or with enzymes spread across many organelles. Some examples of biosynthetic pathways include the synthesis of lipid membrane components and the production of nucleotides. It is often synonymous with anabolism, the process of building up of molecules. Due to the defensive function of latex in plant development, it is understandable that the composition of latex can exhibit complexity and diversity. The specific constituents of latex can significantly differ based on various factors including species, developmental stage, external and internal stimuli and stress conditions. However, it is important to note that latex can be broadly classified as secondary metabolites and proteins, both of which possess bioactive properties. These compounds often exhibit cytotoxic effects and laticifers have evolved as specialized compartments to store and sequester these substances independently from the plant vascular system. This arrangement allows for a rapid and pre-existing defense mechanism in response to herbivory, providing an immediate means of protection. When a plant is damaged, the internal pressure of the latex plays a crucial role in initiating a rapid response. This pressure causes the prompt release of highly concentrated active substances at the site of injury, typically happening within a matter of seconds. This immediate release of active compounds allows for an instantaneous and localized defensive reaction to counteract the damage inflicted on the plant. When considering the composition of latex, it is important to highlight that certain constituents demonstrate a synergistic mode of action, working together in a cooperative manner. For instance, there is evidence of synergistic interactions between terpenes and phenolic compounds, as well as the collaboration of various proteins with defense role against fungi or insects. These cooperative actions have been observed in latex. The content of latex varies greatly depending on the plant species. There are numerous common classes of chemicals present in several types of laticifers. Phenolics, alkaloids, cardenolides, and terpenes are examples of these. These many chemical groups contribute to the specific features and functions of latex in various plants. It is worth noting that the specific types and quantities of these metabolites might vary substantially, reflecting the plant evolutionary adaptations and ecological interactions.

2.2.1 Alkaloids

Alkaloids are basic low-molecular-weight compounds contain nitrogen atoms generally present in the heterocyclic nucleus and possess specific physiological action on human body [21]. Plant alkaloids that are well-known include morphine, quinine, lobeline, sanguinarine, nicotine, and others (Fig. 2.1). The concentration and accumulation of alkaloids in latex-bearing plants depending on numerous factors such as plant parts, climatic and geographical conditions [22]. Latex-producing plants distributed among plant families like Papaveraceae, Apocynaceae, Campanulaceae, Euphorbiaceae and Moraceae are known for higher alkaloidal content. The discovery of the first alkaloid, morphine, dates back to 1804 when it was isolated from the latex of unripe capsule of *Papaver somniferum* belongs to Papaveraceae family [23]. Opium poppy latex is particularly rich in morphine, constituting approximately 5% and 25% of its fresh weight and dry weight respectively, including the poppy seeds [24, 25]. Codeine, papaverine, noscapine and thebaine are some of the other alkaloids contained in opium [26]. Morphine, as a phenanthrene opioid receptor agonist, demonstrates specific binding to activation of M-opioid receptor (MOR) inside the human central nervous system. Morphine primarily exhibits central nervous system depressant effects and act on gastrointestinal tract of vertebrates. However, MOR activation can also lead to associated symptoms such as euphoria, respiratory depression, sedation, dizziness, tolerance and physical dependence [27, 28].

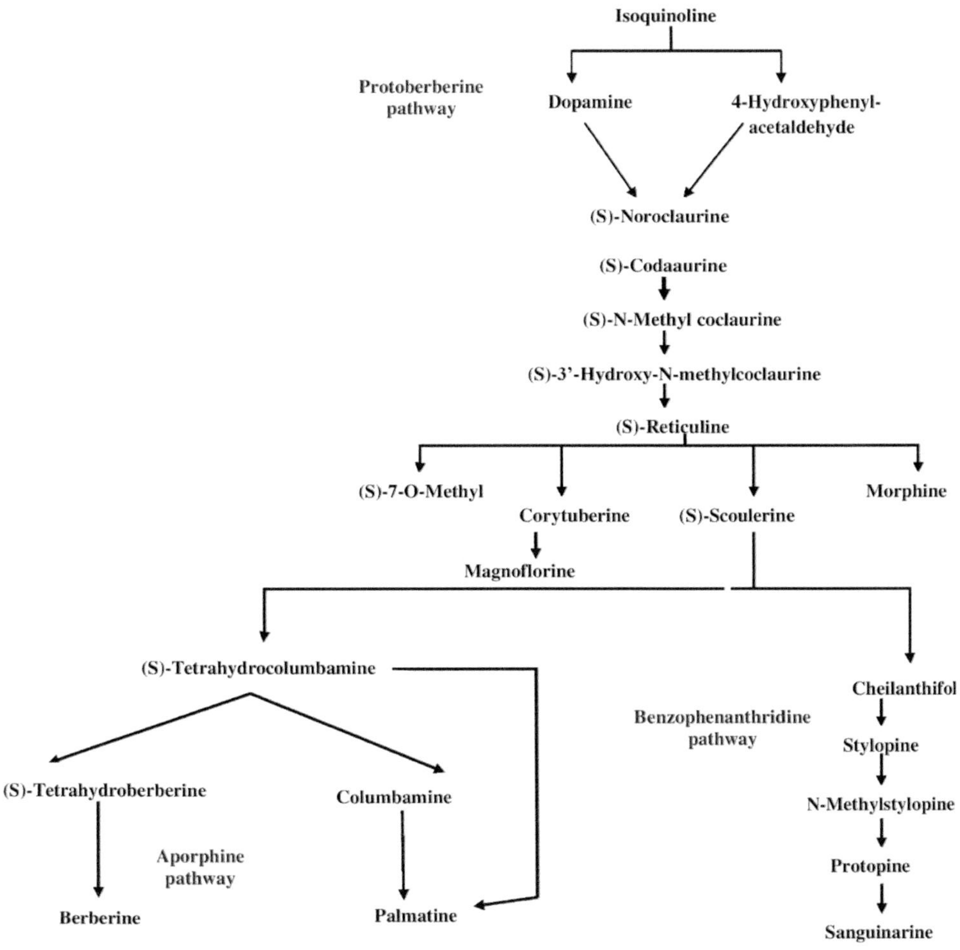

Figure 2.1 Biosynthetic pathway of benzyl isoquinoline alkaloids from latex-bearing plants.

Various plant latex also contains sanguinarine, which is another well-known alkaloid. It is reported that the thick red latex derived from the rhizomes of the bloodroot plant (*Sanguinaria canadensis*) mainly contains sanguinarine, along with other alkaloids like sanguilutine, allocryptopine chelerythrine and protopine. These alkaloids play a protective role for the plant, as sanguinarine exhibits toxicity towards both vertebrates and insects. It interferes with DNA synthesis, choline acetyltransferase function, and neurotransmission [29, 30]. Sanguinarine exhibits inhibitory effects on NF-kB3, which regulates various cellular processes including cell development, cell division and programmed cell death. The inhibition of NF-kB3 by sanguinarine can impact cellular signaling pathways and gene expression, potentially influencing cell growth, differentiation, and survival. This interaction between sanguinarine and NF-kB3 highlights the diverse biological activities of plant alkaloids in modulating cellular processes. The roots of *Papaver somniferum* are high in sanguinarine, where it serves as a deterrent to diseases and herbivorous insects [31]. Sanguinarine has antiviral, anti-inflammatory, antimicrobial and cytotoxic actions [32, 33].

Lobeline is obtained from the leaves and cardinal flowers of *Lobelia cardinalis* of Campanulaceae family (Fig. 2.2) [34]. It is water soluble white, amorphous powder. Lobeline pre-dominantly affects serotonin and dopamine uptake by acting on nicotinic acetylcholine receptors [35]. The ingestion of lobeline can lead to a range of symptoms in vertebrates, such as increased blood pressure, coughing, nausea, vomiting, tremors, diarrhea, mental confusion, weakness and seizures. Reports have indicated that lobeline possesses a narrow therapeutic index, indicating that its margin of safety is very less [36]. In plant latex, alkaloids, along with other secondary metabolites, have multiple functions. They help to speed up plant reproduction and act as defense systems against both biotic and abiotic stressors. These compounds target neurotransmission and/or provide antioxidant activity to facilitate damage repair [37]. Alkaloids are essential in defending plants against herbivores by exhibiting characteristics such as a bitter taste, disturbance of protein processes during digestion or metabolism, and disruption of the central nervous system (CNS) [38]. Moreover, alkaloidal latex serve the crucial function of attracting flower visitors through their stimulating properties. This process significantly raise the plant chances of surviving. These plant alkaloids also exhibit a broad spectrum of biological actions that are advantageous for human therapies. They possess antioxidant, antitumoral, anti-inflammatory and stimulant properties [39].

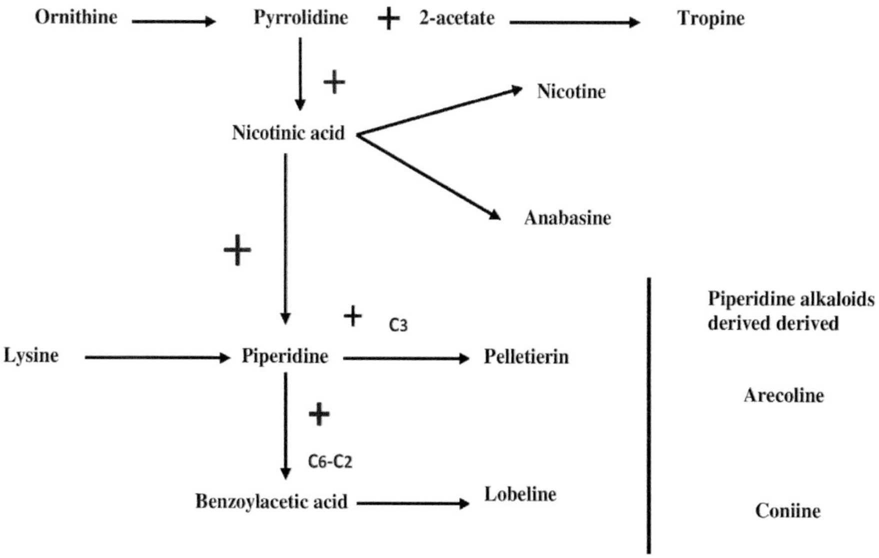

Figure 2.2 Biosynthetic pathway of tropane and piperidine alkaloids from latex-bearing plants (Datura and Lobelia species).

2.2.2 Terpenoids

Terpenoids are the organic compounds distributed in plant latex and are made up of five carbon atom isoprene units. Terpenoids can be monoterpenes, diterpenes, sesquiterpenes and polyterpenes. Rubber is an example of polyisoprene.

Mevalonic acid (MVA) pathway (Fig. 2.3) and 1-deoxy-D-xylulose-5-phosphate (DXP) pathway (Fig. 2.4) are involved in the biosynthesis of sesquiterpenes, triterpenes, monoterpenes, diterpenes and tetraterpenes.

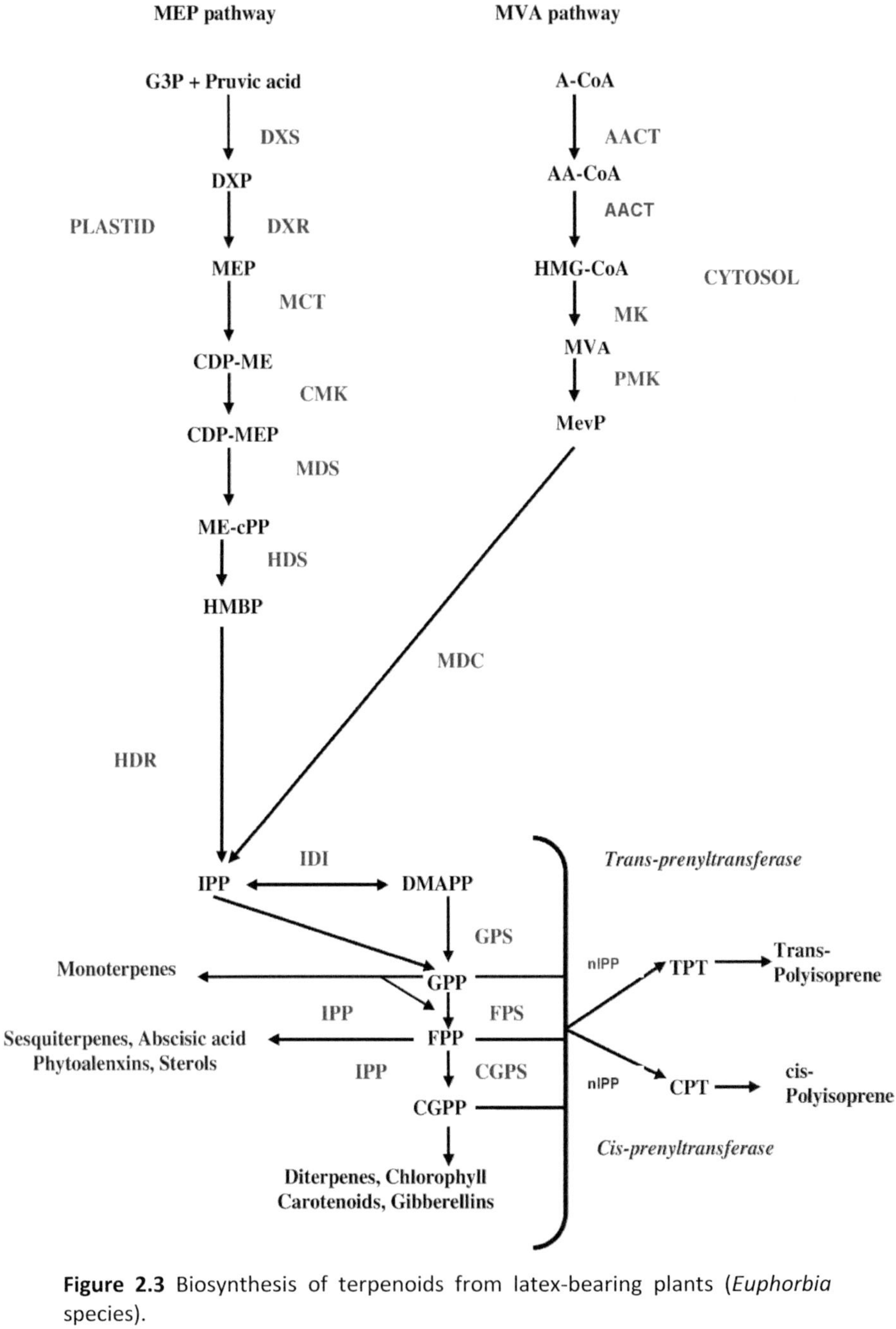

Figure 2.3 Biosynthesis of terpenoids from latex-bearing plants (*Euphorbia* species).

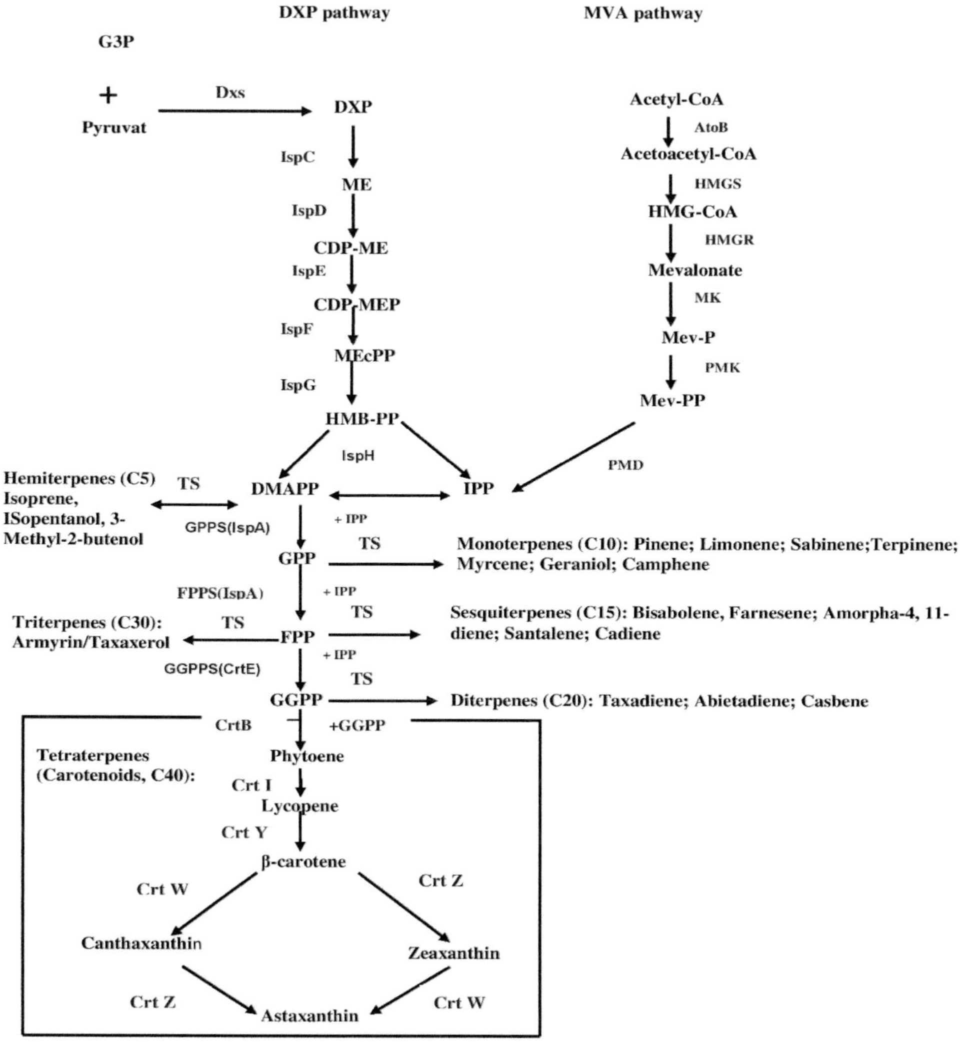

Figure 2.4 Biosynthesis of isoprenoid compounds from latex-bearing plants.

Abbreviations: DXP, 1-deoxy-D-xylulose 5-phosphate; ME, 4-Diphosphocytidyl-2-C-methyl-D-erythritol; CDP-ME, 4-Diphosphocytidyl-2-C methylerythritol; MEcPP, Methylerythritol cyclodiphosphate; HMB-PP, (E)-4-hydroxy-3-methyl-but-2-enyl pyrophosphate; DMAPP, Dimethylallyl pyrophosphate; GPP, Geranyl pyrophosphate; FPP, Farnesyl pyrophosphate; GGPP, Geranylgeranyl pyrophosphate; IPP, Isopentenyl pyrophosphate; HMG-CoA, 3-Hydroxy-3-methylglutaryl-coenzyme A reductase.

Terpenoids are widely distributed in plant species, with more than 50,000 terpenoids are identified so far. Sesquiterpenoids are a specific class of terpenoids that produce latex. They serve as defense mechanisms, acting as anti-feedants to deter herbivores, particularly locusts, through compounds like lactucopicrin and 8-deoxylactucin found in sesquiterpene lactones. Sesquiterpenoids also exhibit antimicrobial properties against plant pathogens such as *Cladosporium* species. *Lactuca sativa* possesses sesquiterpene lactones like lactucin, which are predominantly localized in vacuoles. Another group of terpenoids are diterpenes, with phorbol groups and are obtained from latex of *Euphorbia* species. In summary, terpenoids, including sesquiterpenoids and diterpenes, play an important role in plant defense mechanisms and inducers of specific behaviors in herbivorous insects.

2.2.3 Phenolics

Phenolics are secondary metabolites generated by many plant species. Polyphenols and phenolic acids are the two main classes comprising of polyhydroxy substituted aromatic ring [40]. Flavonoids, lignans, coumarins, anthraquinones, lignin, phenylpropanoids, stilbenes, and tannins are a few examples of phenolic chemicals. These substances may be water soluble, soluble in organic solvents, or entirely insoluble [41].

Studies conducted previously indicated that the number of identified plant phenolic compounds exceeds 8000, with flavonoids representing the largest subgroup within this class [42]. Flavonoids are low molecular weight phenolic compounds with fifteen carbon atoms and the configuration of C6-C3-C6. Flavonoids contain two aromatic rings and two heterocyclic rings with an oxygen atom. Shikimic acid, mevalonic acid and malonic acid pathways are involved in the biosynthesis of flavonoids [43]. The polypropanoid process is used to create flavonoids from phenylalanine molecules (Fig. 2.5). They are further classified as flavan-3-ols, flavonols, flavanones, flavonones and isoflavones [44].

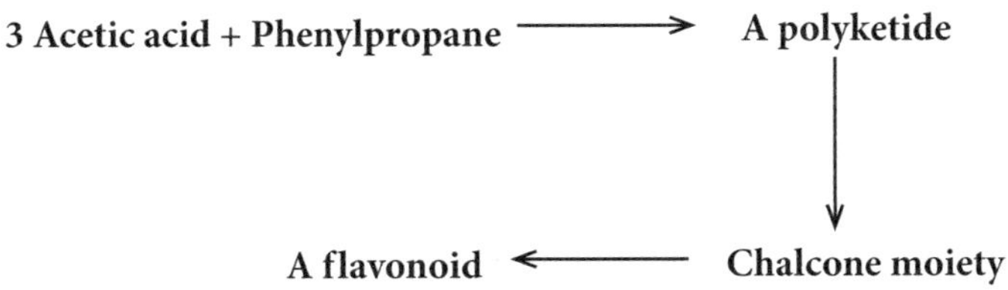

Figure 2.5 Biosynthesis of flavonoid molecules from latex-bearing plants.

Tannins are high molecular weight phenolic compounds where the phenyl ring has been substituted by polyhydroxy groups [45]. On the basis of nature and properties tannins can be categorized as hydrolysable and condensed tannins. Other varieties of tannins include phlorotannins, labiatae tannins, and coffee tannins [46]. Plant latex contains substantial levels of phenolic chemicals. Flavonoids like gallic acid, quercetin, and rutin in *Hevea brasiliensis* have been reported by earlier researchers [47, 48]. *Ipomoea batatas,* a member of the Convolvulaceae family, is a sweet potato whose latex is well recognized for having a high concentration of phenolic substances. Compared to other sections of the sweet potato, including the vine (3%) the latex from the roots has a larger content (10%) [49]. Additionally, studies conducted previously found that weevils are more vulnerable to high concentrations of coumarates [50]. The latex of the Euphorbiaceae family member *Jatropha curcas* contains *flavonoids* and tannins, which aid in larvicidal and antibacterial actions [51]. Plants produces phenolic compounds for a variety of purposes like antioxidants, free radical scavengers, UV protection and maintain homeostatic balance of the body [52]. Phenolic compounds can also function as attractants for pollinators aiding in plant reproduction or as repellents to deter pests [53]. In addition to their defensive role, phenolic compounds, including tannins and phytoalexins, provide mechanical support to latex-producing plants [54]. Furthermore, lignin acts as a structural polymer found in the latex of vascular plants and is a major component of plant cell wall [55].

2.2.4 Proteases

Plant latex contains proteins and enzymes, including proteases, chitinases, and oxidases, as described in recent literature [56]. These proteins and enzymes serve various functions in plant latex, such as signal transduction, oxidative defense, and protection against natural enemies (Fig. 2.6). Among them, proteases, also known as proteinases or peptidases, are highly prevalent and diverse in plant latex. In the early 1900s, extensive studies were conducted to isolate and characterize numerous plant latex proteases, aiming to understand their proteolytic activities. These proteases are capable of hydrolyzing peptide bonds in various protein substrates, including collagen, fibrin, fibrinogen, and others. Proteases can be classified into two main categories: endopeptidases and exopeptidases. Endopeptidases catalyze the hydrolysis of peptide bonds within the protein, while exopeptidases act on the carboxyl or amino ends of the protein [57]. The cysteine and serine protease families are majority of proteases that may be discovered in latex-bearing plants. Serine proteases and cysteine proteases are found in the several plant families. Specific plant latex species have lower prevalences of aspartate and metalloproteases [58].

2.2.5 Cardenolides

Cardenolides, also referred as cardenolide glycosides or cardiac glycosides, are steroid compounds present in plant latex (Fig. 2.7). They contain sugar, lactone moieties, and steroidal nucleus with 5- or 6-membered lactone rings typically positioned at C17 position. Cardenolides are known for their ability to induce cardiac arrest. This toxicity is because of their ability to inhibit Na^+ or K^+ ATPases enzymes. This enzyme is crucial for maintaining cellular electrical potential and controlling ion concentration in both animals and insects. Specifically, cardenolides disrupt neurotransmission by binding to the Na^+-K^+ ATPases enzyme. However, certain insect species are tolerant to cardenolides. This tolerance is attributed to specific amino acid substitutions occurring in the cardenolide-binding sites of their Na^+-K^+ ATPases enzymes.

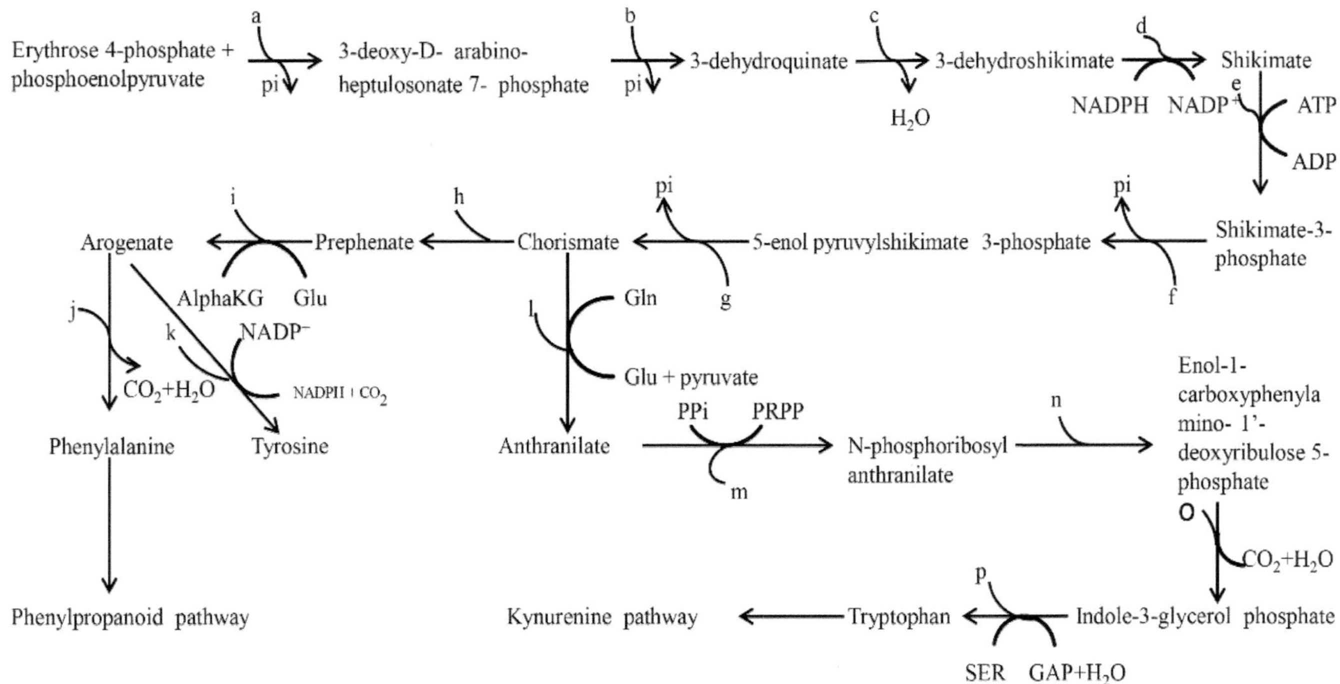

Figure 2.6 Production of amino acids in latex-bearing plants by Shikimic acid pathway. *Abbreviations*: NADPH, Nicotinamide adenine dinucleotide phosphate; NADP, Nicotinamide adenine dinucleotide phosphate; ATP, Adenosine triphosphate; ADP, Adenosine diphosphate; Alpha KG, ketoglutarate; Glu, Glutamic acid; Gln, Glutamine; PRPP, Phosphoribosyl pyrophosphate synthetase 1; GAP, GTPase activating proteins; SER, Smooth endoplasmic reticulum.

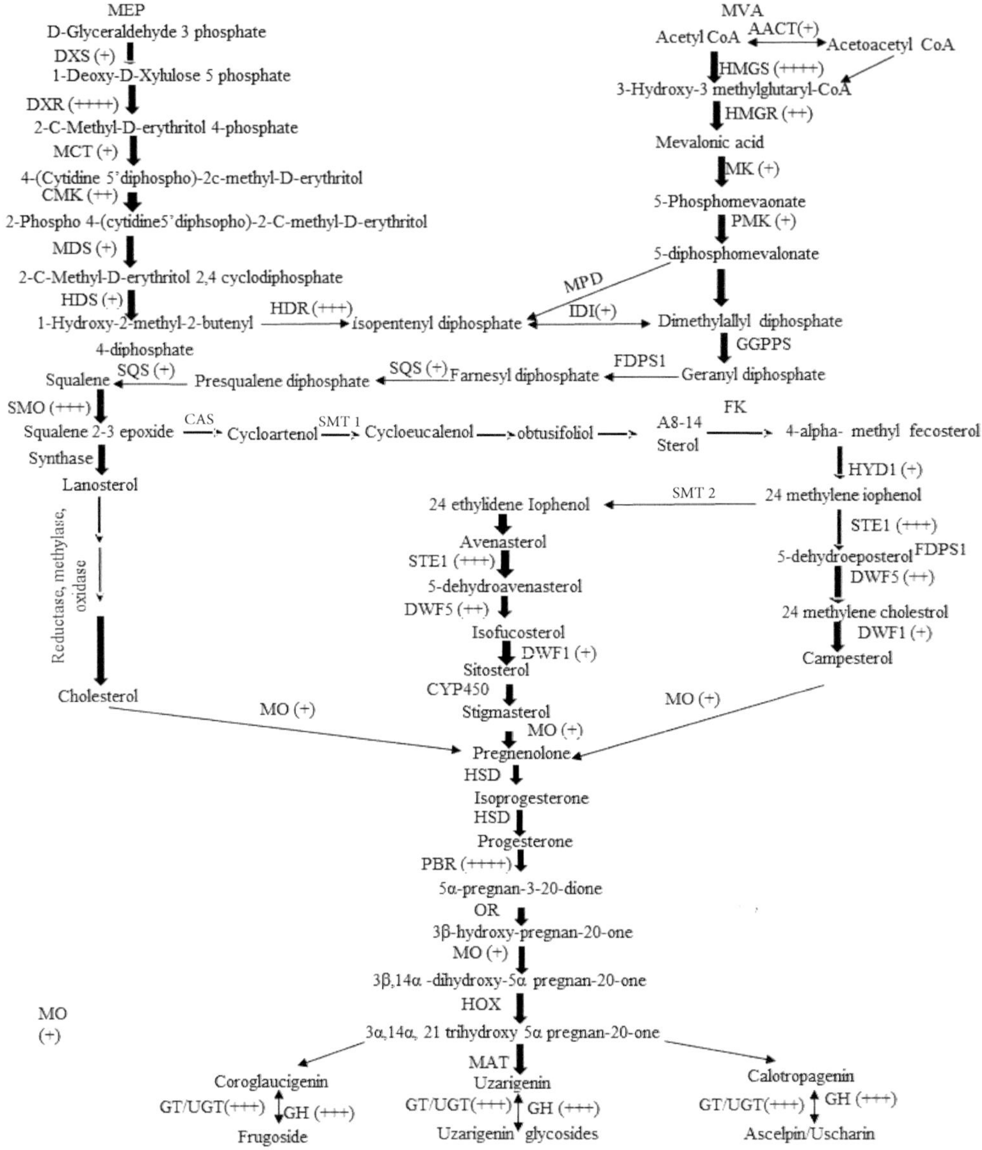

Figure 2.7 Cardenolides biosynthesis in latex-bearing plants, *C. procera* (*Continued*).

Figure 2.7 (Continued) *Abbreviations*: MCT, 2-C-methyl-D-erythritol 4-phosphate cytidyltransferase (ISP); CMK, 4-Diphosphocytidyl-2-C-methyl-D-erythritolkinase (4-diphosphocytidyl-2C-methyl-D-erythritol synthase); MDS, 2-C-methyl-D-erythritol 2,4-cyclodiphosphate synthase; HDS, 4-Hydroxy-3-methylbut-2-en-1-yl diphosphate synthase; HDR, 4-Hydroxy-3-methyl-but-2-enyl diphosphate reductase; AACT, Acetyl-CoA C-acetyltransferase; HMGS, Hydroxymethylglutaryl-CoA synthase; HMGR, Hydroxymethylglutaryl-CoA reductase; MK, Mevalonate kinase; PMK, Phosphomevalonate kinase; MPD, Diphosphomevalonate decarboxylase; IDI, Isopentenyl-diphosphate delta-isomerase; SQS1, Squalene synthase; SMO, Squalene monooxy-genase; CAS1, Cycloartenol synthase 1; SMT1, Sterol 24-C-methyltrans-ferase; CPI1, Cyclopropyl Isomerase; CPI1; Cycloeucalenol cycloisomerase; CYP51G1, Cytochrome P450, family 51, subfamily A (sterol 14-demethylase); FK, Delta14-sterol reductase; HYD1, Cholestenol delta-isomerase; SMT2, Sterol 24-C-methyl transferase-2.

Within the plant latex, cardenolides primarily act against herbivorous insects and pests and have been distributed in the latex of members of Apocynaceae family such as milkweed (*Asclepias* spp.) [59]. Cardenolides are primarily concentrated in the latex. However, they are also moderately accumulated in the leaves of numerous plant species. In the case of *A. californica*, the latex contains lower levels of cardenolides compared to the leaves. Furthermore, Novel nor-cardenolides, known as toxicariosides, have been reported from the latex of *Antiaris toxicaria*, belonging to the family Moraceae [60]. These compounds have historically been utilized as poisons for darts or arrows as reported earlier [56]. Details of biosynthetic pathways of phytoconstituents present in latex is given below.

2.2.6 Factors Affecting Production of Latex

The production of latex in the plant material is affected by several extrinsic and intrinsic factors. Some of the important factors affecting latex production from *H. brasiliensis* are climatic factors, and plantation management. The soil nutrients, nitrogen, potassium, phosphorus, and magnesium, have great effects on latex production. In addition, diseases and pests, tapping system and technology, plant varieties, plant age, and management of rubber plantation affect latex yield.

It has been reported that latex production from *Jatropha curcas* L. (*J. curcas*) is significantly influenced by climatic conditions, crown diameter, fruit production, and plant height. Hence, it is recommended to perform the latex extraction in post-harvest fruit periods. Similarly, morphology of the plant material affects latex production from the members of Euphorbiaceae family (*H. brasiliensis* and *J. curcas*). In *H. brasiliensis* species with articulated laticiferous vessels produce more latex (Mesquita and Oliveira 2010) while latex production is less in *J. curcas* L., plant with articulated anastomosing and non-anastomosing laticifers [60, 61].

There are numerous intrinsic factors like proteins, enzymes, and secondary messengers that affect the production of latex from latex-bearing plants. For example, the low-molecular-weight PR proteins from *Chelidonium majus* and *H. brasiliensis* are induced by phytopathogens and salicylic as well as jasmonic acid which are considered to be the signaling molecules. The genes involved in transcription and protein synthesis were up-regulated in SY107, high latex yielding variety [62–66].

Similarly, biotic elicitation, abiotic elicitation, precursor feeding, callus and suspension culture are the techniques that affect latex production from the plant material.

Temperature affects the growth and production of alkaloids from *Papaver bracteatum* Lindl. and *P. somniferum* L. The in vitro grown plantlets were obtained from seed explants grown on MS basal medium (BM). The best growth was observed at 18.5 and 20°C. The influence of vessel size on the growth of multiple shoots was studied and it was maximum in an automated plant culture system (APCS). The net production of alkaloids was higher in large vessels. The alkaloidal content of *P. somniferum* L (morphine, thebaine and codeine) was higher than *Papaver bracteatum* Lindl [67].

It is reported that cultivation and medicinally important secondary metabolites (tetrahydrocannabinol, cannabidiol, and cannabigerol) accumulation was affected by different light treatments. The effect of light intensity and wavelength on Cannabinoids accumulation from *C. sativa* is presented in Table 2.1 [61].

Table 2.1 Effect of light intensity on cannabinoid production from *C. sativa*

Wavelength of light	Intensity of light	Cannabinoids	Refs
UV-B	6.7 and 13.4 kJ m^{-2}	Tetrahydrocannabinol	[68]
Full-spectrum LEDs	900 μmol m^{-2} s^{-1}	No effect	[69]
Solar radiation (1,200 m ASL)	—	Cannabidiol	[70]
Blue & Red (450 + 650 nm)	90 μmol $m^{-2}s^{-1}$	Tetrahydrocannabinol, cannabigerolic acid	[71]
450 + 630 nm	450 μmol $m^{-2}s^{-1}$	Cannabidiol, tetrahydrocannabinol	[72]
540 + 670 nm	450 μmol $m^{-2}s^{-1}$	Cannabidiol, tetrahydrocannabinol, Cannabigerol, tetrahydrocannabivarin	[72]
440 + 660 nm	50–500 μmol $m^{-2}s^{-1}$	Cannabigerolic acid, tetrahydrocannabinol	[73]
440 + 660 nm	50–500 μmol $m^{-2}s^{-1}$	Sesquiterpene alcohol, i.e., cis-nerolidol	[73]
440 + 530 + 660 nm	50–500 μmol $m^{-2}s^{-1}$	Monoterpenes: α-pinene, limonene, myrcene Sesquiterpene alcohol: cis-nerolidol Terpene alcohol: Linalool	[73]
1,200 m ASL	130 m ASL	Monoterpenes: α-pinene, limonene, myrcene	[70]
450 + 650 nm	180–200 μmol $m^{-2}s^{-1}$	Terpenes	[71]

Note: KJ, kilo joules, μMol, micro mol.

2.3 Biotic Interactions Contribution to the Production of Latex

Elicitors are substances that when used in small amounts in a biological system may enhance secondary metabolites production and this process is known as elicitation. Elicitor affect several physiochemical processes and phytoalexin accumulation. On the basis of their nature the elicitors may be abiotic or biotic, and depending on their 'origin,' the elicitors may be exogenous and endogenous.

Biotic elicitors are of biological origin and are released by plants or microorganisms. They are directly secreted by microorganisms as enzymes or cell wall fragments formed as a result of microbial activity on plant cell walls [74].

Exogenous biotic elicitors are formed as a result of microbial metabolism or by interaction between phytoenzymes and microbial cells (e.g., chitin and glucanes).

Endogenous biotic elicitors are formed as a result of degradation of plant cell wall as well as their intracellular molecules by the microbial enzymes. These are jasmonic acid and salicylic acid derivatives [75, 76].

The effect of biotic elicitors like *Chaetomium globosum, Aspergillus niveoglaucus, Paecilomyces lilacinus* and *Trichoderma harzianum* fungus on the cell suspension culture of *Papaver somniferum* has been studied. The optimized concentration of biotic elicitors was added to suspension culture of *Papaver somniferum*. The mixture was incubated for a period of time. The samples were analyzed for growth and sanguinarine content as determined by liquid chromatography techniques. Maximum biomass accumulation and sanguinarine production were also obtained with *T. harzianum* culture. It was further increased with the combination of *T. harzianum* culture and shikimic acid. This combination elicits genes, like cheilanthifoline synthase, stylopine synthase, tetrahydroprotoberberine cis-*N*-methyltransferase and protopine 6-hydroxylase (P6H), involved in the biosynthesis of sanguinarine. Agitation speed, impeller type, media volume, inoculum weight and culture duration affect sanguinarine production [27].

It is reported that *Papaver armeniacum* hairy root culture elicited by methyl jasmonate and salicylic acid affects the production of morphinan metabolites. This may be due to the upregulation of *SalAT*, *COR*, and *T6ODM* genes involved in their biosynthesis, respectively. Similarly, addition of 100 micromoles of salicylic acid to the hairy root culture of *Papaver armeniacum* resulted in a maximum accumulation of papaverine and noscapine. This may be due to the upregulation of *DBOX* and *BBE* genes such as phenylalanine and methyl jasmonate [77].

A combination of a precursor and elicitors affects the production of secondary metabolites from cell suspension culture of *C. sativa.* This increase expression of genes linked to elicitors like phenylalanine and tyrosine. After treatment, the activity and expression of phenylalanine ammonia-lyase and tyrosine aminotransferase gene increased. As a result, the accumulation of phenolic compounds increased up to 82% compared with untreated cells [78].

The content of papain enzyme is increased using MS media supplemented with *S. cerevisiae*. The papain content in papaya callus was maximum by adding 25 ppm of *S. cerevisiae* to MS media [79]. The effect of arachidonic acid, salicylic acid, methyl jasmonate and harpin protein on accumulation of carotenoid and polyphenolic compounds from *Lactuca sativa* was studied. After treatment with these elicitors, the phenolic and flavonoid content were increased in green and red lettuce while carotenoids were increased in red lettuce. The best results were reported with methyl jasmonate, which induced lycopene beta cyclase gene expression [80].

2.3.1 Plant Growth Regulators

Organic substances that are actively present in plants and control the growth and development of the cells, tissues, and organs of plants when used in small amount are known as plant growth regulators or plant hormones or phytohormones. Additionally, they take part in the precursor biosynthesis, which is the process by which primary and secondary plant metabolites are formed.

Plant growth hormones are classified into many classes that have special role in plant tissue culture such as auxins, cytokinins, gibberellic acid, abscisic acid, ethylene [81]. Phytohormones act as chemical messengers for intercellular communication and are working in a synergistic way to stimulate plant growth and development. They work as secondary messengers and elicit metabolic pathways that leads to biosynthesis of secondary metabolites. The description of individual plant hormones and their effect is given below.

2.3.2 Auxins

Auxins are the derivative of acetic acid. The main plant auxin is indole-3-acetic acid (IAA). Other examples are indole butyric acid (IBA), naphthoxy acetic acid (NAA), 2,4-dichlorophenoxy acetic acid (2,4-D), and 2,4,5-trichlorophenoxyacetic acid (2,4,5-T). The effect of auxin on growth and differentiation of latex-producing cells was reported earlier [82]. Reports indicated that auxin response factor genes (HbARF2 and HbARF16.3) are the negative regulators for trunk growth of *H. brasiliensis* and affect latex production. Similarly, auxin improves latex yield from *H. brasiliensis* by regulating the action of two aquaporins genes, i.e., HbPIP2 and HbTIP1 [83].

2.3.3 Cytokinins

Cytokinins are the derivatives of purine. Some of the important cytokinins are kinetin, zeatin, benzyl amino purine, isopentyladenine (iP). They are the antagonist of auxins. Suri and Ramawat (1995) [84] studied the effect of cytokinin on laticifer differentiation. He reported that kinetin is more effective for differentiation of this structure than other cytokinins. The impact of cytokinins on cannabinoid content from the latex of *Cannabis sativa* has been studied. It is observed that cannabinoid content is not significantly improved by a combination of optimized ratio of auxin and cytokinin. However, it significantly affect plant morphology of plant [85].

2.3.4 Gibberellins

Gibberellins are compounds derived from gibberellic acid. There are more than 80 gibberellins found in plants that are essential for their growth and development, with GA3–GA7 being the most important. Rubber tree seedlings benefit from this as they sprout and grow [86]. Similarly, gibberellic acid (GA_3), abscisic acid and light improves germination of *Parthenium argentatum* [87]. Gibberellic acid improves morphine production from *Papaver somniferum* L. latex. The activity of the enzyme 1-deoxy-D-xylulose 5-phosphate synthase is reduced in both male and female plants by gibberellic acid (GA3). Consequently, the levels of chlorophylls, carotenoids, tetrahydrocannabinol and an elevation in the levels of α-tocopherol has been reported. Additionally, 3-hydroxy-3-methylglutaryl coenzyme A reductase (HMGR), an enzyme, becomes more active when GA3 is present. Elevated level of HMGR stimulate the accumulation of steroidal moieties such as stigmasterol and β-sitosterol in both male and female plants, whereas campestrol is accumulated only in male plants [88].

2.3.5 Abscisic Acid (ABA)

Chemically, abscisic acid (ABA) is a terpenoid involved in seed germination, protein synthesis and regulating stress. It promotes seed development and stomatal closure and helps in somatic embryogenesis during maturation and germination. Latex production from *H. brasiliensis* tree treated with abscisic acid leads to a significant increase in latex yield which has been regulated by abscisic acid signaling pathways. In *H. brasiliensis* the HbSRPP gene involved in latex production is regulated by abscisic acid. Similarly, TbSRPP gene involved in latex production from *Taraxacum brevicorniculatum* was also regulated by abscisic acid [83, 89].

2.3.6 Ethylene

Ethylene is a gaseous hydrocarbon with the chemical formula $H_2C{=}CH_2$. Ethylene significantly affects the development of roots and shoots. It may influence plant growth and development, cell expansion, fruit ripening and promote abscission of leaves. The production of latex from *Hevea brasiliensis* was affected by ethylene [90]. The effect of ethylene on gene involved in latex production, regeneration of laticifers and coagulation of latex is reported by earlier researchers. Hevein, a lectin like gene involved in latex coagulation while glutamine synthetase, an enzyme involved in protein metabolism and regeneration of laticifers. Ethylene stimulates latex production from *Hevea brasiliensis* by upregulating the activity of glutamine synthetase and mRNA level [91, 92].

2.3.7 Jasmonic Acid and Methyl Jasmonate

In response to biotic and abiotic stress, plant growth, development and defense mechanisms are regulated by jasmonic acid (JA) and its derivatives, which are regarded as secondary messengers [93]. The optimized concentration of methyl jasmonate caused a significant increase in the thebaine content in the roots of *Papaver bracteatum* Lindl. This is due to over expression of berberine bridge enzyme (BBE) and norcoclaurine synthase (NCS) genes [94]. It is also observed that methyl jasmonate significantly increase cannabinoid production from *C. sativa* by regulating the expression of cannabinoid synthase genes [95].

2.3.8 Brassinolides

Brassinolide is a plant hormone isolated from rapeseed, *Brassica napus*. It promotes stem elongation and cell division. Brassinolide decreases laticifers by upregulation of genes involved in xylem development in high wood yield *H. brasiliensis* [96].

2.3.9 Salicylic Acid

Salicylic acid is considered as a plant growth regulator. It is reported that phytohormones SA and MeJA can enhance the growth and biomass production of *C. sativa.* This leads to concentration-dependent accumulation of cannabidiol and tetrahydrocannabinol in leaves and inflorescences of *C. sativa* [97].

Production of benzylisoquinoline alkaloids from the hairy root culture of *Papaver armeniacum* L. was elicited by salicylic acid. It is observed that SA resulted in the maximum accumulation of papaverine and noscapine by upregulating dihydrobenzo phenanthridine oxidase (DBOX) and berberine bridge enzyme family (BBE) gene. *Thevetia peruviana* cell suspension culture experienced an accumulation of phenolic compounds due to the combined effects of salicylic acid and jasmonic acid [98].

2.3.10 Systemin

Systemin is the only peptide possessing hormone-like activity in plants [99]. Systemin affects secondary messenger JA defensive genes [100].

Examples of phytohormones and their structure are given in Table 2.2.

Table 2.2 Phytohormones and their structures

No.	Phytohormones	Structures
1.	Indole-3-acetic acid (IAA)	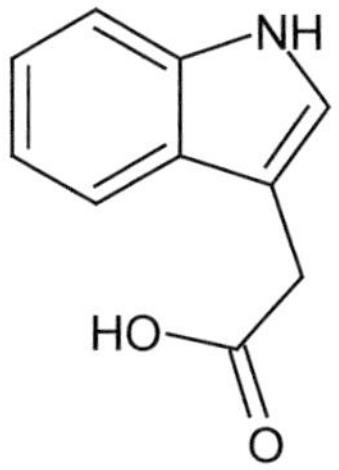
2.	Indole-3-butyric (IBA)	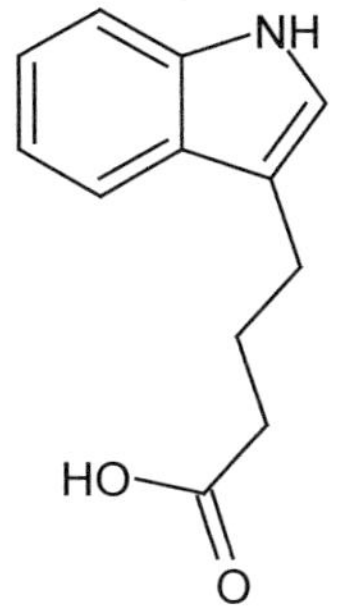

(*Continued*)

Table 2.2 (*Continued*)

No.	Phytohormones	Structures
3.	1-Naphthalene acetic acid (NAA)	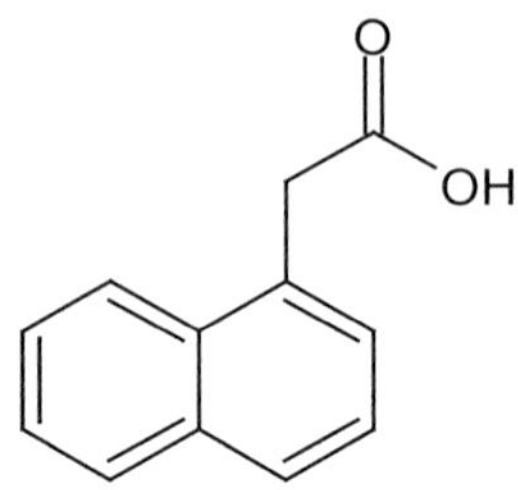
4.	2,4-Dichlorophenoxyacetic acid (2,4-D)	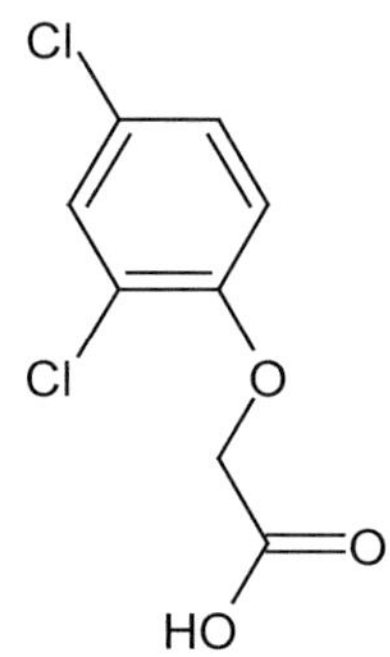
5.	2,4,5-Trichlorophenoxyacetic acid (2,4,5-T)	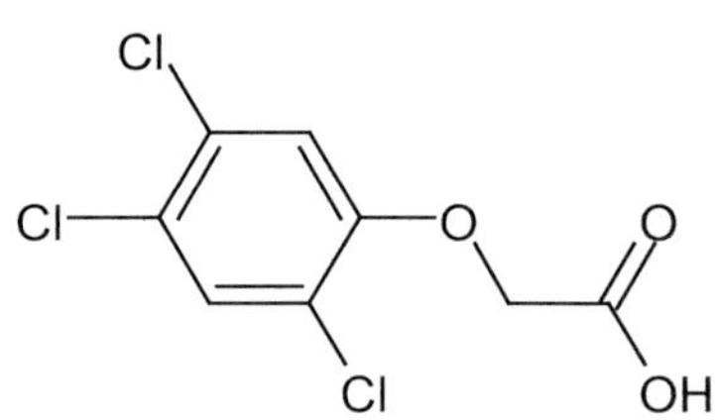
6.	6-Benzylaminopurine (BAP)	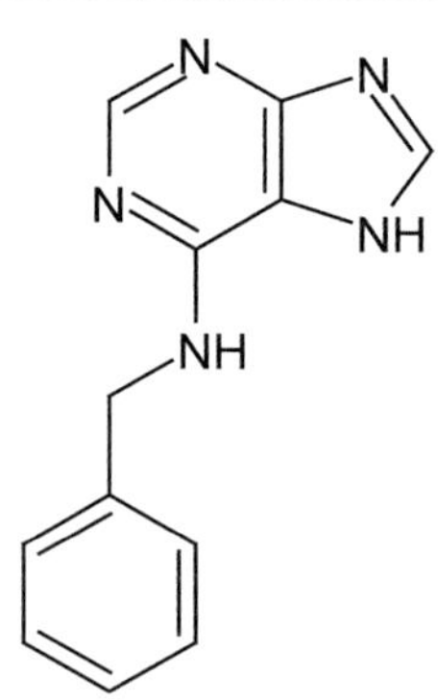

No.	Phytohormones	Structures
7.	Kinetin (Kn)	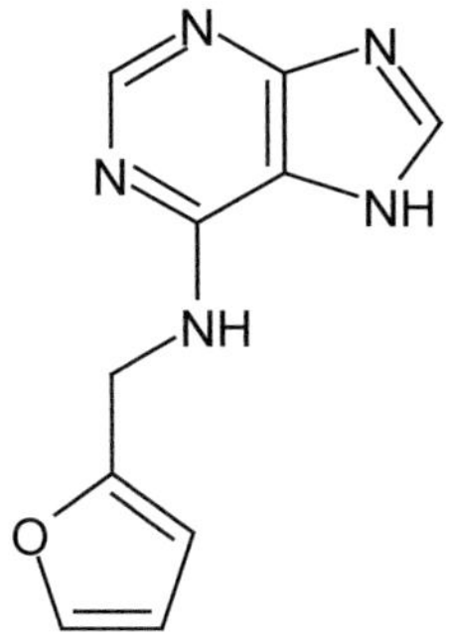
8.	Zeatin	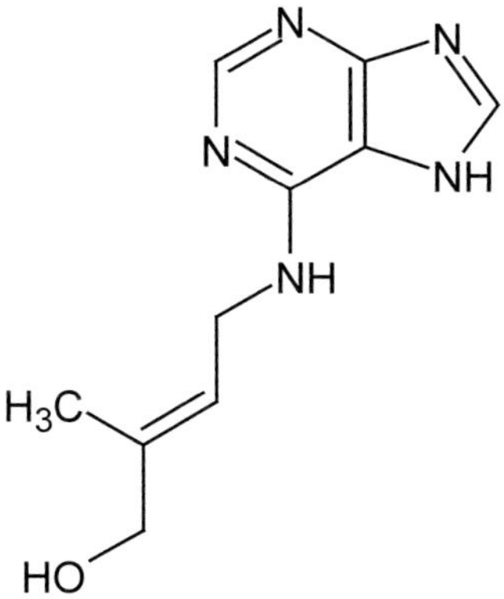
9.	Benzyl adenine	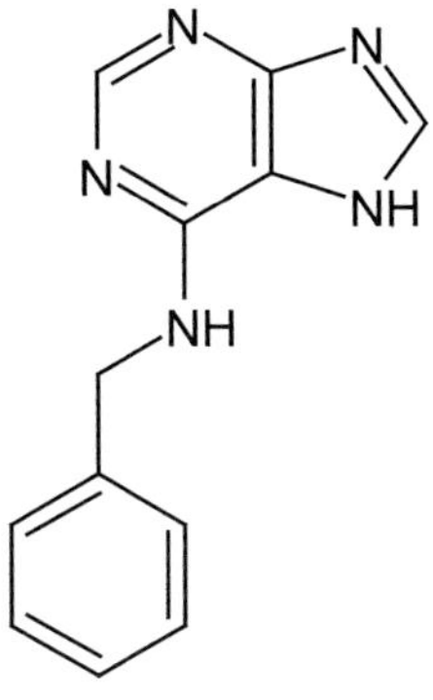

(*Continued*)

Table 2.2 (*Continued*)

No.	Phytohormones	Structures
10.	Gibberellic acid (GA_3)	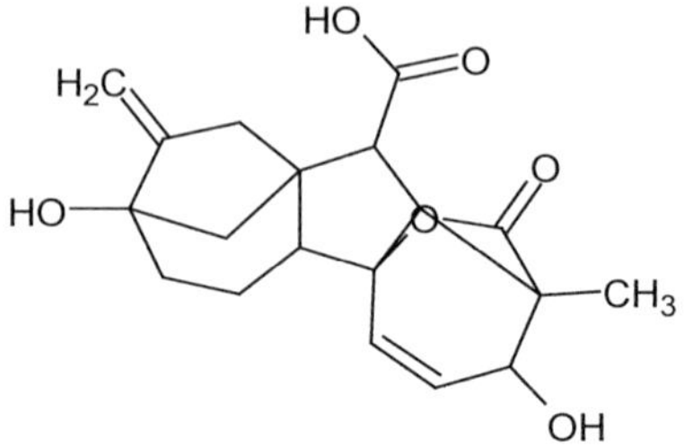
11.	Abscisic acid	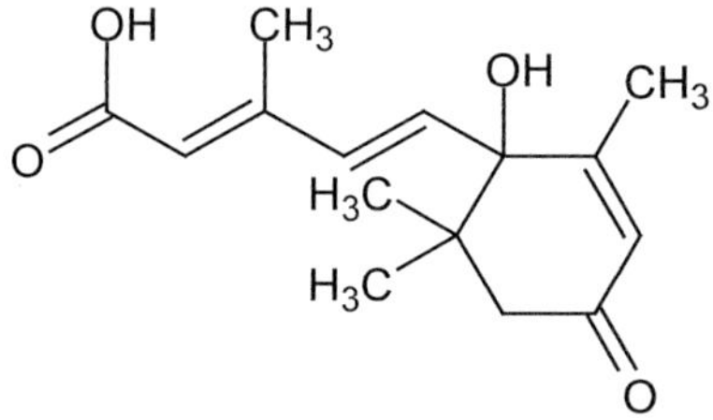
12.	Brassinolides	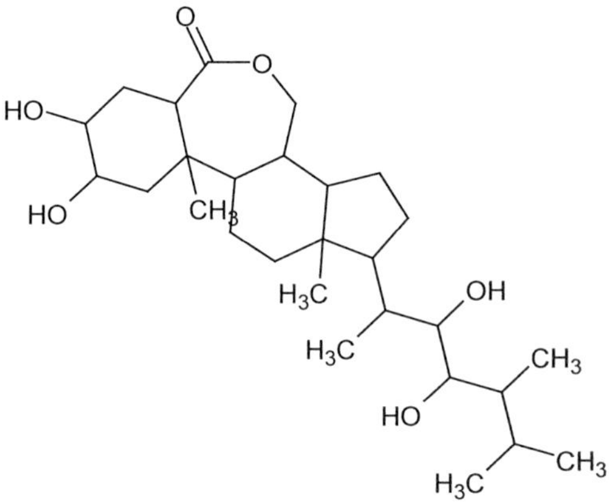
13.	Jasmonic acid	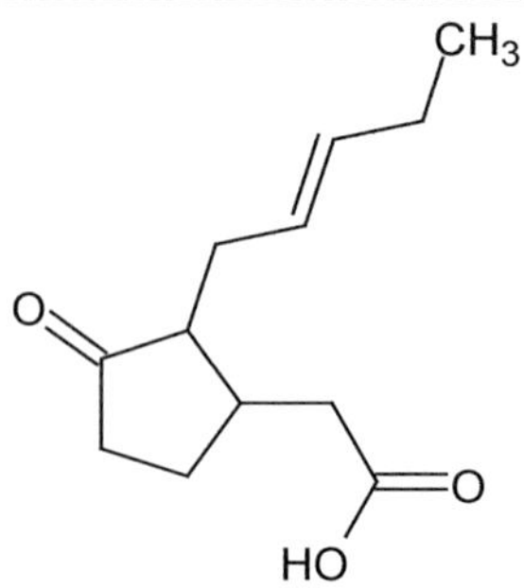

No.	Phytohormones	Structures
14.	Salicylic acid	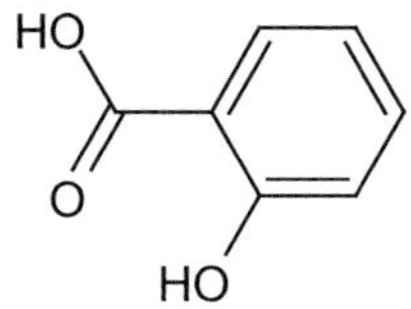
15.	Systemin	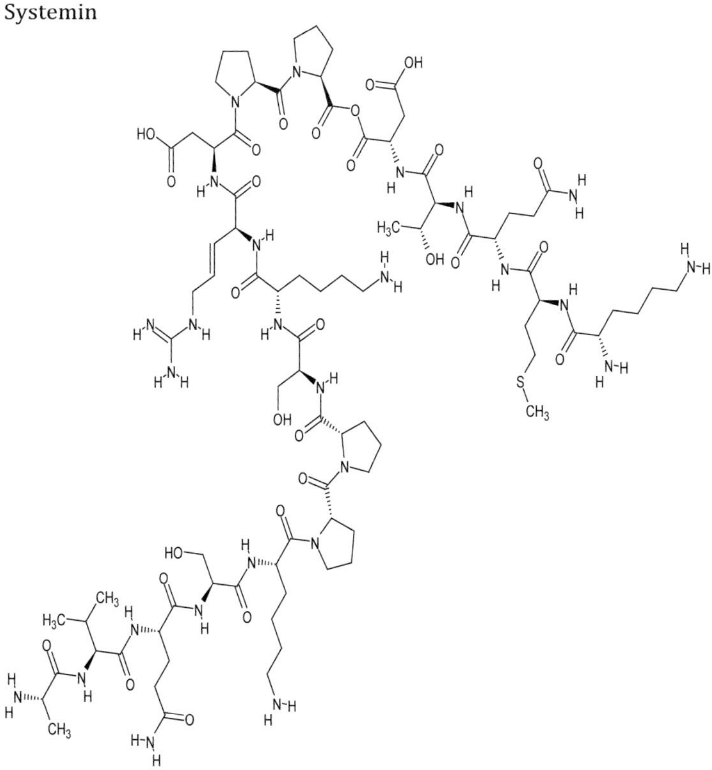

2.4 Abiotic Interactions Contribution to the Production of Latex

The term abiotic elicitors refers to substances that are nonbiological in nature. They include ions, pH, UV light, heavy metals, xenobiochemicals, fungicides and herbicides [74]. Elicitation using abiotic elicitors like hydrogen peroxide (H_2O_2), salicylic acid (SA), and carbon dioxide (CO_2) was performed by adding optimized concentration of abiotic elicitor preparation to cell suspension culture of *Papaver somniferum* and the cell culture will be analyzed for growth and secondary metabolite content. It was found that optimized concentration of abiotic elicitors enhanced the production of therapeutically important metabolites sanguinarine. The optimized concentration of abiotic elicitors was added to the cell suspension culture of *Papaver somniferum*. The mixture was incubated for a time period. The cultured flasks were sampled and analyzed for growth and sanguinarine content as determined by liquid chromatography techniques. It was found that sanguinarine content was significantly increased in presence of abiotic elicitors used with respect to untreated cell suspension culture. Hydrogen peroxide and carbon dioxide caused low level of sanguinarine accumulation confirms the downregulation of sanguinarine pathway while maximum sanguinarine content was reported with salicylic acid. Cannabinoids accumulation in *C. sativa* leaves and inflorescences has been observed to be enhanced by low dose of salicylic acid and methyl jasmonate. Also, 0.1 mM concentration of methyl jasmonate and salicylic acid has been reported to induce maximum accumulation of THCA and CBDA. Similarly other elicitors like benzothiadiazole (BTH), chitosan, or hexanoic acid could produce similar results [97]. The accumulation of benzophenanthridine alkaloids in *Chelidonium majus* L. improved by the upregulation of cheilanthifoline synthase, stylopine synthase and tetra-hydroprotoberberine *N*-methyltransferase genes after methyl jasmonate and salicylic acid elicitation. The upregulation of these genes resulted in the accumulation of chelidonine and sanguinarine content [101].

Treatment of *H. brasiliensis* bark with ethephon increased latex volume and latex regeneration. This also increased lutoids stability and duration of latex outflow. Over-excitation of ethephon caused dilution of latex, loss in nutritional components like nucleic acids, proteins and loss of latex outflow [102, 103].

Jatropha curcas L. callus cultures were examined for the effects of an abiotic elicitor (Ultraviolet-B). The accumulation of flavonoids, vitexin, isovitexin, and apigenin was seen upon irradiation of callus culture [104].

2.5 Biotechnological Techniques for Latex Bearing Plants

2.5.1 Micropropagation

Micropropagation (in vitro propagation) is an artificial technique used for the clonal propagation of medicinal plants under controlled conditions. Micropropagated plantlets are genetically similar and are used for the propagation of latex-bearing medicinal plants [105].

Plant regeneration is one of the major applications of plant tissue culture and it is mainly used for the regeneration of rare endangered medicinal plants and plants of higher commercial and therapeutic potential. Regeneration of plant material can be achieved by organogenesis and somatic embryogenesis which is the genesis of different organs and somatic embryos, respectively. Both processes are affected by phytohormones. The process of plant regeneration is based on totipotency. The totipotent cells may form somatic embryos via the process of embryogenesis, which leads to the regeneration of whole plants [106]. Explants may be seeds, fruits, shoots, leaves, stems of plant species. However, each plant species and explants are most required system of plant regeneration. The physiological and morphological properties of the explants have significant role for responding to plant tissue culture methodology. For example, immature explants generally respond better than mature explants.

Plant tissue culture is an in vitro technique and is applicable for the production of rare endangered and disease free medicinal plants. Also, this technique is used for germplasm conservation and secondary metabolites production. Figure 2.8 describes plant regeneration, which is predicated on the totipotency concept.

Secondary metabolites are produced and medicinal plants that produce latex are propagated using the plant tissue culture technique [107, 108]. By reducing the generation of reactive oxygen species and ethylene, changing gene expression, and enhancing antioxidant enzyme activities, it has been reported that adding nanoparticles to the culture media enhances callus formation, root initiation, shoot initiation, organogenesis, and somatic embryogenesis [109-114]. The presence of phytohormones and media composition affect callus formation, organogenesis and morphological structure of the cells or tissues. Higher amount of auxins promotes root proliferation while higher amount of cytokinin promotes shoot proliferation (Fig. 2.9) [115].

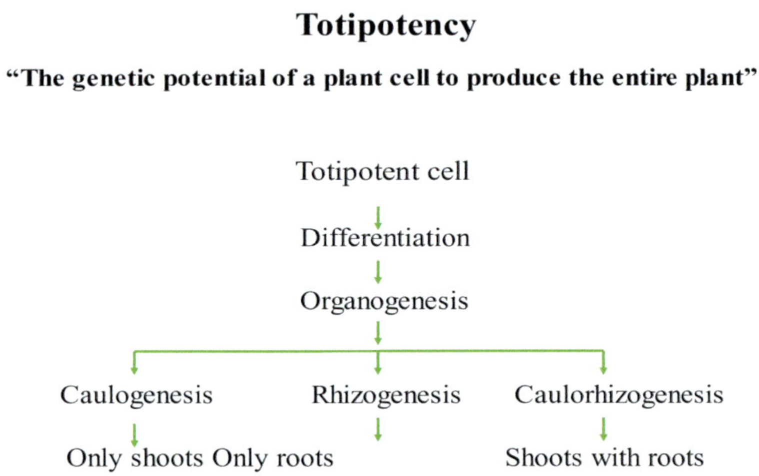

Figure 2.8 Plant regeneration and totipotency.

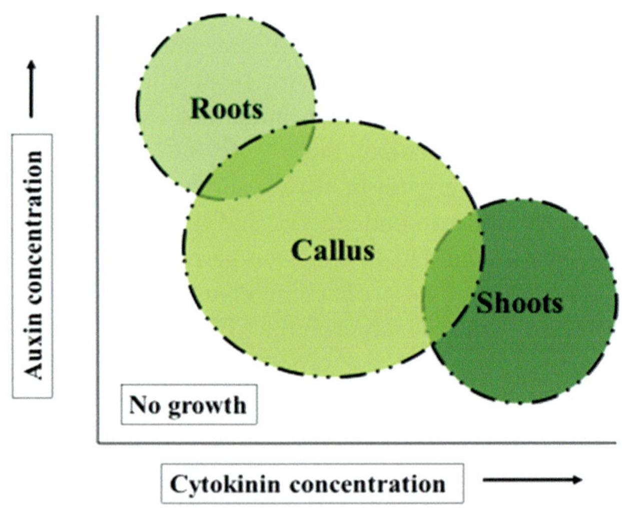

Figure 2.9 Effect of plant hormones on callus formation and organogenesis.

In vitro propagation protocol of *Hevea brasiliensis* was developed for mass production of rubber planting materials. MS medium containing kinetin significantly increased shoot and root formation [116].

The effect of culture medium, sucrose concentration and nano silver on in vitro propagation of *H. brasiliensis* is also being studied. It has been found that the highest shooting with maximum survival rate has been reported with Enjalric and Carron medium supplemented with 6 g/L sucrose and 4 ppm of nano silver [117]. In MS media supplemented with 4 ppm of BA, *Ficus elastica* shoot regeneration reached its peak. Significant protein increase was also seen in the regenerated shoots. With MS media treated with 3 ppm of IBA and potassium nitrate, the plantlets including shoots could be rooted with ease. The plantlets were moved to compost after being satisfactorily acclimated [118].

For cannabis recently photoautotrophic micropropagation method has been developed for large-scale propagation of plantlets because of limitations associated with conventional micropropagation methods. Photoautotrophic micropropagation uses passive ventilation and scales up in vitro clonal propagation and commercial production of plantlets and cannabidiol [119]. For the development of micropropagation protocol of *Papaver bracteatum*, the seedling was inoculated on MS media supplemented with NAA and BA. It was observed that full-strength MS media inhibited callus induction and proliferation, but 1/2 MS was suitable for callus induction. Similarly, temperature affects callus growth and it was good when the temperature was lower than 25°C. Shoot regeneration was taken place in MS medium containing NAA, BA and casein hydrolysate. With half-strength MS medium and no growth regulators added, the roots were induced from shoots. For the purpose of hardening and acclimatization, the rooted plantlets were moved to soil [120].

The micropropagation protocol of latex-bearing medicinal plants, *Chelidonium majus* L, and phenolic content among different tissues grown under in vitro conditions was reported. The results indicated that maximum shoots formation was obtained from the medium supplemented with 0.5 ppm of TDZ while higher concentration of TDZ decreases shoot multiplication.

Table 2.3 Summary of micropropagation protocol of *Cannabis sativa*

Techniques	Steps	Explants used	Culture media	Outcome	References
Direct regeneration	Shoot induction	Axillary buds	MS + BAP	Shooting	[146]
	Rooting		MS + IAA	Rooting	
Direct regeneration	Culture initiation	Seeds	MS		[147]
	Regeneration	Shoot tips	MS + BAP+ NAA	Auxiliary bud multiplication	
	Rooting		MS + IAA	Rooting	
Direct regeneration	Culture initiation	Seeds	MS		[148]
	Shoot initiation	Cotyledons	MS+ TDZ	Shooting	
	Rooting		MS+IBA	Rooting	
Indirect regeneration	Culture initiation	Seeds	MS		[149]
	Callus induction	Leaves	MS+NAA	Callus formation	
	Rooting		MS+IAA	Rooting	
Indirect regeneration	Culture initiation	Seeds	DARIA		[150]
	Callus induction	Stems	DARIA+Kn+NAA	Callus initiation	
	Rooting	Cotyledons	DARIA+IAA		
Indirect regeneration	Culture initiation	Seeds	MS		[151]
	Callus induction	Cotyledons	MS+BA	Callus initiation	
	Shooting		MS+BA	Shoot regeneration	
	Rooting		MS+NAA	Root regeneration	

Shoot length was affected by BAP and it was maximum with 2 ppm of BAP. The maximum rooting was reported with MS medium supplemented with 2 ppm of NAA or 1 ppm of IAA. The results of hardening showed that 85% of the plant material was acclimatized and survived. It was also reported that total phenol content of in vitro grown leaf was higher than in the stem and root of *Chelidonium majus* L [121]. Ex vitro propagation of *H. brasiliensis* by stem cutting is one of the methods of mass propagation of rubber plant material. This method protect genetic traits leads to the generation of genetically uniform plant material. Also, this method address phenological and intraclonal problems linked with rubber propagation. Brown and green stem cutting was used for propagation where the brown stem survived. Stem cuttings treated with naphthalene-acetic acid significantly developed higher number and lengths of roots and shoots [122]. The protocol for vegetative greenhouse propagation of *C. sativa* was developed using vegetative cutting. The propagule developed showed higher accumulation of cannabinoid content. The THC content of vegetative propagules was higher than in seed propagules [123]. Also, clonal propagation of leafy cuttings, rooting of leafy cuttings and seed germination are another method reported for the propagation of *C. sativa* [124]. For propagation of *P. somniferum*, seedling transplantation and sowing methods were used. The seedling transplantation method is faster than the sowing method. The highest yield of capsule and seed were obtained from seedling plants in the month of October whereas morphine and codeine content were not significantly affected [125]. Summary of micropropagation protocol of *Cannabis sativa* is presented in Table 2.3.

2.5.2 Callus Culture

The callus refers to unspecialized, unorganized dividing group of cells. The callus is produced when explants (cells) are incubated in a suitable medium and aseptic environment. In practice, callus culture involves callus induction and organogenesis (organ production). The callus culture is often initiated in many kinds of media combinations such as Murashige and Skoog medium. The culture consists of a combination of macronutrients, micronutrients, vitamins, amino acid, sugar content and agar.

Callus formation from latex-bearing plants *H. brasiliensis* and *Asclepias* species is given in Fig. 2.10. The protocol for the callus culture of *C. procera* was given previously [84].

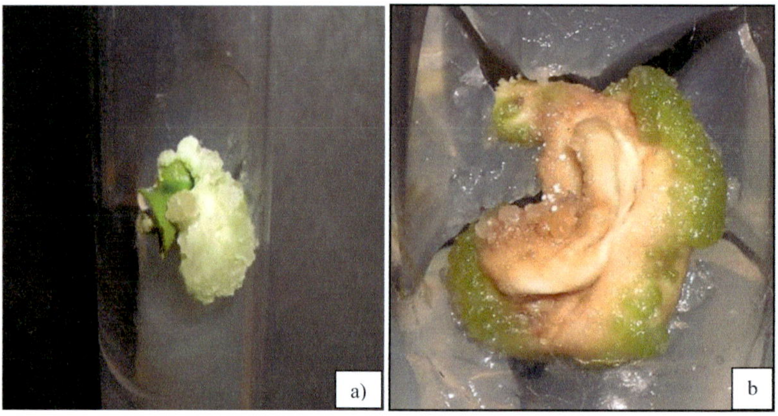

Figure 2.10 Callus culture of latex-bearing plants (a) *H. brasiliensis* and (b) *Asclepias* species.

The effect of polyethylene glycol and tyrosine on the accumulation of morphinian alkaloids from the callus culture of *P. somniferum* has been reported. It has been observed that MS medium supplemented with 20 ppm of PEG and 60 ppm of tyrosine yields the highest percentage of morphine and papaverine while MS media supplemented with 20 ppm of PEG and 40 ppm of tyrosine yields the highest percentage of codeine. The yield of morphinian alkaloids is minimal when the medium is free of polyethylene glycol and tyrosine.

The effect of sucrose concentration on callus initiation and callus proliferation from *Thevetia peruviana* was studied. It was found that compared with the control group (3% sucrose), callus induction did not differ significantly with the group supplemented with 2.5% sucrose while the group supplemented with 2% sucrose showed a significant reduction in callus induction. However, fresh weight of callus was significantly different among all groups and it was maximum with the control group supplemented with 3% sucrose. The group treated with 2% sucrose showed the lowest proliferation of callus with a friable appearance.

Callus initiation from *Papaver bracteatum* Lindl. was shown to be optimal on MS medium supplemented with ascorbic acid, kinetin and 2,4-dichlorophenoxyacetic acid. In this research, calli of *P. bracteatum* were induced at a higher efficiency. A combination of 2,4-dichlorophenoxyacetic acid, kinetin, and ascorbic acid were added to the MS basic media to provide the optimal callus induction. The concentrations of 2,4-D and ascorbic acid were found critical to callus induction and proliferation. Full-strength MS medium inhibits callus induction while half-strength MS medium induces callus induction. Studies using MS media containing benzylamine purine and naphthalene-acetic acid similarly demonstrated shoot multiplication from the established callus culture [126].

2.5.3 Cell Suspension Culture

Cell suspension culture is also known as cell culture. In this culture, aggregates of small cells are suspended in an agitated liquid medium and multiply. The single-cell cultures have properties of plant cells and provide a tremendous prospect to investigate the potentialities of plant cells.

Aggregates of tiny cells are formed into a liquid media by inoculating friable callus and agitating it thereafter. In favorable conditions, these produced cells grow continuously and divide very rapidly, ultimately resulting in a cell suspension culture. The cultures are sub-cultured into fresh nutrient medium and continuously repeated after a particular time period. Cell suspension culture of latex-bearing plants (*Asclepias* species) is given in Fig. 2.11 [115]. The cell suspension culture protocol of latex-bearing plants *C. gigantea* Linn has been established [115]. The suspension culture was established by getting friable callus using hypocotyl explants in MS media supplemented with auxin and cytokinin. The suspension culture can be further utilized for plant regeneration, cardenolides production as well as for callus culture.

Figure 2.11 Suspension culture of latex-bearing plants.

The effects of different parameters on the growth and accumulation of oleandrin from *Nerium oleander* were reported. The highest oleandrin yield was obtained in MS medium containing two-third lower nitrogen concentration and two-third higher phosphate concentration along with sucrose, coconut water, IBA and BA. Higher speed of shaking and low pH–induced oleandrin accumulation. Addition of chemicals like progesterone and cholesterol improved the accumulation of oleandrin [127].

2.5.4 Single-Cell Culture

The technique of obtaining individual cells that have been seeded on nutritional or baseline medium under controlled aseptic conditions is known as single-cell culture. The single-cell culture is isolated from intact living cells and these cell cultures are transferred on the nutrient medium for the nourishment of cells. Single cells are obtained from different type of tissues and organs of plants, callus culture and cell suspension. Single cells are isolated from the whole plant tissue such as leaf, stem and root by mechanical or enzymatic means. The production of benzylisoquinoline alkaloids from *P. somniferum* latex involves sieve elements and laticifers. Some enzymes like hydroxylase and reductase are involved in the biosynthesis of morphine and sanguinarine from *P. somniferum* latex. These enzymes are distributed in the parietal region of sieve elements, laticifers and companion cell [22]. Also, two major latex proteins (MLPs) and protein of PR10 family members from *P. somniferum* latex trigger the biosynthesis of morphine [128].

2.5.5 Protoplast Culture

Protoplast means cells without cell walls. There are various methods for the protoplast culture such as the hanging-drop cultures method, microculture chamber method, and soft agar matrix method. After culture protoplast can grow, divide, differentiate into the whole plant. This process includes repeated subculturing of the cells.

The protoplast culture technique is a significant way to produce a huge number of cells which are used in various studies such as plant regeneration from protoplast culture, hybridization of plants, cell-to-cell cloning, genetic manipulation, etc.

The protoplast culture of plant materials requires different culture medium supplemented with different concentration of nutritional elements. For example, Smaller concentration of iron and zinc while higher concentration of calcium favors protoplast culture. Similarly, higher auxin/kinetin ratio is required for cell division and a higher concentration of kinetin/auxin ratio is necessary for regeneration. This method is used for the regeneration of *H. brasiliensis* plant where the callus cell has been digested with enzymes to get the protoplast [129].

Isolation, transformation and transient gene expression of protoplast have been studied in *Cannabis sativa* with low tetrahydrocannabinol content. The yield and viability of isolated protoplast is good. It can be transformed with an auxin-responsive reporter gene and the reaction involving treatment with indole-3-acetic acid. This method can be used as a model system for cultivation and secondary metabolite accumulation from *C. sativa* [130].

2.5.6 Plant Seeds Culture

Plant seed culture is the technique in which mature and immature seeds are cultured in vitro to produce seedlings or plants. It is the best method for raising the sterile plantlet. This method is used for the production of pharmacologically active anti-inflammatory proteins from *Calotropis procera* plant latex [131].

2.5.7 Embryo Culture

The embryo culture is also termed as zygotic culture. The embryo culture technique involves the separation of a seed embryo from the fruits of plants for in vitro growth.

The embryo explants used for plant tissue culture may be mature and immature. Mature embryos are produced by ripe seeds and immature embryos are obtained from unripe/hybrid seeds used for embryo culture. Thus, the embryo cultures are capable of forming completely viable plants. This culture method is used to develop *Hevea brasiliensis* plants [132, 133].

2.5.8 Organ Culture (Root, Shoot and Meristem Culture)

In organ culture any organ of a plant such as root, shoot, leaf and flower is used as an explant for in vitro plant growth and development. Any explants can be cultured on a simple medium. For example, lateral roots or root tip of lateral roots can be used as explants for in vitro *root* cultures. However, for in vitro shoot culture, the shoot tips from auxiliary or adventitious buds can be cultured.

Organ culture provides remarkable outcomes on the patterns of growth, differentiation and development. This culture method can be used for clonal propagation. Figure 2.12 describes the shoot and root culture of latex-bearing *Euphorbia* species.

Abbasi [134] established the organogenesis protocol for latex-bearing plants *C. procera*. The effect of various concentration of auxin and cytokinin on root and shoot formation was studied. It was found that the highest callusing was recorded with 2,4-D, shoot regeneration with a combination of BA and NAA while root formation was obtained with IBA.

Figure 2.12 Shoot and Root culture of latex-bearing plants (*Euphorbia* species).

2.5.9 Somatic Embryogenesis

Somatic embryogenesis is a method of plant propagation which is used in plant conservation. In somatic embryogenesis, embryo-like structures are developed into plants like zygotic embryos. This somatic embryo is formed by direct or indirect methods. In direct somatic embryogenesis, the embryo is obtained directly from a small group of cells without the production of callus. In indirect somatic embryogenesis, callus is first produced from the explants.

It has been reported that somatic embryogenesis affects the production of benzylisoquinoline alkaloids (Morphine) from hypocotyl and cotyledons of *Papaver somniferum album* and *Papaver orientale splendidissimum.* Somatic embryogenesis has been done in both solid and suspension culture using culture media containing auxin and cytokinin [135].

2.5.10 Hairy Root Culture

The hairy root culture is an example of genetic transformation through microbial vector, i.e., *Agrobacterium rhizogenes,* the contributing agent of hairy root disease.

The hairy root culture showed biosynthetic capacity equivalent to the intact plant roots. Therefore, this is an alternate protocol for the root biomass production and to obtain plant root-derived compounds. The hairy root culture method is the most promising approach for the production of bioactive compounds particularly when they are accumulated in roots. This technique is used for the production of alkaloids and proteins from latex-bearing plants *Papaver somniferum, Cannabis sativa* and *Chelidonium majus L* [19]. The transformed roots of *Calotropis procera* are used for the production of terpenes [136]. Studies conducted previously have reported *Agrobacterium rhizogenes*-mediated hairy root induction, plant regeneration and secondary metabolites production from *Ficus carica* L [137]. The effect of mechanical wound and elicitation on the secondary metabolite accumulation from hairy root culture of *Calotropis gigantea* was reported. All the elicitors induced the production and accumulation of cardenolides. Chitosan showed better results than methyl jasmonate, yeast extract and mechanical wound [138].

2.5.11 Precursor Feeding Study

A plant produces primary and secondary metabolites in response to stress by using various biosynthetic pathways. In biosynthetic pathways, numerous compounds and enzymes are involved. The addition of such organic compounds may enhance secondary metabolites production and are known as precursors. The addition of precursors in an optimized concentration may enhance the yield of desired compounds. The selection of precursor depends on their impact on the biosynthetic route as well as their availability and cost. Sometimes depending upon the conditions, researchers can utilize other intermediates as well for enhancing the yield of secondary metabolites. For example, amino acids and other compounds have been added to cell suspension culture media for the production of alkaloids, terpenoids, flavonoids, etc. In the introductory part of the chapter, the basic biosynthetic pathways of latex-producing plants have been discussed. Chemically latex is of diverse nature like alkaloids, flavonoids, phenolics. Their precursors for the biosynthesis will be different. For example, tyrosine is the precursor to benzylisoquinoline alkaloids like opium alkaloids [139]. Similarly, pyruvate and D-glyceraldehyde-3-phosphate are the precursors for the biosynthesis of rubber latex. Tryptophan and geraniol were selected as precursors for the biosynthesis of monoterpene indole alkaloids [140], which are present in *Tabernaemontana catharinensis* latex [140]. The precursors are added to the cell suspension culture at optimized concentration and the culture analyzed for growth and secondary metabolite content.

The effect of precursors like tyrosine and shikimic acid on the sanguinarine content of suspension culture of latex-bearing plant, *Papaver somniferum*, has been studied. The precursors were added in different concentration and it was optimized. The flasks were sampled from 10th day to 40th day and analyzed for growth and sanguinarine content as determined by liquid chromatography techniques (HPLC). In comparison to an untreated cell suspension culture, it was discovered that the presence of precursors dramatically enhanced the sanguinarine concentration two to three times.. The activity and gene expression linked with tyrosine and phenylalanine increased upon precursor and elicitor treatment [78].

2.5.12 Micro and Macronutrient

Some of the micro- and macronutrients essential for plant nourishment and their desirable functions [108] are listed in Table 2.4. Macronutrients are the primary nutrients that provide energy to the plant materials and are required in higher quantity (>10 mM) to maintain their growth and development while the minor nutrients or trace elements are required in lesser quantity (<10 mM) and are required for growth and metabolism.

Examples of Macronutrients: Nitrogen (N), phosphorus (P), potassium (K), calcium (Ca), magnesium (Mg), sulphur (S), oxygen (O), carbon (C) and hydrogen (H).

Examples of Micronutrients: Boron (B), iron (Fe), chlorine (Cl), manganese (Mn), copper (Cu), zinc (Zn), molybdenum (Mo), and nickel (Ni).

The effect of various elements on latex production has been studied. It has been observed that micronutrients like molybdenum (Mo) increase rubber latex yield by reducing the shrinkage of tapping line and tapping panel dryness. Similarly, inorganic phosphorous and thiol group also improve latex yield from *H. brasiliensis* [141–143]. The effect of macronutrients like nitrogen, phosphorous, potassium and magnesium on the growth of rubber tree and latex yield was also reported. Nitrogen-phosphorous-Potassium (NPK) treatment did not alter cannabidiol and tetrahydrocannabinol level while phosphorous decreased THC and cannabidiol content in the inflorescence leaves. The presence of NPK did not alter the accumulation of cannabigerol in leaves, while the accumulation of cannabigerol in flowers significantly increased. NPK lowered cannabinol levels in both flowers and leaves [143].

Table 2.4 Elements and its role

S.No.	Elements	Functions
1.	Nitrogen	• Constituent of proteins, enzymes and nucleic acids • Essential for growth and morphogenesis in tissue cultures • Regulate photosynthesis and metabolism
2.	Potassium	• It regulates pH and osmotic environment in the cell • Principal inorganic cation of cell • It affects water intake • It provides resistance to fungal diseases and insect pests
3.	Calcium	• Enzyme cofactor in plant tissue culture • Required for membrane function • Essential for cell wall synthesis • Required for cell signaling pathways
4.	Magnesium	• Enzyme cofactor in plant tissue culture • Essential element of chlorophyll biosynthesis • In plants, plays role in somatic embryogenesis
5.	Phosphate	• Constituent of nucleic acids, lipid like phospholipids and co-enzymes • It has participated in energy transfer processes • It is the constituent of intermediates in respiration and photosynthesis • In tissue culture, it induces shoot formation and multiplication • Necessary for root development • Important for flowering, fruiting, and seed production

S.No.	Elements	Functions
6.	Sulphur	• Constituent of some amino acids and cofactors • Essential for growth and protein synthesis in cell suspension cultures
7.	Chloride	• Regulate for plant growth and development • Important for photosynthesis process
8.	Iron	• It is the component of cytochrome system and plays an important role in electron transfer reactions • Required for root growth • It is required for photosynthesis and chlorophyll formation • It influences nitrogen reduction and fixation
9.	Cobalt	• Constituent of some vitamins and nucleic acid synthesis • Essential for growth and morphogenesis of plant tissues
10.	Copper	• Constitutes of Enzyme cofactor • Involved in electron transfer reactions
11.	Zinc	• Constitute of metallo-enzyme • Biosynthesis of proteins, nucleic acids and chlorophyll • It balances the action of phytohormones • It influences root, seed and fruit formation
12.	Molybdenum	• Enzyme cofactor • Constituent of nitrate reductase enzyme
13.	Boron	• It is important for cell wall formation, flowering and fruiting • Regulate amino acid formation and crop reproduction
14.	Manganese	• Active role in photosynthesis • Affecting chloroplast production • It is required for enzyme activation and germination
15.	Nickel	• It is the component of urease enzyme • It influences shoot tip culture

It has been studied that callus cultures of *Calotropis procera* can be enhanced by reducing calcium, and adding Ca^{+2} ion channel blockers [148].

References

1. L AB. Rubber gutta percha and chicle. *Phytochem.* 1973; 2: 310–43.
2. Hagel JM, Yeung EC, Facchini PJ. Got milk? The secret life of laticifers. *Trends Plant Sci.* 2008; 13(12): 631–9.
3. Pickard WF. Laticifers and secretory ducts: two other tube systems in plants. *New Phytol.* 2008; 177(4): 877–88.
4. Thomas R, Sah NK, Sharma PB. Therapeutic biology of *Jatropha curcas*: a mini review. *Curr Pharm Biotechnol.* 2008; 9(4): 315–24.
5. Yagami T, Sato M, Nakamura A, Shono M. One of the rubber latex allergens has lysozyme activity. In *International Archives of Allergy and Immunology*, Vol. 104, Switzerland; 1994. p. 307.
6. Domsalla A, Melzig MF. Occurrence and properties of proteases in plant latices. *Planta Med.* 2008; 74(7): 699–711.
7. Yagami T, Sato M, Nakamura A, Komiyama T, Kitagawa K, Akasawa A, et al. Plant defense-related enzymes as latex antigens. *J Allergy Clin Immunol.* 1998; 101(3): 379–85.
8. Sequeira BJ, Vital MJS, Pohlit AM, Pararols IC, Caúper GSB. Antibacterial and antifungal activity of extracts and exudates of the Amazonian medicinal tree Himatanthus articulatus (Vahl) Woodson (common name: sucuba). *Mem Inst Oswaldo Cruz.* 2009; 104(4): 659–61.
9. Kitajima S, Kamei K, Taketani S, Yamaguchi M, Kawai F, Komatsu A, et al. Two chitinase-like proteins abundantly accumulated in latex of mulberry show insecticidal activity. *BMC Biochem.* 2010; 11: 6.
10. Mesquita ML de, Desrivot J, Bories C, Fournet A, Paula JE de, Grellier P, et al. Antileishmanial and trypanocidal activity of Brazilian Cerrado plants. *Mem Inst Oswaldo Cruz.* 2005; 100(7): 783–7.
11. Metcalfe CR. Distribution of latex in the plant kingdom. *Econ Bot.* 1967; 21(2): 115–27.
12. Farrell BD. A4 - Dussourd, DE. A4 - Mitter, C. BDA-F. Escalation of plant defense: do latex and resin canals spur plant diversification. *Am Nat.* 1991; 138(4): 881–900.
13. Lewinsohn T. The geographical distribution of plant latex. *Chemoecology.* 1991; 2: 64–8.
14. Coley PD, Aide TM. Comparison of herbivory and plant defenses in temperate and tropical broad-leaf forests. In: Price PW, Lewinsohn

TM, Fernandes GW, Benson WW (eds). *Plant-Animal Interactions: Evolutionary Ecology in the Tropical and Temperate Regions*, John Wiley & Sons, Inc. 1991, pp. 25–49.

15. Buranov AU, Elmuradov BJ. Extraction and characterization of latex and natural rubber from rubber-bearing plants. *J Agric Food Chem.* 2010; 58(2): 734–43.
16. Tabassum N, Hamdani M. Plants used to treat skin diseases. *Pharmacogn Rev.* 2014; 8(15): 52–60.
17. Schiff PL. Opium and Its Alkaloids. *Am J Pharm Educ.* 2004; 66, 186–194.
18. Covolan V, Innocentini-Mei L, Rossi C. Chemical modifications on polystyrene latex: preparation and characterization for use in immunological applications. *Polym Adv Technol.* 1997; 8: 44–50.
19. Gracz-Bernaciak J, Mazur O, Nawrot R. Functional studies of plant latex as a rich source of bioactive compounds: focus on proteins and alkaloids. *Int J Mol Sci.* 2021; 22(22): 12427.
20. Agrawal AA, Konno K. Latex: A model for understanding mechanisms, ecology, and evolution of plant defense against herbivory. *Annu Rev Ecol Evol Syst.* 2009; 40: 311–31.
21. Zaman K, Adibah M, Azzreena MA. Plant toxins : alkaloids and their toxicities. *GSC Biol Pharm Sci.* 2019; 06(02): 21–9.
22. Bird DA, Franceschi VR, Facchini PJ. A tale of three cell types: alkaloid biosynthesis is localized to sieve elements in opium poppy. *Plant Cell.* 2003; 15(11): 2626–35.
23. Carlin MG, Dean JR, Ames JM. Opium alkaloids in harvested and thermally processed poppy seeds. *Front Chem.* 2020; 8: 737.
24. Weid M, Ziegler J, Kutchan TM. The roles of latex and the vascular bundle in morphine biosynthesis in the opium poppy, Papaver somniferum. *Proc Natl Acad Sci U S A.* 2004; 101(38): 13957–62.
25. Onoyovwe A, Hagel JM, Chen X, Khan MF, Schriemer DC, Facchini PJ. Morphine biosynthesis in opium poppy involves two cell types: sieve elements and laticifers. *Plant Cell.* 2013; 25(10): 4110–22.
26. Croaker A, King GJ, Pyne JH, Anoopkumar-Dukie S, Liu L. Sanguinaria canadensis: traditional medicine, phytochemical composition, biological activities and current uses. *Int J Mol Sci.* 2016; 17(9): 1414.
27. Verma P, Khan SA, Mathur AK, Ghosh S, Shanker K, Kalra A. Improved sanguinarine production via biotic and abiotic elicitations and precursor feeding in cell suspensions of latex-less variety

of Papaver somniferum with their gene expression studies and upscaling in bioreactor. *Protoplasma*. 2014; 251(6): 1359–71.

28. Facchini PJ, Hagel JM, Liscombe DK, Loukanina N, MacLeod BP, Samanani N, et al. Opium poppy: blueprint for an alkaloid factory. *Phytochem Rev*. 2007; 6(1): 97–124.
29. Trujillo-Villanueva K, Rubio-Piña J, Monforte-González M, Ramírez-Benítez E, Vázquez-Flota F. The sequential exposure to jasmonate, salicylic acid and yeast extract promotes sanguinarine accumulation in Argemone mexicana cell cultures. *Biotechnol Lett*. 2012; 34(2): 379–85.
30. Oppel CB, Dussourd DE, Garimella U. Visualizing a plant defense and insect counterploy: alkaloid distribution in Lobelia leaves trenched by a plusiine caterpillar. *J Chem Ecol*. 2009; 35(6): 625–34.
31. Felpin F-X, Lebreton J. History, Chemistry and biology of alkaloids from Lobelia inflata. *Tetrahedron*. 2004; 60.
32. Tamargo J, Le Heuzey J-Y, Mabo P. Narrow therapeutic index drugs: a clinical pharmacological consideration to flecainide. *Eur J Clin Pharmacol*. 2015; 71(5): 549–67.
33. Matsuura HN, Fett-Neto AG. The major indole alkaloid N,β-D-glucopyranosyl vincosamide from leaves of Psychotria leiocarpa Cham. & Schltdl. is not an antifeedant but shows broad antioxidant activity. *Nat Prod Res*. 2013; 27(4–5): 402–11.
34. Mithöfer A, Boland W. Plant defense against herbivores: chemical aspects. *Annu Rev Plant Biol*. 2012; 63: 431–50.
35. Thawabteh A, Juma S, Bader M, Karaman D, Scrano L, Bufo SA, et al. The biological activity of natural alkaloids against herbivores, cancerous cells and pathogens. *Toxins (Basel)*. 2019; 656: 11(11).
36. Sulaiman CT, Balachandran I. Total phenolics and total flavonoids in selected Indian medicinal plants. *Indian J Pharm Sci*. 2012; 74(3): 258–60.
37. Tungmunnithum D, Thongboonyou A, Pholboon A, Yangsabai A. Flavonoids and other phenolic compounds from medicinal plants for pharmaceutical and medical aspects: an overview. *Med (Basel, Switzerland)*. 2018; 5(3): 93.
38. Sieniawska E. Activities of tannins--from in vitro studies to clinical trials. *Nat Prod Commun*. 2015; 10(11): 1877–84.

39. Bochkov DV, Sysolyatin SV, Kalashnikov AI, Surmacheva IA. Shikimic acid: review of its analytical, isolation, and purification techniques from plant and microbial sources. *J Chem Biol.* 2012; 5(1): 5–17.
40. Dai J, Mumper RJ. Plant phenolics: extraction, analysis and their antioxidant and anticancer properties. *Molecules.* 2010; 15(10): 7313–52.
41. Okuda T, Ito H. Tannins of constant structure in medicinal and food plants—hydrolyzable tannins and polyphenols related to tannins. *Molecules.* 2011; 16: 2191–217.
42. Kandungan P, Getah P. Determination of polyphenol contents in Hevea brasiliensis and rubber-processing effluent. 2018; 22(2): 185–96.
43. Kim MY, Lee BW, Lee H-U, Lee YY, Kim MH, Lee JY, et al. Phenolic compounds and antioxidant activity in sweet potato after heat treatment. *J Sci Food Agric.* 2019; 99(15): 6833–40.
44. Snook, Data EM, Kays S. Characterization and quantification of hexadecyl-, octadecyl-, and eicosyl-esters of p-coumaric acids in the vine and root latex of sweetpotato [Ipomoea batatas (L.) Lam.]. *J Agric Food Chem.* 1994; 42: 2589–95.
45. Rampadarath S, Puchooa D, Jeewon R. Original article. *Asian Pac J Trop Biomed.* 2016; 6(10): 858–65.
46. Lin D, Xiao M, Zhao J, Li Z, Xing B, Li X, et al. An overview of plant phenolic compounds and their importance in human nutrition and management of type 2 diabetes. *Molecules.* 2016; 1374: 21(10).
47. Abdul-Hafeez E, Karamova N, Ilinskaya O. Antioxidant activity and total phenolic compound content of certain medicinal plants. *Int J Biosci.* 2014; 5: 213–22.
48. Babenko LM, Smirnov OE, Romanenko KO, Trunova OK. Rreview. 2019; 91: 5–18.
49. Jeandet P. Phytoalexins: current progress and future prospects. *Molecules.* 2015; 20: 2770–4.
50. Glasser WG. About making lignin great again-some lessons from the past. *Front Chem.* 2019; 7: 565.
51. Boncan DAT, Tsang SSK, Li C, Lee IHT, Lam H-M, Chan T-F, et al. Terpenes and terpenoids in plants: interactions with environment and insects. *Int J Mol Sci.* 2020; 7382: 21(19).
52. Pichersky E, Raguso RA. Why do plants produce so many terpenoid compounds? *New Phytol.* 2018; 220(3): 692–702.

53. Bensch K, Braun U, Groenewald JZ, Crous PW. The genus Cladosporium. *Stud Mycol.* 2012; 72(1): 1–401.

54. Chadwick M, Trewin H, Gawthrop F, Wagstaff C. Sesquiterpenoids lactones: benefits to plants and people. *Int J Mol Sci.* 2013; 14(6): 12780–805.

55. Dussourd DE. Chemical stimulants of leaf-trenching by cabbage loopers: natural products, neurotransmitters, insecticides, and drugs. *J Chem Ecol.* 2003; 29(9): 2023–47.

56. Carter CA, Gray EA, Schneider TL, Lovett CM, Scott L, Messer AC, et al. Toxicarioside B and toxicarioside C. New cardenolides isolated from Antiaris toxicaria latex-derived dart poison. *Tetrahedron.* 1997; 53(50): 16959–68.

57. Konno K. Plant latex and other exudates as plant defense systems: roles of various defense chemicals and proteins contained therein. *Phytochemistry.* 2011; 72(13): 1510–30.

58. Mazorra-Manzano MA, Ramírez-Suarez JC, Yada RY. Plant proteases for bioactive peptides release: a review. *Crit Rev Food Sci Nutr.* 2018; 58(13): 2147–63.

59. Züst T, Petschenka G, Hastings AP, Agrawal AA. Toxicity of Milkweed leaves and latex: chromatographic quantification versus biological activity of cardenolides in 16 *Asclepias* species. *J Chem Ecol.* 2019; 45(1): 50–60.

60. Mimura K, Teh JL, Okayama H, Shiraishi K, Kua L-F, Koh V, et al. PD-L1 expression is mainly regulated by interferon gamma associated with JAK-STAT pathway in gastric cancer. *Cancer Sci.* 2018; 109(1): 43–53.

61. Desaulniers Brousseau V, Wu B-S, MacPherson S, Morello V, Lefsrud M. Cannabinoids and terpenes: how production of photo-protectants can be manipulated to enhance *Cannabis sativa* L. phytochemistry. *Front Plant Sci.* 2021; 12: 620021.

62. Nawrot R, Barylski J, Lippmann R, Altschmied L, Mock H-P. Combination of transcriptomic and proteomic approaches helps to unravel the protein composition of Chelidonium majus L. milky sap. *Planta.* 2016; 244(5): 1055–64.

63. D'Amato A, Bachi A, Fasoli E, Boschetti E, Peltre G, Sénéchal H, et al. In-depth exploration of Hevea brasiliensis latex proteome and "hidden allergens" via combinatorial peptide ligand libraries. *J Proteomics.* 2010; 73(7): 1368–80.

64. Tang C, Xiao X, Li H, Fan Y, Yang J, Qi J, et al. Comparative analysis of latex transcriptome reveals putative molecular mechanisms underlying super productivity of Hevea brasiliensis. *PLoS One.* 2013; 8(9): e75307.

65. Lamb C, Dixon RA. The oxidative burst in plant disease resistance. *Annu Rev Plant Physiol Plant Mol Biol.* 1997; 48: 251–75.

66. Hilder VA, Gatehouse AMR, Sheerman SE, Barker RF, Boulter D. A novel mechanism of insect resistance engineered into tobacco. *Nature.* 1987; 330(6144): 160–3.

67. Tisserat B, Berhow M. Production of pharmaceuticals from papaver cultivars in vitro. *Eng Life Sci.* 2009; 9(3): 190–6.

68. Lydon J, Teramura AH, Coffman CB. UV-B radiation effects on photosynthesis, growth and cannabinoid production of two *Cannabis sativa* chemotypes. *Photochem Photobiol.* 1987; 46(2): 201–6.

69. Westmoreland FM, Bugbee B. Sustainable Cannabis Nutrition: elevated root-zone phosphorus significantly increases leachate P and does not improve yield or quality. *Front Plant Sci.* 2022; 13: 1–11.

70. Giupponi L, Leoni V, Pavlovic R, Giorgi A. Influence of altitude on phytochemical composition of Hemp inflorescence: a metabolomic approach. *Molecules.* 2020; 25(6): 1381.

71. Namdar D, Charuvi D, Ajjampura V, Mazuz M, Ion A, Kamara I, et al. LED lighting affects the composition and biological activity of *Cannabis sativa* secondary metabolites. *Ind Crops Prod.* 2019; 177–85.

72. Magagnini G, Grassi G, Kotiranta S. The effect of light spectrum on the morphology and cannabinoid content of *Cannabis sativa* L. *Med Cannabis Cannabinoids.* 2018; 1(1): 19–27.

73. Hawley D. The influence of spectral quality of light on plant secondary metabolism and photosynthetic acclimation to light quality. 2018.

74. Namdeo AG. Phcog rev: review article plant cell elicitation for production of secondary metabolites: a review. *Pharm Rev.* 2007; 1(1): 69–79.

75. Ramirez-Estrada K, Vidal-Limon H, Hidalgo D, Moyano E, Golenioswki M, Cusidó RM, et al. Elicitation, an effective strategy for the biotechnological production of bioactive high-added value compounds in plant cell factories. *Molecules.* 2016; 21(2): 182.

76. Ochoa-Meza LC, Quintana-Obregón EA, Vargas-Arispuro I, Falcón-Rodríguez AB, Aispuro-Hernández E, Virgen-Ortiz JJ, et al.

Oligosaccharins as elicitors of defense responses in wheat. *Polymers (Basel)*. 2021; 13(18): 3105.

77. Naeini MS, Naghavi MR, Bihamta MR, Sabokdast M, Salehi M. Production of some benzylisoquinoline alkaloids in Papaver armeniacum L. hairy root cultures elicited with salicylic acid and methyl jasmonate. *Vitr Cell Dev Biol*. 2021; 57(2): 261–271.
78. Gabotti D, Locatelli F, Cusano E, Baldoni E, Genga A, Pucci L, et al. Cell suspensions of *Cannabis sativa* (var. Futura): effect of elicitation on metabolite content and antioxidant activity. *Molecules*. 2019; 24(22): 4056.
79. Suryaningsih DR, Prakoeswa SA, Eryanto A. Analysis of growth and enzyme contents of papain callus papaya (*Carica papaya* L.) through tissue culture engineering with *Saccharomyces cerevisiae* elicitors on MS and VW media. In *International Conference on Science, Technology & Environment* (ICoSTE) 2020.
80. Moreno-Escamilla JO, Jimeńez-Hernández FE, Alvarez-Parrilla E, De La Rosa LA, MartÍnez-Ruiz NDR, González-Fernández R, et al. Effect of elicitation on polyphenol and carotenoid metabolism in butterhead lettuce (*Lactuca sativa* var. capitata). *ACS Omega*. 2020; 5(20): 11535–46.
81. Gamborg OL, Miller RA, Ojima K. Nutrient requirements of suspension cultures of soybean root cells. *Exp Cell Res*. 1968; 50(1): 151–8.
82. Suri SS, Ramawat KG. In vitro hormonal regulation of laticifer differentiation in *Calotropis procera*. *Annals of botany*. 1995; 5(5): 477–80.
83. Tungngoen K, Viboonjun U, Kongsawadworakul P, Katsuhara M, Julien J-L, Sakr S, et al. Hormonal treatment of the bark of rubber trees (Hevea brasiliensis) increases latex yield through latex dilution in relation with the differential expression of two aquaporin genes. *J Plant Physiol*. 2011; 168(3): 253–62.
84. Suri SS, Ramawat KG. In vitro hormonal regulation of laticifer differentiation in *Calotropis procera*. *Ann Bot*. 1995; 75(5): 477–80.
85. Burgel L, Hartung J, Schibano D, Graeff-Hönninger S. Impact of different phytohormones on morphology, yield and cannabinoid content of *Cannabis sativa* L. *Plants (Basel, Switzerland)*. 2020; 9(6): 725.
86. Sappalani GI, Cabahug LM, Valleser VC. Impact of gibberellic acid and organic growth media on seed germination and seedling development of rubber (Hevea brasiliensis). *Int J Hortic Sci Technol*. 2021; 9(2): 165–74.

87. Dissanayake P, George DL, Gupta ML. Effect of light, gibberellic acid and abscisic acid on germination of guayule (Parthenium argentatum Gray) seed. *Ind Crops Prod.* 2010; 32(2): 111–7.

88. Mansouri H, Asrar Z, Mehrabani M. Effects of gibberellic acid on primary terpenoids and delta-tetrahydrocannabinol in *Cannabis sativa* at flowering stage. *J Integr Plant Biol.* 2009; 51(6): 553–61.

89. Fricke J, Hillebrand A, Twyman RM, Prüfer D, Schulze Gronover C. Abscisic acid-dependent regulation of small rubber particle protein gene expression in Taraxacum brevicorniculatum is mediated by TbbZIP1. *Plant Cell Physiol.* 2013; 54(4): 448–64.

90. Liu J-P, Zhuang Y-F, Guo X-L, Li Y-J. Molecular mechanism of ethylene stimulation of latex yield in rubber tree (Hevea brasiliensis) revealed by de novo sequencing and transcriptome analysis. *BMC Genomics.* 2016; 17: 257.

91. Pujade-Renaud V, Clement A, Perrot-Rechenmann C, Prevot JC, Chrestin H, Jacob JL, et al. Ethylene-induced increase in glutamine synthetase activity and mRNA levels in Hevea brasiliensis latex cells. *Plant Physiol.* 1994; 105(1): 127–32.

92. Broekaert I, Lee HI, Kush A, Chua NH, Raikhel N. Wound-induced accumulation of mRNA containing a hevein sequence in laticifers of rubber tree (Hevea brasiliensis). *Proc Natl Acad Sci U S A.* 1990; 87(19): 7633–7.

93. Yang J, Duan G, Li C, Liu L, Han G, Zhang Y, et al. The crosstalks between jasmonic acid and other plant hormone signaling highlight the involvement of jasmonic acid as a core component in plant response to biotic and abiotic stresses. *Front Plant Sci.* 2019; 10(October): 1–12.

94. Tarkesh Esfahani S, Karimzadeh G, Naghavi MR, Vrieling K. Altered gene expression and root thebaine production in polyploidized and methyl jasmonate-elicited Papaver bracteatum Lindl. *Plant Physiol Biochem.* 2021; 158: 334–41.

95. Apicella PV., Sands LB, Ma Y, Berkowitz GA. Delineating genetic regulation of cannabinoid biosynthesis during female flower development in *Cannabis sativa. Plant Direct.* 2022; 6(6): 1–8.

96. Arreewichit P, Sae-Lim P, Nirapathpongporn K, Viboonjun U, Kongsawadworakul P, Narangajavana J. Opposite physiological effects upon jasmonic acid and brassinosteroid treatment on laticifer proliferation and co-occurrence of differential expression of genes involved in vascular development in rubber tree. *Physiol Mol Biol Plants.* 2019; 25(5): 1283–99.

97. Garrido J, Rico S, Corral C, Sánchez C, Vidal N, Martínez-Quesada JJ, et al. Exogenous application of stress-related signaling molecules affect growth and cannabinoid accumulation in medical cannabis (*Cannabis sativa* L.). *Front Plant Sci.* 2022; 13: 1082554.

98. Mendoza D, Cuaspud O, Arias JP, Ruiz O, Arias M. Effect of salicylic acid and methyl jasmonate in the production of phenolic compounds in plant cell suspension cultures of Thevetia peruviana. *Biotechnol Reports.* 2018; 19: e00273.

99. Dzhavakhiya VG, Ozeretskovskaya OL, Zinovyeva SV. Immun response. In Dyakov YT, Dzhavakhiya VG, Korpela T, editors. *Comprehensive and Molecular Phytopathology,* Chapter 10. Amsterdam: Elsevier; 2007. pp. 265–314. (Studies in Plant Science).

100. Ryan CA, Pearce G. Systemins: A functionally defined family of peptide signals that regulate defensive genes in Solanaceae species. *Proc Natl Acad Sci.* 2003; 100(suppl_2): 14577–80.

101. Hashemi SM, Naghavi MR, Bakhshandeh E, Ghorbani M, Priyanatha C, Zandi P. Effects of abiotic elicitors on expression and accumulation of three candidate benzophenanthridine alkaloids in cultured greater celandine cells. *Molecules.* 2021; 26(5): 1395.

102. Coupé M, Chrestin H. Physico-chemical and biochemical mechanisms of hormonal (ethylene) stimulation. In *Phys. Rub. tree latex* 2018 (pp. 295–319). CRC Press.

103. Wang X, Wang D, Sun Y, Yang Q, Chang L, Wang L, et al. Comprehensive proteomics analysis of laticifer latex reveals new insights into ethylene stimulation of natural rubber production. *Sci Rep.* 2015; 5: 1–19.

104. Bascos EM, Ungson L. Ultraviolet-B (UV-B) radiation as an elicitor of flavonoid production in callus cultures of Jatropha (*Jatropha curcas* L.). *Philipp Agric.* 2012; 95: 335–43.

105. Tideman J, Hawker JS. In vitro propagation of latex-producing plants. *Ann Bot.* 1982; 49(2): 273–9.

106. Gamborg OL. Plant tissue culture. biotechnology. *Milestones.* 2002; 38: 84–92.

107. Nazarbekova SP, Nazarbek UB, Characteristics PIP. Physico-chemical characteristics and methods of purification of sodium chloride. *Ind Tech Eng.* 2022; 2(43): 38–43.

108. Moreno PR, van der Heijden R, Verpoorte R. Effect of terpenoid precursor feeding and elicitation on formation of indole alkaloids

in cell suspension cultures of Catharanthus roseus. *Plant Cell Rep.* 1993; 12(12): 702–5.

109. Hesami M, Baiton A, Alizadeh M, Pepe M, Torkamaneh D, Jones AMP. Advances and perspectives in tissue culture and genetic engineering of Cannabis. *Int J Mol Sci.* 2021; 22(11): 5671.

110. Hesami M, Tohidfar M, Alizadeh M, Daneshvar MH. Effects of sodium nitroprusside on callus browning of Ficus religiosa: an important medicinal plant. *J For Res.* 2020; 31(3): 789–96.

111. Hesami M, Daneshvar MH, Yoosefzadeh-Najafabadi M. An efficient in vitro shoot regeneration through direct organogenesis from seedling-derived petiole and leaf segments and acclimatization of Ficus religiosa. *J For Res.* 2019; 30(3): 807–15.

112. Hesami M, Pepe M, Alizadeh M, Rakei A, Baiton A, Phineas Jones AM. Recent advances in cannabis biotechnology. *Ind Crops Prod.* 2020; 158: 113026.

113. Hesami M, Naderi R, Tohidfar M, Yoosefzadeh-Najafabadi M. Application of adaptive neuro-fuzzy inference system-non-dominated sorting genetic algorithm-II (ANFIS-NSGAII) for modeling and optimizing somatic embryogenesis of chrysanthemum. *Front Plant Sci.* 2019; 10: 1–12.

114. Hesami M, Naderi R, Tohidfar M. Introducing a hybrid artificial intelligence method for high-throughput modeling and optimizing plant tissue culture processes: the establishment of a new embryogenesis medium for chrysanthemum, as a case study. *Appl Microbiol Biotechnol.* 2020; 104(23): 10249–63.

115. Roy AT, Koutoulis A, De DN. Cell suspension culture and plant regeneration in the latex-producing plant, Calotropis gigantea (Linn.) R. Br. *Plant Cell Tissue Organ Cult.* 2000; 63(1): 15–22.

116. Antwi-Wiredu A, Amiteye S, Gyinae Diawuoh R, Kofi Asumeng A, Y. P. Klu G. In vitro propagation of rubber tree (Hevea Brasiliensis) using shoot-tip and nodal cutting explants. *Int J Adv Sci Res Eng.* 2018; 4(6): 38–50.

117. Moradpour M, Aziz MA. Establishment of in vitro culture of rubber (Hevea brasiliensis) from field-derived explants: effective role of silver nanoparticles in reducing contamination and browning. *J Nanomed Nanotechnol.* 2016; 7(3).

118. Abdullah RM, AL-Mallah MK. Stimulation of stem callus and regeneration of ficus elastica "Decora" and from apical tips in vitro. 2013; 7(2): 5–12.

119. Zarei A, Behdarvandi B, Dinani ET, Maccarone J. *Cannabis sativa* L. photoautotrophic micropropagation: a powerful tool for industrial scale in vitro propagation. *Vitr Cell Dev Biol.* 2021; 57(6): 932–941.

120. Ilahi I, Ghauri EG. Regeneration in cultures of Papaver bracteatum as influenced by growth hormones and temperature. *Plant Cell Tissue Organ Cult.* 1994; 38(1): 81–3.

121. Ghasemi M, arouiee hossein, Azadi P, Ali Ahmadi A. Optimization of micropropagation of Chelidonium majus L. and study on total phenol content in different organs in in vitro conditions. *J Hortic Sci.* 2020; 34(1): 145–59.

122. Antwi-Wiredu A, Amiteye S, Diawuoh RG, Klu GYP. Ex vitro propagation of rubber tree (Hevea Brasiliensis) using stem cuttings. *Int J Environ Agric Biotechnol.* 2018; 3(3): 846–54.

123. Coffman CB, Gentner WA. Greenhouse propagation of *Cannabis Sativa* L. by vegetative cuttings. *Econ Bot.* 1979; 33(2): 124–7.

124. Casillas A. Propagation of *Cannabis sativa* for commercial production©. In: Acta Horticulturae. International Society for Horticultural Science (ISHS), Leuven, Belgium; 2017. pp. 157–8.

125. Kara N, Baydar H. The influence of sowing and planting seedlings at different dates in autumn on the yield and quality of the opium poppy (*Papaver somniferum* L.). *J Appl Res Med Aromat Plants.* 2021; 21: 100290.

126. Siwach P, Grover K, Gill AR. The influence of plant growth regulators, explant nature and sucrose concentration on in vitro callus growth of Thevetia peruviana Schum, *Asian J Biotechnol.* 2011; 3: 280–292.

127. Ibrahim A, Khalifa S, Khafagi I, Youssef D, Khan I, Mesbah M. Enhancement of oleandrin production in suspension cultures of Nerium oleander by combined optimization of medium composition and substrate feeding. *Plant Biosyst – An Int J Deal with all Asp Plant Biol.* 2009; 143(1): 97–103.

128. Ozber N, Carr SC, Morris JS, Liang S, Watkins JL, Caldo KM, et al. Alkaloid binding to opium poppy major latex proteins triggers structural modification and functional aggregation. *Nat Commun.* 2022; 13(1): 6768.

129. Sushamakumari S, Asokan MP, Anthony P, Lowe KC, Power JB, Davey MR. Plant regeneration from embryogenic cell suspension-derived protoplasts of rubber. *Plant Cell Tissue Organ Cult.* 2000; 61(1): 81–5.

130. Beard KM, Boling AWH, Bargmann BOR. Protoplast isolation, transient transformation, and flow-cytometric analysis of reporter-gene activation in *Cannabis sativa* L. *Ind Crops Prod.* 2021; 164: 113360.

131. Teixeira FM, Ramos MV, Soares AA, Oliveira RSB, Almeida-Filho LCP, Oliveira JS, et al. In vitro tissue culture of the medicinal shrub *Calotropis procera* to produce pharmacologically active proteins from plant latex. *Process Biochem.* 2011; 46(5): 1118–24.

132. Dickson IA, Okere A, Elizabeth J, Mary O. In-vitro culture of Hevea brasiliensis (rubber tree) embryo. 2019; 6(2): 1–5.

133. Varshney A, Tangirala SJ. Efficient plant regeneration from immature embryo cultures of *Jatropha curcas,* a biodiesel plant. *Plant Biotechnol Rep.* 2010; 4: 139–48.

134. Abbasi H. Indirect organogenesis in milkweed (*Calotropis procera*) from mature zygotic embryo explants. *Biotechnologia.* 2017; 98(3): 189–94.

135. Kassem MA, Jacquin A. Somatic embryogenesis, rhizogenesis, and morphinan alkaloids production in two species of opium poppy. *J Biomed Biotechnol.* 2001; 1(2): 70–8.

136. Adabavazeh F, Pourseyedi S, Nadernejad N, Razavizadeh R. Hairy root induction in *Calotropis procera* and optimization of its phytochemical characteristics by elicitors. *Plant Cell, Tissue Organ Cult.* 2023; 1–14.

137. Amani S, Mohebodini M, Khademvatan S, Jafari M. *Agrobacterium rhizogenes*-mediated hairy root induction and plant regeneration from transgenic roots in *Ficus carica* L. *J. Plant Mol. Bre.* 2019; 7(1): 10–21.

138. Sun J, Xiao J, Wang X, Yuan X, Zhao B. Improved cardenolide production in Calotropis gigantea hairy roots using mechanical wounding and elicitation. *Biotechnol Lett.* 2012; 34(3): 563–9.

139. Lichman BR. The scaffold-forming steps of plant alkaloidbio-synthesis. *Nat Prod Rep.* 2021; 38(1): 103–29.

140. Plantakum P. Organic growth factor requirements of tobacco tissue cultures. *Phys Plantarum.* 1965; 18: 100–27.

141. Gao L, Sun Y, Wu M, Wang D, Wei J, Wu B, et al. Physiological and proteomic analyses of molybdenum- and ethylene-responsive mechanisms in rubber latex. *Front Plant Sci.* 2018; 9: 621.

142. Tiva LK, Lacote R, Chan C, Sopheaveasna M, Gohet E. Effect of fertilization on latex yield potential and physiological parameters of clone PB 217 in Cambodia. *CRRI IRRDB Int Rubber Conf Siem Reap.* 21–22 November 2016. 2016: 409–14.

143. Vrignon-Brenas S, Gay F, Ricard S, Snoeck D, Perron T, Mareschal L, et al. Nutrient management of immature rubber plantations. A review. *Agron Sustain Dev.* 2019; 39(1): 11.

144. Bernstein N, Gorelick J, Zerahia R, Koch S. Impact of N, P, K, and humic acid supplementation on the chemical profile of medical cannabis (*Cannabis sativa* L). *Front Plant Sci.* 2019; 10: 436.

145. Suri SS, Ramawat KG. Extracellular calcium deprivation stimulates laticifer differentiation in callus cultures of *Calotropis procera. Ann Bot.* 1997; 79(4): 371–4.

146. Lata H, Chandra S, Khan I, ElSohly MA. Thidiazuron-induced high-frequency direct shoot organogenesis of *Cannabis sativa* L. *Vitr Cell Dev Biol - Plant.* 2009; 45(1): 12–9.

147. Wang R, He L, Xia B, Tong J, Li N, Peng F. A micropropagation system for cloning of hemp (*Cannabis Sativa* L.) by shoot tip culture. *Pakistan J Bot.* 2009; 41: 603–8.

148. Chaohua C, Gonggu Z, Lining Z, Chunsheng G, Qing T, Jianhua C, et al. A rapid shoot regeneration protocol from the cotyledons of hemp (*Cannabis sativa* L.). *Ind Crops Prod.* 2016; 83: 61–5.

149. Ślusarkiewicz-Jarzina A, Ponitka A, Kaczmarek Z. Influence of cultivar, explant source and plant growth regulator on callus induction and plant regeneration of *Cannabis sativa* L. *Acta Biol Cracoviensia Ser Bot.* 2005; 47(2): 145–51.

150. Wielgus K, Luwanska A, Lassocinski W, Kaczmarek Z. Estimation of *Cannabis sativa* L. Tissue culture conditions essential for callus induction and plant regeneration. *J Nat Fibers.* 2008; 5(3): 199–207.

151. Movahedi M, Torabi S. The effect of different concentrations of TDZ and BA on in vitro regeneration of Iranian cannabis (*Cannabis sativa*) using cotyledon and epicotyl explants. *J Plant Mol Breed.* 2015; 3: 20–7.

Chapter 3

Chemistry of Latex

Arya Lakshmi Marisetti,[a] Ajay Sharma,[a] Abhishek Tiwari,[b] Varsha Tiwari,[b] Saurabh Dahiya,[c] Vineet Mittal,[d] Saurabh Bhatia,[e,f] Ahmed Al-Harrasi,[e] and Laleh Khodaie[g,h]

[a]*Department of Pharmacognosy & Phytochemistry, School of Pharmaceutical Sciences, Delhi Pharmaceutical Sciences and Research University, Mehrauli-Badarpur Road, New Delhi, India*

[b]*Faculty of Pharmacy, Pharmacy Academy, IFTM University, Lodhipur Rajput, Moradabad, Uttar Pradesh, India*

[c]*Department of Pharmaceutical Chemistry and Quality Assurance, Delhi Pharmaceutical Sciences and Research University, Mehrauli-Badarpur Road, New Delhi, India*

[d]*Department of Pharmaceutical Sciences, Maharshi Dayanand University, Rohtak, Haryana, India*

[e]*Natural & Medical Sciences Research Center, University of Nizwa, Birkat Al Mauz, Nizwa, Oman*

[f]*School of Health Sciences, University of Petroleum and Energy Studies, Dehradun, Uttarakhand, India*

[g]*Department of Traditional Pharmacy, Faculty of Traditional Medicine, Tabriz University of Medical Sciences, Tabriz, Iran*

[h]*Department of Pharmacognosy, Faculty of Pharmacy, Tabriz University of Medical Sciences, Tabriz, Iran*

ajaysharmapharma1979@gmail.com

Introduction to Plant-Based Latex: Phytochemistry, Ethnopharmacology, Biological, and Pharmacological Activities
Edited by Laleh Khodaie, Ajay Sharma, Saurabh Bhatia, and Ahmed Al-Harrasi

ISBN 978-981-5129-12-0 (Hardcover), 978-1-003-57071-4 (eBook)
www.jennystanford.com

3.1 Introduction

Latex is a milky fluid secreted by the specialized secretory cells (Laticifers) as a part of defense mechanisms in response to stress. The elemental composition of the latex is characterized by the presence of basic elements with a molecular formula C_3H_3N. Chemically latex is complex in nature and composed of a diverse variety of phytochemicals like alkaloids (with indole, steroid, terpenoid, isoquinoline, iboga and phenanthroindolizidine moieties), carbohydrates, fatty acids, furanocoumarins, glycosides (cardenolides), mucilages, organic acids, phenolics, resins, saponins, sterols, terpenoids (sesquiterpenes, diterpenes, triterpenes), chitinases, glucosidases, hevein, lipases, oxidases, papain, peptidases, peroxidases (POD), plasmin, proteases and phosphatase thrombin are few of the enzymes obtained from plant bearing latex. The amount of these phytoconstituents may vary among the plant species. Moreover, with advanced sophisticated analytical tools like chromatographic and spectroscopic techniques the quantitative and qualitative analysis of latex-based phytoconstituents becomes quite easier and accurate [1–3].

3.2 Physicochemical Properties of Latex

It include both the physical and chemical characteristics and are used for the standardization of latex bearing plants. Physicochemical evaluation aids in the standardization of latex bearing medicinal plants and it ensures quality, safety and efficacy [4]. The physicochemical properties include ash value, organoleptic characteristics (color, odor, taste, texture), solubility, melting point, density, pH, moisture content, UV luminescence, viscosity, surface tension, tensile strength, extractive value (alcohol and water-soluble extractives), swelling index, foaming index, thickness, specific gravity, analysis of crude protein, lipid percentage, carbohydrate percentage, water absorption, dry rubber content (DRC), total solids content (TSC) and elemental analysis etc. The optical properties such as birefringence, extinction angles, isotropy, transparency and pleochroism are also investigated for latex bearing plants.

3.2.1 Organoleptic Characteristics

Color, odor, taste and texture are some of the organoleptic characteristics used for the identification of latex and latex bearing medicinal plants. The organoleptic characteristics of latex varies according to geographical locations, mixes, age and taphonomic circumstances. For example, *Papaver somniferum* latex showed milky white, *Chelidonium majus* showed milky yellow, *Ficus septica* showed pale yellow, *Ficus benghalensis* showed orange and *Croton lecheri* showed scarlet appearance [1].

3.2.2 Solubility of Latex

The capacity of a substance to dissolve in a certain solvent serves as the definition of solubility. The latex monomers have a greater tendency to dissolve and polymerize at certain temperature. A latex is a colloidal dispersion where the insoluble colloidal particles are scattered throughout the liquid. Some of the solvents that greatly affect the solubility of latex are hexane, toluene, and xylene. The effect of solubility parameters on the diffusion properties of NR/NBR (natural rubber/nitrile rubber blends) was reported earlier [5–7].

3.2.3 Melting Point

Melting point can be used to distinguish latex, especially if it is pure and homogeneous. The most widely used natural latex from rubber tree melt at 180°C. It is required to add plasticizers or other excipients to maintain physical properties of the commercially available latex formulations. Pure polyisoprene, the primary polymer in natural latex rubber, has a very low melting point of only 25°C. At this temperature, natural latex rubber does not melt. Differential scanning calorimetry (DSC) investigate changes in the thermal properties of latex over time for *Ficus carica, Hivea brasiliensis* and *Artocarpus heterophyllus* [8, 9].

3.2.4 Density

Density is the mass per unit volume expressed in kilograms/cubic meters. The firmness of latex can be related to its density and has been widely utilized to differentiate between different varieties. The density measurement is appropriate for pure substances, while the presence of adulteration affects it. Specific gravity and density measurement can be done by densitometer using the test materials (1% w/v) for latex bearing plants [10]. The density is calculated as given in the following equation by dividing the weight by the volume [4].

$$\text{Density of the samples} = \frac{(W_1 - W)}{V} \times 100,$$

where, W_1 is the weight of bottle with sample, W is the weight of the bottle when it is empty, and V is the volume of the sample.

3.2.5 pH

Fresh latex material has a pH of 7. The pH may vary on long-term storage and goes up to 10 if latex is stored for a period of 10 h. For the best agglomeration of the latex upon freezing, the pH must be maintained within a narrow pH range. If the pH is too high, the latex is not adequately destabilized for effective agglomeration upon freezing and if the pH is too low, too much pre-coagulum is formed in both the pH adjustment stage and the subsequent agglomeration process. The pH of the original latex should be adjusted to a range of 7.8 to 8.5 for optimum agglomeration. Beyond the optimum pH, stability of latex gets affected. Moreover, pH affects the catalytic properties of lipase enzyme from dried crude latex of *Euphorbia unispina* and *Ficus trichopoda.* The enzymatic activity was maximum at pH 5 (11).

3.2.6 Moisture Content

The moisture content is responsible for the degradation of crude latex drugs either by chemical alteration or by micro-biological deterioration. Hence, the moisture content of drugs must be identified and managed. The moisture content is determined by heating the latex sample to a constant weight in an oven at 105°C. The percentage ratio of the weight change to the weight of original sample is used to calculate moisture content.

For example, *Hevea brasiliensis* latex at various percentages of moisture content varied from 42.47% to 77.57% [12].

$$\%\text{Moisture content} = \frac{(W_1 - W_2)}{W_1} \times 100,$$

where W_1 represents the actual weight of the samples and W_2 represents their weight post drying.

3.2.7 Ultraviolet Light

Certain medications fluoresce when exposed to ultraviolet radiation, which aids in the identification and differentiation of latex drugs. Some species are quite difficult to differentiate, especially in powdered form. However, inspection under UV light reveals such distinct changes in fluorescence that the varieties can be easily differentiated from one another. Fluorescence analysis of *Euphorbia heterophylla* under UV light is reported and the sample showed different color in day light and UV light [13].

3.2.8 Viscosity and Thixotropy

The samples viscosity and thixotropy are determined using an instrument called a Rheometer. Important information for characterizing various types of latex can be obtained via rheological experiments that establish viscoelastic data by displaying flow curves and utilizing theoretical models. Because of the pseudo-plastic behavior, the viscosity of the natural latex rubber samples decreased with increasing shear rate. As the shear rate increases, viscosity decreases, defining the boundary of the Newtonian fluid zone. The viscosity and thixotropy behavior of *Hevea brasiliensis* samples are determined by this method [14, 15].

3.2.9 Surface Tension via Contact Angle

The optical contact angle and surface tension are one of important physicochemical parameters of latex and are used to calculate a surface and a droplet surface angle with regard to airflow at various times after the sample droplets were put on a flat surface. Both the contact angle and surface tension phenomena of natural rubber bio membranes affect cell response and cell adhesion. Hydrophilic surface favors these phenomena while hydrophobic surface disfavors these events [16, 17].

3.2.10 Ash Value

Ash value determines the quality and purity of the powdered crude drug containing latex. Ashing phenomena remove all organic material from the crude drug which may hinder the process of analytical determinations. Hence, for latex, acid-insoluble ash would be preferred. When the ash is treated with dilute HCl, calcium oxalate is converted to calcium oxide or calcium carbonate, which is acid soluble. The remaining ash, which is referred to as the acid-insoluble ash, is weighed. By doing so, we can identify the presence of excessive amounts of earthy material from latex of roots, rhizomes, and leaves. Water-soluble ash is used to find materials that have been depleted by water. Sulfated ash is created by adding H_2SO_4 to obtain sulfate salts, and the percentage of ash is estimated using the air-dried medication as a reference. This is done at a temperature greater than 600°C. For example, the value of latex from *H. brasiliensis* ranges from 8% to 30% and is determined by the following formula:

$$\%\ \text{Ash content} = \frac{(C_1 - C_2)}{C_1} \times 100,$$

where C_1 denotes the dry weight of the ash and C_2 denotes the weight of the ash upon igniting.

The ash values *Ficus bengalensis* are determined as per the method reported [17, 18].

3.2.11 Extractive Values

Extractive values will be determined by using different solvents and are used to assess the quality, safety and efficacy of extracts and to detect adulteration. Depending upon the nature of phytoconstituents present in latex a wide variety of solvent systems is used. Alcohol-soluble extractive is used if latex-bearing plant materials contain tannins, glycosides, mucilage and other water-soluble constituents. For drugs containing volatile constituents and fats, an ether-soluble extractive is used. The extractive value of *Ficus bengalensis* is determined as per the method reported and is an indication of the nature of phytoconstituents present [17]. The extractive value is determined as per WHO guidelines.

3.2.12 Determination of Alcohol Soluble Extractive

This value is determined by macerating the desired amount of drug material with a suitable solvent, i.e., alcohol. The mixture is shaken frequently initially for a period of 6 h and then kept for a period of 1–2 days. It is filtered and the filtrate is weighed after drying. In case of *Ficus bengalensis* it is determined as per the method reported and is an indication of the nature of phytoconstituents present [17].

Percentage of alcohol-soluble extractive

$$= \frac{\text{Weight of the residue}}{\text{Weight of the sample taken}} \times 100$$

3.2.13 Determination of Water-Soluble Extractive

Water soluble extractive is used for drugs containing tannins, glycosides, resins and other water soluble constituents.

Weighed amount of latex-containing powdered material is added with desired amount of water and Keiselghur and then the mixture is filtered. The filtrate is weighed after drying. It may be determined for *Ficus bengalensis* as per the method reported earlier [17].

Percentage of water-soluble extractive value

$$= \frac{\text{Weight of the residue}}{\text{Weight of the sample taken}} \times 100$$

3.2.14 Swelling Index

The volume in milliliters occupied by the swelling of 1 g of latex under specific circumstances is known as the swelling index. Its determination is based on the inclusion of water or another designated swelling agent. Cross-link swelling index of rubber is 2.75.

$$\text{Swelling index} = \frac{\text{Final weight} - \text{Initial weight}}{\text{Final weight}}$$

The swelling index of *Calotropis procera* is 1.6 cm. It is an indication of the presence of mucilaginous material [19]. The swelling index is also used to determine the cross-link density of the natural rubber latex films.

3.2.15 Foaming Index

An aqueous decoction of latex is evaluated according to its foaming capacity using a foaming index. A sufficient quantity of powdered latex is mixed with water and boiled for a period as mentioned in the protocol. The decoction is poured into test tubes and makes up the volume. The test containers are then shaken for few seconds and left to stand for few minutes and foam height is measured.

The foaming index is calculated as

$$\text{Foaming index} = \frac{1000}{a}$$

a = volume (ml) of decoction in test tube where foam of 1 cm height is observed.

The foaming index of *Calotropis procera* is less than 100, this is an indication of the presence of saponins [19].

3.2.16 Analysis of Crude Protein

The crude content of the latex samples is performed using Kjeldahl's method. In this method, a weighed amount of the test sample can be taken in a flask and a weighed amount of saturated H_2SO_4 (98.9%) is taken in flask and mixed it with occasional stirring. A mixture of $NaSO_4$ and $CuSO_4$ (10:1 ratio) is added to the flask. The mixture is heated for some time and then cooled to initiate the digestion of the sample. To the digested sample, sufficient amount of boric acid along with a few drops of indicator is added followed by the addition of NaOH. Thereafter, the flask is titrated till the last point. Protein content is determined by the following formula:

$$\% \text{ Protein} = \frac{1.4 \times \text{DF} \times 6.25}{\text{Weight of sample}}$$

where DF is dilution factor

The crude protein content of *Ficus trichopoda, Carica papaya* and *Euphorbia unispina* was reported earlier [11].

3.2.17 Calculation of Lipid Percentage

The Association of Official Analytical Chemists (AOAC) gravimetric technique is used for the determination of lipid content of samples. In this method, the samples will be kept in a clean round bottom flask (W_1) and a sufficient amount of nonpolar solvent (hexane or petroleum ether) is added to it. Soxhlet apparatus is used for this purpose in which the solvent is heated up and the vapors are condensed using cold-water circulation. This process is carried out from 6 to 24 h depending upon the samples and requirements [7]. The oil remaining is allowed to cool and then weighed (W_2). The % lipids content is calculated using the following equation:

$$\% \text{ Crude lipid content} = \frac{W_2 - W_1}{\text{Weight of sample}} \times 100$$

The lipid content of *Carica papaya* dried latex showed the presence of crude fat (approximately 5%) [20].

3.2.18 Carbohydrate Content

It is estimated by phenol-sulfuric acid method. In this method, the latex material is mixed with phenol and concentrated sulfuric acid. The mixture is incubated for 30 min. Carbohydrates content is determined from the calibration curve of glucose and the results are calculated as g/100 g fresh weight of the sample.

Alternatively, carbohydrates and fiber contents of latex-bearing medicinal plants are calculated using the following formula [21, 22]:

$$\text{Carbohydrates value} = 100 - (\%\ \text{lipid} + \%\ \text{moisture} + \%\ \text{ash} + \%\ \text{protein})$$

3.2.19 Absorption of Water

In this method, water-swellable rubber strips are immersed in distilled water at room temperature. The samples are removed periodically at specified time intervals. The dry and fresh weight of each sample is recorded [23].

$$\%\text{Water absorption} = \frac{\text{Wet weight} - \text{Dry weight}}{\text{Wet weight}} \times 100$$

3.2.20 Dry Rubber Content

The percentage of rubber hydrocarbon in the latex is known as the DRC and is calculated as per the protocol given by ISO 126 [11].

3.2.21 Total Solid Content

The percentage of solid particles in the latex is indicated by the term total solid content (TSC). According to ISO 1244, the TSC is the weight percentage of all the solids (rubber and non-rubber) that are present in latex [15].

The weight percentage of natural rubber solids (NRS), is represented by the difference between the TSC and DRC values. The total solid content of latex decreases with time of collection. Hence, when higher dry DRC is required, it is recommended not to mix the latex collected during the early stage with the latex collected during the late stage.

3.2.22 Elemental Analysis

It is reported that certain elements like phosphorus (P), calcium (Ca), potassium (K), magnesium (Mg), and sodium (Na) significantly affect metabolism of plant latex. Moreover, these elements affect oxidative degradation of natural rubber latex which can be regulated in the presence of certain molecules like proteins, vitamins and amino acids. Hence, the determination of these elements in natural rubber latex is important [24–26]. Elemental analysis of *Plumeria alba, Calotropis procera, Euphorbia neriifolia, Nerium indicum, Mimusops elengi* and Cannabis has also been reported [2, 26].

3.2.23 Solubility of Plant Latex Material

It is the capacity of a substance to dissolve in a specific solvent and is used to identify distinct types of molecules present in latex [5]. The solubility phenomenon is based on the principle of Like Dissolves Like and it is affected by the nature of the molecules, intermolecular attraction, polarity, dispersion, and other factors as well. Solubility was traditionally used to distinguish between latex, gums, and resins [10].

Hevea brasiliensis tree produces NRL which is a colloidal polymer (1,4-cis-isoprene). It includes polar and non-polar substances including proteins, fatty acids, phospholipids, carbohydrates, etc. Due to its exceptional qualities, including high flexibility and excellent film production, the latex mixing process is used in a variety of applications [27, 28]. In this process, the solubility of different compounds plays an important role.

3.3 Chemical Composition of Plant Latex

Latex containing plants has the molecular formula C_3H_3N and it is composed of carbon, hydrogen and nitrogen.

Secondary metabolites like alkaloids, glycosides, terpenoids, resins, saponins and numerous proteolytic enzymes are the most chemical compounds obtained from latex bearing medicinal plants [2, 3]. Figure 3.1 shows major components isolated from latex bearing medicinal plants.

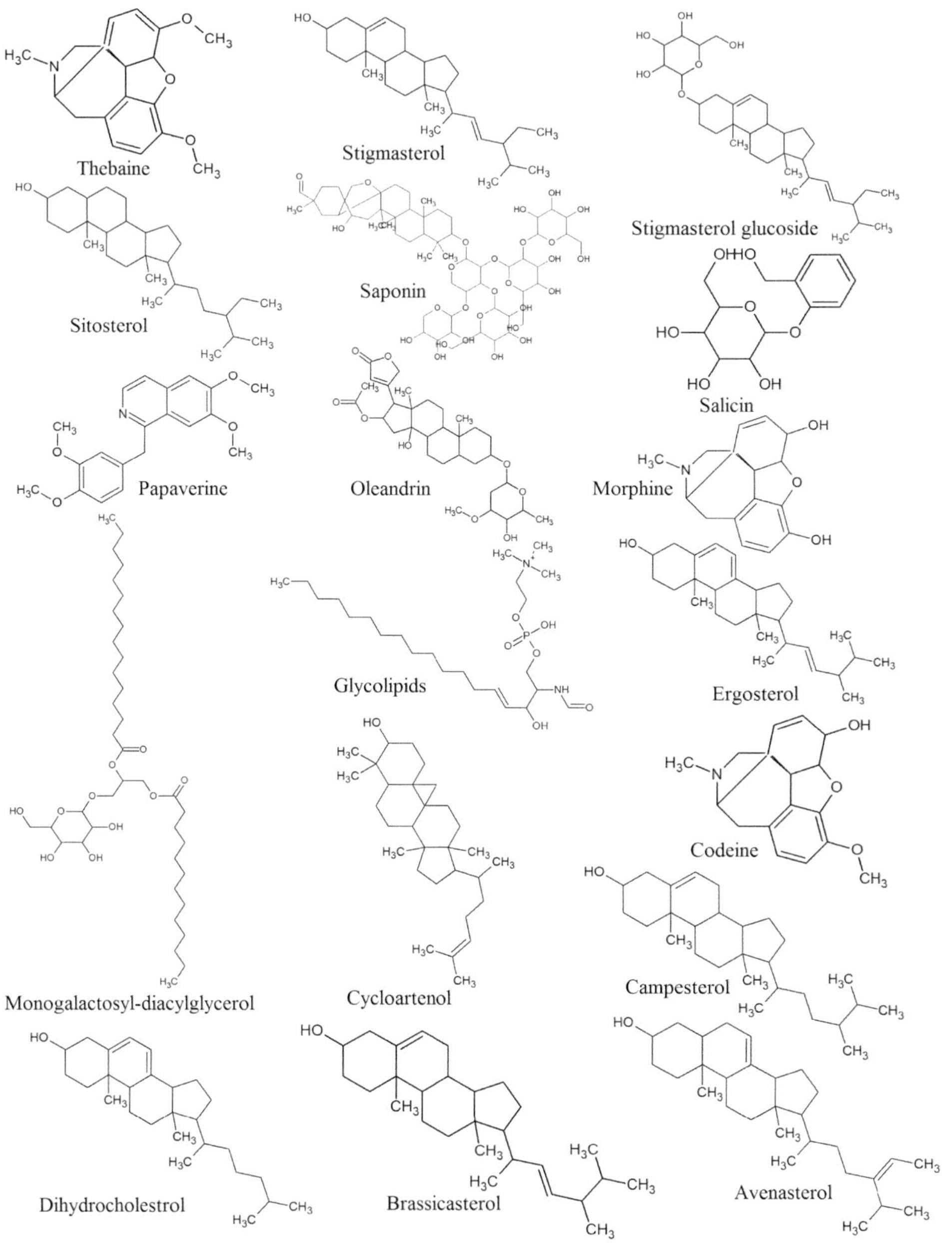

Figure 3.1 Phytoconstituents from latex bearing medicinal plants.

3.4 Classification of Main Chemical Components

Secondary metabolites and proteins can be identified as major latex components. Notably, several latex constituents have been shown to have a synergistic mechanism of action against insects or fungi as in combination of terpenes and phenolics. Secondary metabolites such as alkaloids, cardenolides, terpenoids, phenolics, proteases, protease inhibitors, proteins, chitinases, oxidases and others are often found in a variety of laticifers [29–33].

3.4.1 Alkaloids

Alkaloids are basic, low molecular weight compounds containing one or more nitrogen generally in the heterocyclic nucleus [34]. The various functional groups such as hydroxyl, methoxy, acetyl and methyl on the heterocyclic nucleus affect physicochemical and biological properties of latex. They are frequently present in the latex sap of certain plants. For example, Opium poppy (*Papaver somniferum*), Greater celandine (*Chelidonium majus*), Lobelia (*Lobelia cardinalis*), Mulberry trees (Morus spp.) and Ficus species. The chemistry of latex alkaloids focuses on their structural features, biosynthesis and medicinal properties. The concentration and accumulation of alkaloids in latex producing plants are influenced by exogenous and endogenous factors like plant parts, climate and geographical location [35]. The characteristics of alkaloids from latex bearing plants along with their chemical nature are depicted in Table 3.1.

3.4.2 Cardenolides

Cardenolides are cardiac glycosides which are steroidal in nature. A sugar, a steroid backbone, and a lactone ring make up these molecules. They are frequently detected in the sap of certain plant species latex. Cardenolides protect plants against herbivorous insects.

Table 3.1 Characteristics of alkaloids from latex-bearing plants

No.	Name	Biological source	Family	Chemical constituents	Pharmacological uses	Nature of chemical compound	Structure of main constituents present in latex	Refs.
1.	Greater celandine	*Chelidonium majus* L.	Papaveraceae	Coptisine, chelidonine, berberine	Anti-viral, antitumor, treatment of warts, papillae, condylomas	Isoquinoline alkaloid	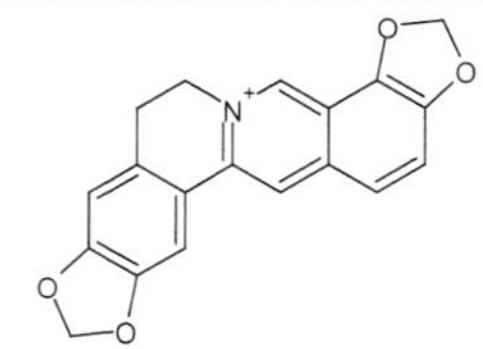 Coptisine	[36, 37]
2.	Bois lait	*Tabernaemontana citrifolia* L.	Apocynaceae	Ibogane, aspidospermane, corynane	Anthelmintic, anti-infective	Indole alkaloids	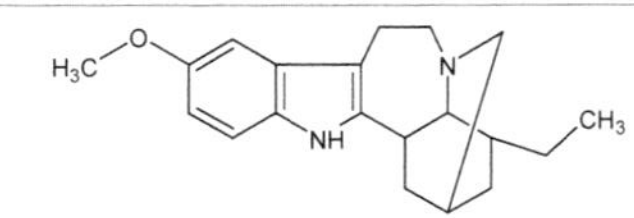 Ibogane	[38, 39]
3.	Pinwheel flower, East Indian rosebay	*Tabernaemontana divaricata* (L.) R. Br.	Apocynaceae	Ibogamine-18-carboxylic acid, 21-hydroxy-methyl ester, taberdines, 16α-hydroxyl-ibogamine	Abortifacient, antidiabetic, anticancer, anti-Alzheimer, antifungal, anti-epileptic, anti-mania, antioxidant, anti-infective	Iboga alkaloids	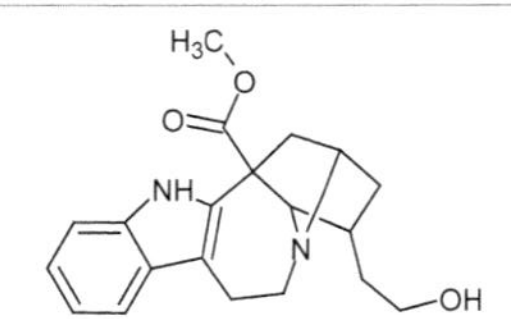 Ibogamine-18-carboxylic acid, 21-hydroxy-methyl ester	[40–50]

No.	Name	Biological source	Family	Chemical constituents	Pharmacological uses	Nature of chemical compound	Structure of main constituents present in latex	Refs.
4.	Buddha belly plant	*Jatropha podagrica* Hook.	Euphorbiaceae	Podacycline A, B, jatrogricaine A	Antihypertensive, antibacterial		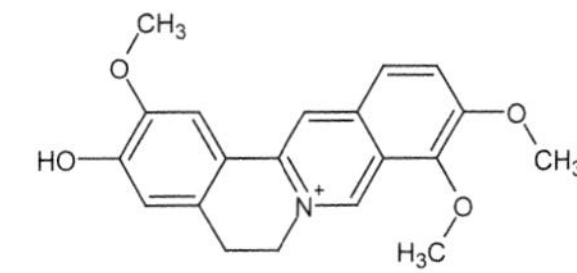 Jatrorrhizine	[51–55]
5.	Sadabahar	*Catharanthus roseus* L.	Apocynaceae	Vindoline, catechin, caffeic acid, vinblastine, catharanthine, vincristine, ajmalicine, reserpine, serpentine	Anti-oxidant, antidiabetic, anticancer, antibacterial	Indole alkaloids	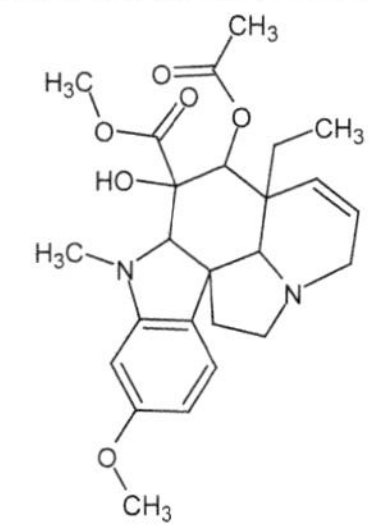 vindoline	[56–58]
6.	Arkaparni	*Tylophora indica* (Burm.f.) Merr.	*Asclepiadaceae*	Tylophorine, tylophorinine, tylophorinidine, kaempferol, quercetin, α-amyrin, β-amyrin, β-sitosterol	Antiasthmatic, antidiarrheal, anticancer, antiarthritic, antiepileptic, anti-inflammatory, immunomodulator	Phenanthro-indolizidine alkaloids	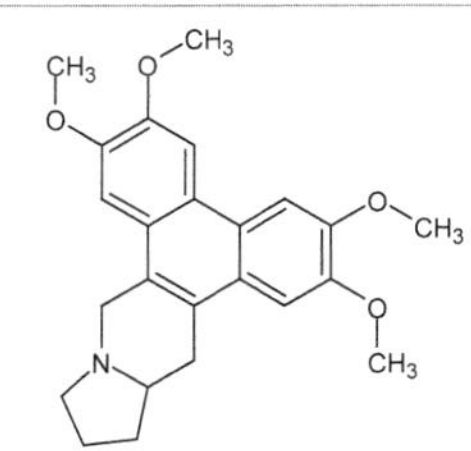 Tylophorine	[59–64]

(*Continued*)

Table 3.1 (*Continued*)

No.	Name	Biological source	Family	Chemical constituents	Pharmacological uses	Nature of chemical compound	Structure of main constituents present in latex	Refs.
7.	Sesudu	*Euphorbia antiquorum* L.	Euphorbiaceae	Atisine, quorumolides, rhizophorin B, ent-atisine, ent-atisane-3α,16α,17-triol	Antidiabetic, anti-inflammatory	Diterpenoid alkaloid	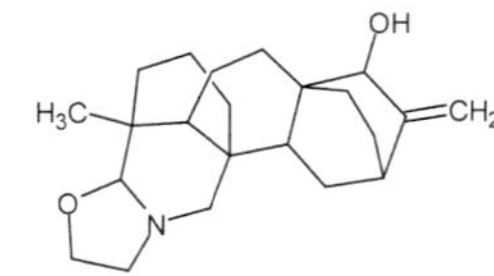 Atisine	[65–72]
8.	Kutaja	*Holarrhena pubescens*	Apocynaceae	Holanamine, conessine, conessimine, conimine, conessidine, holarrhimine, conarrhimine	Anti-malarial, anticancer, antidiabetic	Steroidal alkaloid	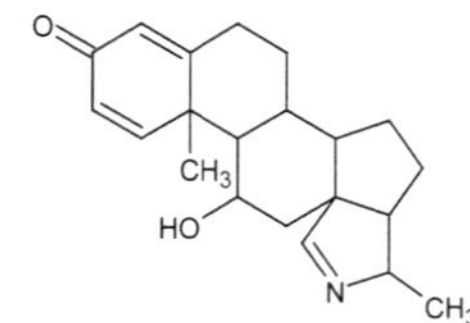 Holanamine	[73–78]
9.	Belly-ache bush	*Jatropha gossypifolia L.* var. elegans Muell. Arg	Euphorbiaceae	Jatrophenone, isogadain, jatrorrhizine	Analgesic, neurological disorders, anti-diarrheal, antidiabetic, antifertility	Isoquinoline alkaloid	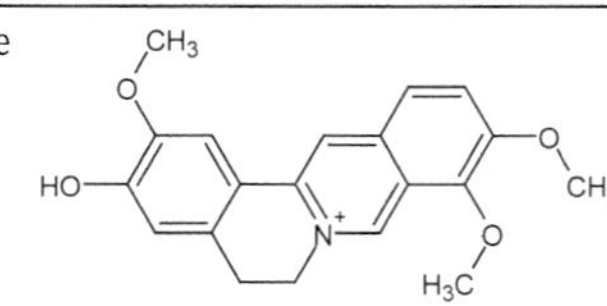 Jatrorrhizine	[79–85]

No.	Name	Biological source	Family	Chemical constituents	Pharmacological uses	Nature of chemical compound	Structure of main constituents present in latex	Refs.
10.	Black-board tree	*Alstonia scholaris* L.	Apocynaceae	Echitamine, alstonine, pleiocarpamine, macralstonine, lupeol	Antimicrobial, antiinflammatory, anticancer	Indole alkaloid	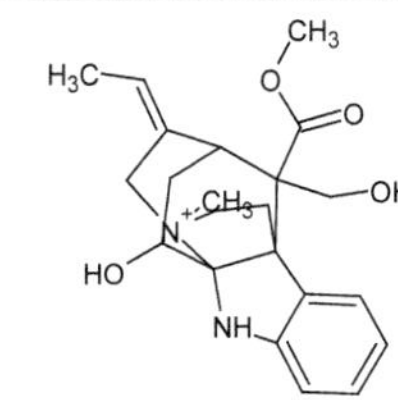 Echitamine	[86–91]
11.	Janthaala mara	*Alstonia macrophylla Wall*	Apocynaceae	Alstonerine, strictamine, macralstonine, thungfaine, sweroside, naresuanoside, rauvomitin, brucine, triterpenoids	Antimicrobial	Indole alkaloids	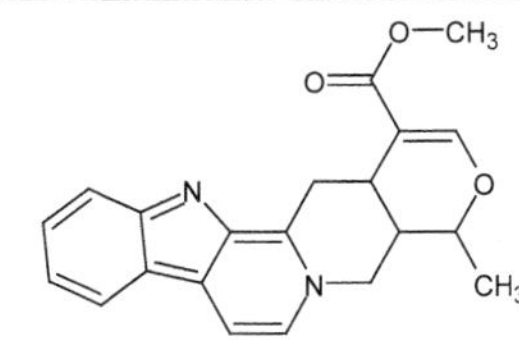 Alstonerine	[92–99]

The steroidal backbone composed of tetracyclic ring (A-D) system with common functional groups include hydroxyl (–OH) and unsaturated bonds. They have a lactone ring at C-17 position which can be five-membered (-cardenolides) or six-membered (-cardenolides). On the steroidal backbone several substituents and functional groups such as hydroxyl (–OH) and unsaturated bonds might be present. Cardenolides operate as Na^+/K^+-ATPase inhibitors, which are essential for maintaining cell electrical potential, regulating cellular volume and transmembrane transport. Cardenolides are extremely hazardous to most animals due to their inhibitory action on these enzymes. These chemicals are mostly found in plant groups like Apocynaceae and Asclepiadaceae with *Digitalis purpurea* and Asclepias species as examples.

The information regarding precursors, biosynthetic pathways and biotransformation of cardenolide is given below

A. **Precursor Molecules:** The biosynthesis of latex cardenolides involves the conversion of simple precursors such as cholesterol or sitosterol into cardenolides intermediates.
B. **Enzymatic Transformations:** Enzymatic reactions including hydroxylation, oxidation, glycosylation, and lactonization, significantly affect biosynthesis of latex cardenolides. Cytochrome P450 monooxygenases and glycosyltransferases are among the enzymes involved in these transformations.
C. **Pathways:** Different plant species may employ specific biosynthetic pathways to produce latex cardenolides. Examples of these pathways include mevalonate pathway and the deoxyxylulose phosphate pathway.

Chemical Names of Selected Latex Cardenolides

A. ***Digitalis purpurea* Cardenolides:** Digitoxin, Digoxin, Gitoxin.
B. **Asclepias Cardenolides:** Asclepin, Calotropin, Frugoside.
C. ***Antiaris toxicaria*:** *Antiaris toxicaria* Lesch., (Moraceae family) is rich with cardenolides content. Traditionally it is used as a poison on hunting darts by South Asian tribes. Characteristics of cardenolides from latex-bearing plants are given in Table 3.2.

Table 3.2 Characteristics of cardenolides from latex-bearing plants

No.	Name	Biological source	Family	Chemical constituents	Pharma-cological uses	Nature of chemical compound	Structure of main constituents present in latex	Refs.
1.	Aak, madar	*Calotropis procera*	Apocynaceae	Afroside, 15β-hydroxycalactin, calactoprocin, calotoxin, calotropin	Anti-inflammatory, treatment of snake bite, scabies, boils, migraine, leprosy, ascites	Cardenolides	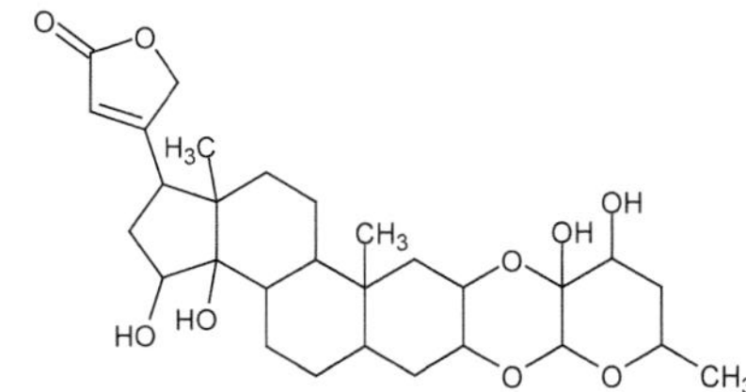 Afroside	[100–103]
2.	Upas tree, Sacking tree	*Antiaris toxicaria* Lesch.	Moraceae	Toxicariosides A, Toxicarioside F, G, H	Anticancer	Cardenolides	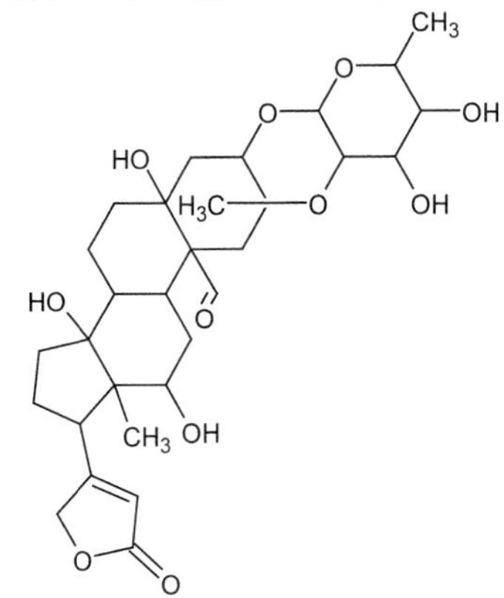 Toxicaroside A	[104, 105]

Table 3.2 (*Continued*)

No.	Name	Biological source	Family	Chemical constituents	Pharma-cological uses	Nature of chemical compound	Structure of main constituents present in latex	Refs.
3.	Tropical milkweed	*Asclepias curassavica*	Apocynaceae	Coroglaucigenin, corotoxigenin	Analgesic, anti-inflammatory	Cardenolides	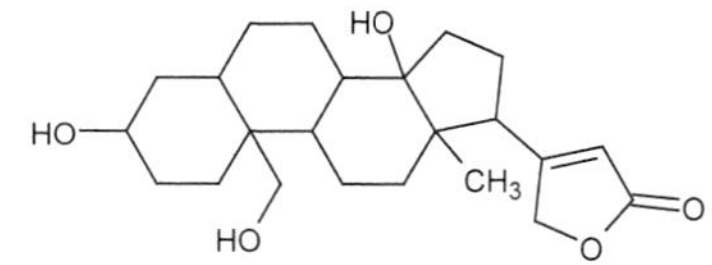 Coroglaucigenin	[106–108]
4.	Crown Flower	*Calotropis gigantea* L.	Apocynaceae	Caloside G, uscharidin, calotropin	Anticancer, antimigraine	Cardenolides	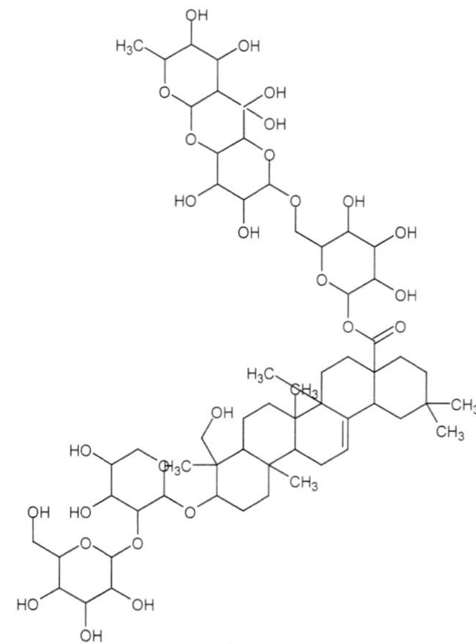 Caloside G	[109–114]

3.4.3 Terpenoids

Terpenoids comprising of five carbon (C5) atom isoprene units (cis-1,4-polyisoprene). This chemical is present in a number of plant families (8) and genera (300) [115]. Terpenoids serve a variety of purposes in plants such as helping in pollination, providing protection and taking part in primary metabolism. The latex of Asteraceae plant *Lactuca sativa* L. constitutes many sesquiterpene lactones (SLs) along with lettucenin A [116]. In addition to lettucenin A, sesquiterpene lactones show antifungal activity [116]. Certain members of Euphorbiaceae family such as *Euphorbia biglandulosa* Desf., contains higher amount of diterpenes, phorbols and their derivatives. These molecules improve cell adhesion and have antitumor properties. Moreover, these molecules produce skin irritation and are toxic to animal and insects [117, 118]. Table 3.3 presents the characteristics of terpenoids from latex bearing plants.

3.4.4 Phenolics

Phenolic compounds (e.g., tannins, lignans, coumarins, and flavonoids which are synthesized through shikimic acid pathway and regulates physiological and biological processes [187]. Though little research has been done on phenolics existence in latex. One research indicated that the latex of *Hevea brasiliensis* contains phenolic acids and flavonoids whereas *Ipomoea batatas* (L.) Lam. (Convolvulaceae), contains greater quantities of p-coumaric acids [188].

Fresh vine and root latex of sweet potato had 3% and 10% p-coumarate esters respectively and are protective in nature [189, 190]. In addition, it is well known that the latex of Rhus species contains urushiol and it can cause strong skin irritations. The variability in reactions to phenolics at the individual and ecological level may be largely explained by oxidative activation [188]. Table 3.4 presents the characteristics of phenolics from latex bearing plants.

Table 3.3 Characteristics of terpenoids from latex-bearing plants

No.	Name	Biological source	Family	Chemical constituents	Pharma-cological uses	Nature of chemical compound	Structure of main constituents present in latex	Refs.
1.	Sandbox Tree	*Hura crepitans*	Euphor-biaceae	Huratoxin, prohuratoxin, 20-OH-octadecanoyl huratoxin	Piscicidal, antiviral, anti-tumor	Terpenes	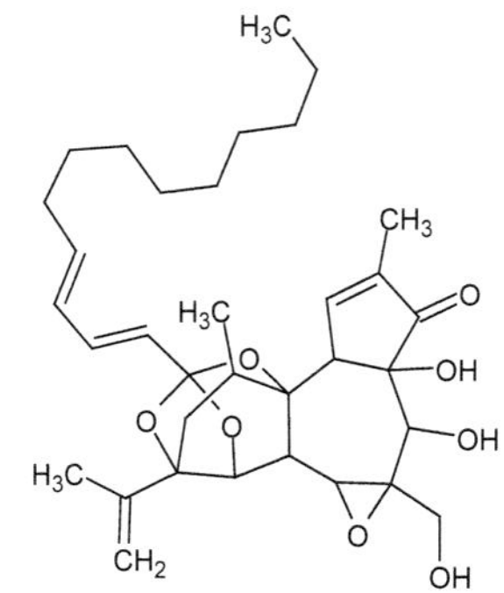 Huratoxin	[119, 120]
2.	Dravantī	*Jatropha glandulifera* Roxb.	Euphor-biaceae	Calenduladiol, 2α-OH-jatrophone, 2β-OH-jatrophone	Wound healing, anti-inflammatory, antimicrobial, antioxidant, antithrombotic, antitumor, antiplasmodial	Triterpenoid	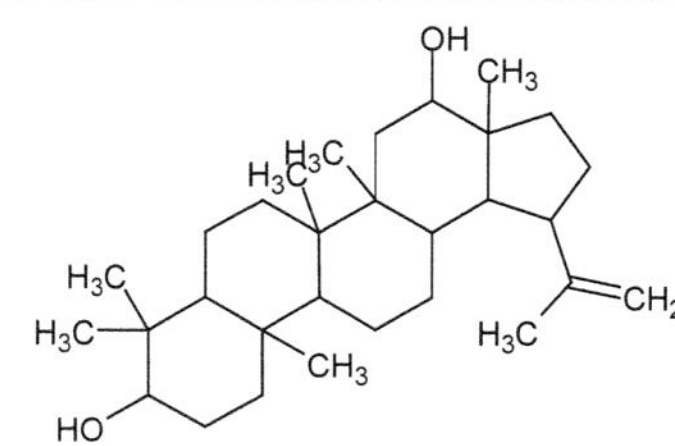 Calenduladiol	[121–123]

No.	Name	Biological source	Family	Chemical constituents	Pharma-cological uses	Nature of chemical compound	Structure of main constituents present in latex	Refs.
3.	Indian Bat tree	*Ficus amplissima* J.E.Sm.	Moraceae	Oleanolic acid, betulinic acid, lupeol, β-sitosterol, catechin, gallic acid	Antidiabetic, anti-inflammatory, wound healing, antioxidant	Pentacyclic triterpenoids	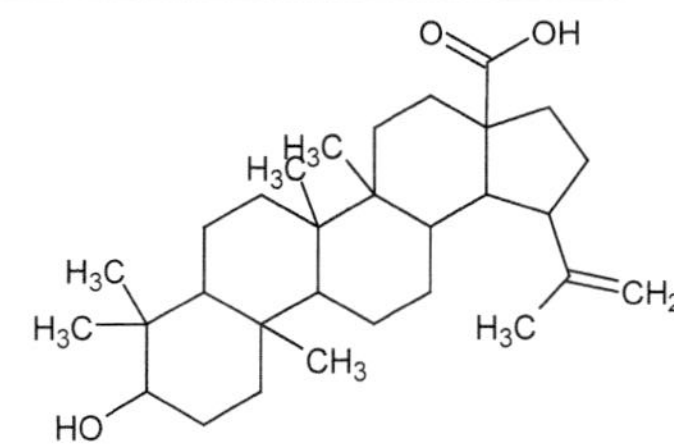 Betulinic acid	[124, 125]
4.	Bully tree, sawo, and marmalade plum	*Manikara zapota* L.P. van Royan	Sapotaceae	Taraxerol methyl ether, spinasterol, 6-OH-flavanone, dihydrok-aempferol, taraxerol, taraxerone	Antidiarrhoeal, antinociceptive, antimicrobial, anthelmintic, antidiabetic, antilipidemic	Triterpenoids	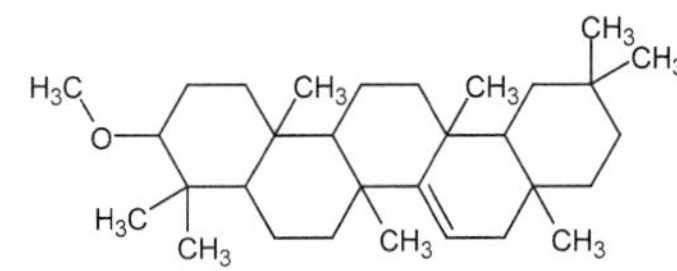 Taraxerol methyl ether	[126–130]
5.	Indian rock fig	*Ficus arnottiana* (Miq.)	Moraceae	Oleanolic acid, betulinic acid, lupeol, β-sitosterol, catechin, gallic acid	Antitumor, anti-inflammatory, antiviral	Pentacyclic triterpenoids	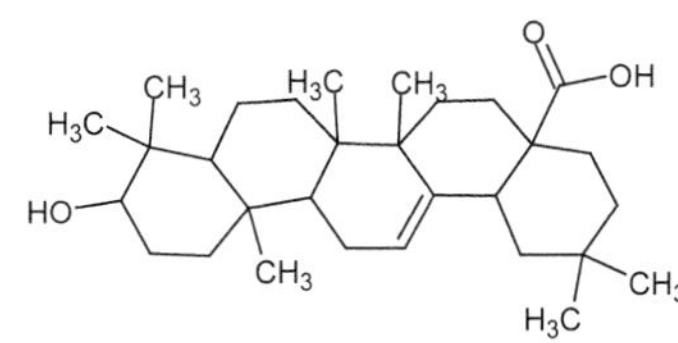 Oleanolic Acid	[131, 132]

(*Continued*)

Table 3.3 (*Continued*)

No.	Name	Biological source	Family	Chemical constituents	Pharma-cological uses	Nature of chemical compound	Structure of main constituents present in latex	Refs.
6.	Spurge	*Euphorbia indica* Lamk.	Euphor-biaceae	Taraxerol	Antiviral, antimicrobial	Triterpenes	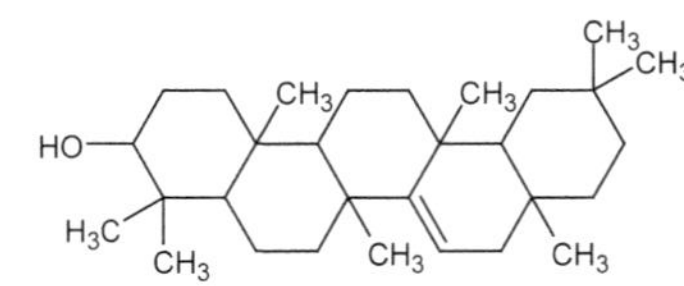 Taraxerol	[133]
7.	Prostrate Spurge	*Euphorbia granulate* Forsk	Euphor-biaceae	12-deoxy-ingenol, 20-deoxy-ingenol, resiniferotoxin, 16-hydroxy phorbol,12-deoxy-phorbol	Antiviral, antimicrobial	Diterpenes	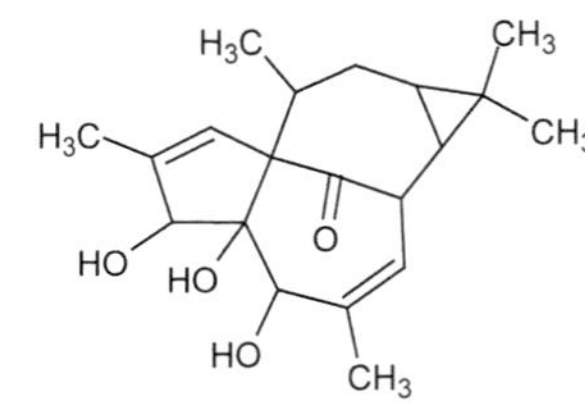 20-Deoxyingenol	[134–137]
8.	Physic nut	*Jatropha curcas* L.	Euphor-biaceae	12-deoxy-16-OH-phorbol, preussomerin, phorbol ester	Wound healing, anti-inflammatory, antibacterial, anti-fungal	Diterpenes	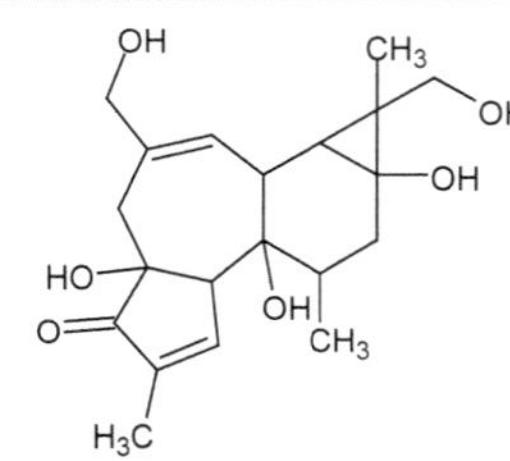 12-Deoxy-16-hydroxyphorbol	[138–144]

No.	Name	Biological source	Family	Chemical constituents	Pharma-cological uses	Nature of chemical compound	Structure of main constituents present in latex	Refs.
9.	Clark's Spurge	*Euphorbia heynena* Spreng.	Euphor-biaceae	Eupneonoids, lathyrane, jatrophane, ingenane, ingenol, ingol	Anti-tumor, antibacterial	Diterpenes	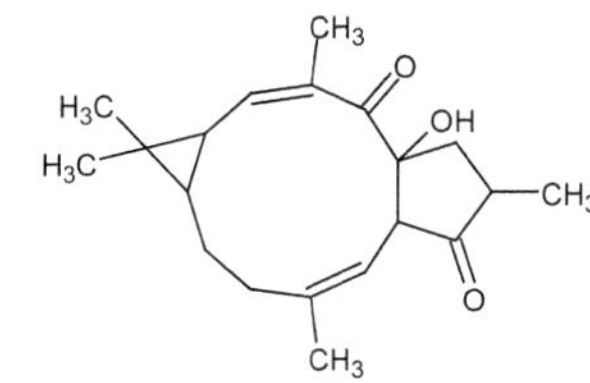 Lathyrane	[145–147]
10.	Common fig	*Ficus carica* L.	Moraceae	Lupeol acetate, lupeol palmitate, germacrene-D, δ-cadinene, triacontane	Anti-tumor, anthelmintic, antioxidant, antibacterial	Triterpenes	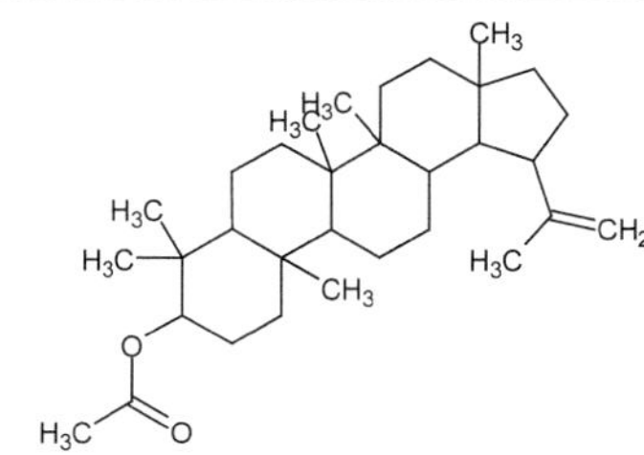 Lupeol acetate	[148–152]

(*Continued*)

Table 3.3 (*Continued*)

No.	Name	Biological source	Family	Chemical constituents	Pharma-cological uses	Nature of chemical compound	Structure of main constituents present in latex	Refs.
11.	Dragon Spurge	*Euphorbia dracunculoides* Lamk	Euphor-biaceae	Euphor-draculoates 8-hydroperoxy tigliane, euphorantin A eupulcherol A	Antioxidant, anti-inflammatory, antiarthritic	Terpeniods	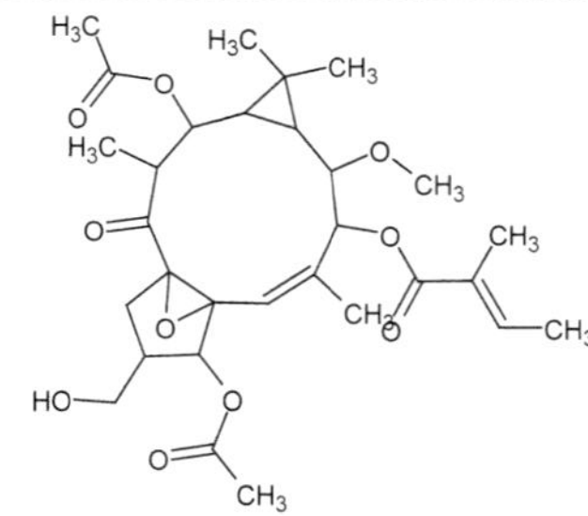 Euphorantin A	[153–158]
12.	Lettuce	*Lactuca sativa* L.	Asteraceae	Lactucopicrin oxalate, lactucin, phytoalexin, lettucenin A	Antifungal, antimicrobial	Sesquiterpenes	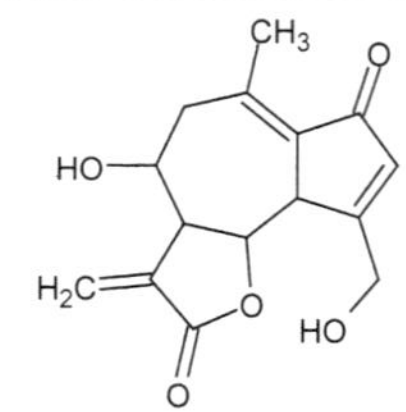 Lactucin	[116, 117, 159, 160]

No.	Name	Biological source	Family	Chemical constituents	Pharma-cological uses	Nature of chemical compound	Structure of main constituents present in latex	Refs.
13.	Petty spurge, radium weed	*Euphorbia peplus* L.	Euphor-biaceae	Ingenol-mebutate	Anticancer	Diterpenes	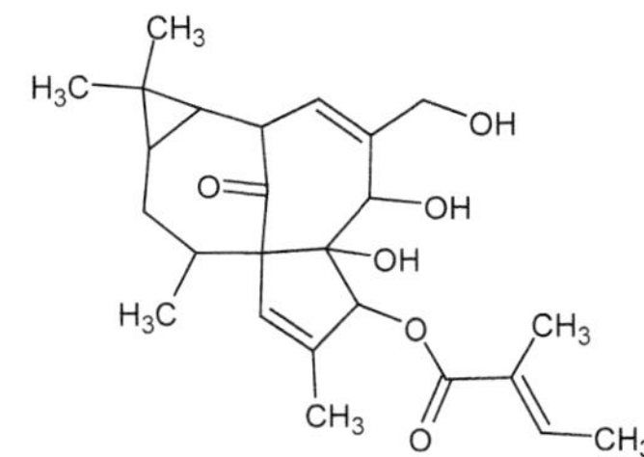 Ingenolmebutate	[161, 162]
14.	Pencil cactus	*Euphorbia tirucalli* L	Euphor-biaceae	Tigliane, ingenane	Anticancer, antiasthmatic, ulcer protective, antimutagenic, anti-inflammatory	Diterpenoid	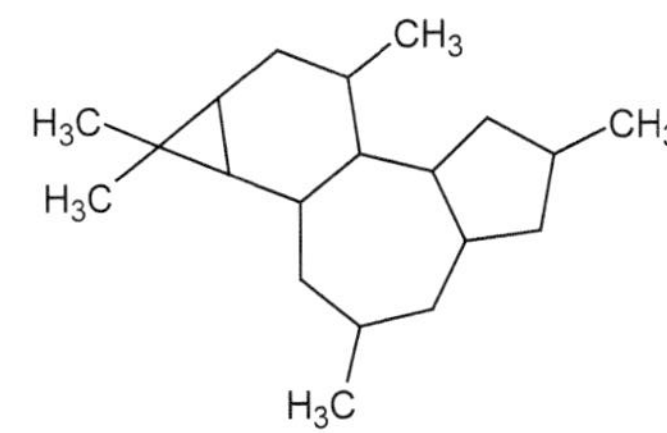 Tigliane	[163–168]

(*Continued*)

Table 3.3 (*Continued*)

No.	Name	Biological source	Family	Chemical constituents	Pharma-cological uses	Nature of chemical compound	Structure of main constituents present in latex	Refs.
15.	Australian cowplant	*Gymnema sylvestre* (Retz.)	Apo-cynaceae	Gurmarin, gymnemic acid, gymnema saponins	Antidiabetic, antioxidant, antibiotic, anti-inflammatory, antiviral, gastroprotective, hepatoprotective, anticancer	Triterpenoid glycosides	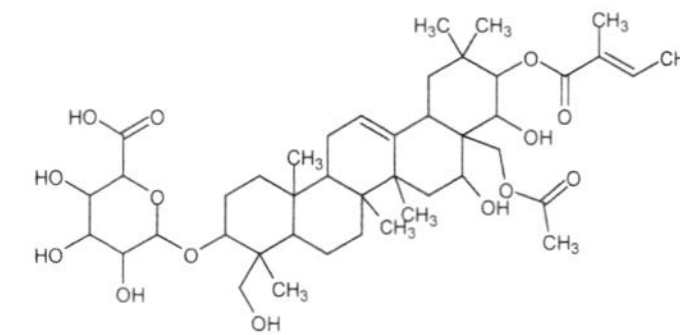 Gymnemic acid	[169–173]
16.	Scentless Cowplant	*Gymnem a tingens* (Roxb.) Wt. & Am.	Asclepia-daceae	Gymnefuranols, gymnetinosides, sequinoside K	Hepatoprotective, antidiabetic	Triterpenoid saponins	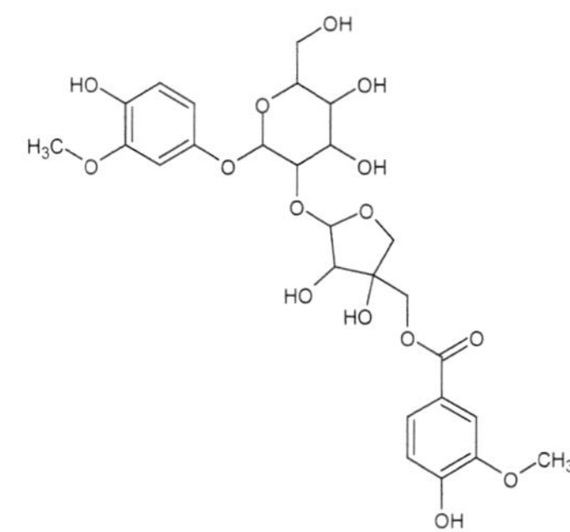 Sequinoside K	[174–176]

No.	Name	Biological source	Family	Chemical constituents	Pharma-cological uses	Nature of chemical compound	Structure of main constituents present in latex	Refs.
17.	Jivanti, Jivaniya	*Leptadenia reticulata* (Retz.) Wt. & Am.	Asclepi-daceae	Flavonoids, terpenoids and phenolic acids	Antibacterial, antioxidant, anti-inflammatory, antidiabetic	Triterpenes	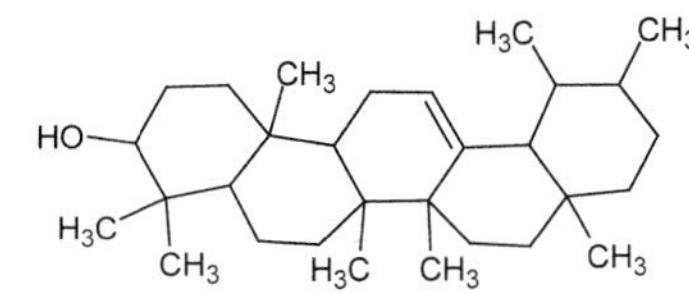 α-Amyrin	[177–181]
18.	Hairy Spurge	*Euphorbia hirta* L.	Euphor-biaceae	Euphorhirtins A,D, euphornin	Immunomodulator, antidiarrhoeal, antibacterial	Diterpene	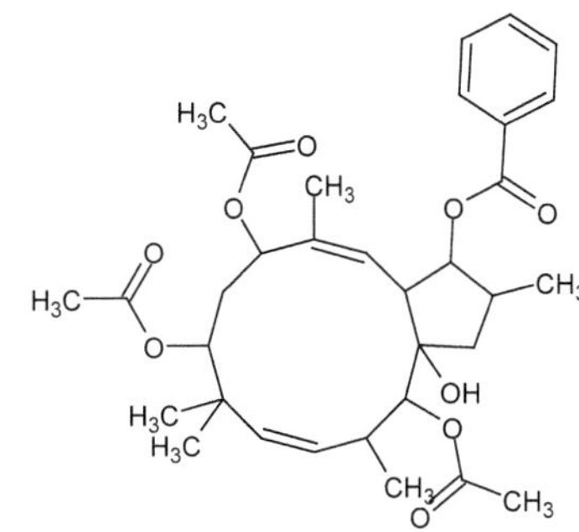 Euphornin	[182–186]

(*Continued*)

Table 3.4 Characteristics of phenolics from latex-bearing plants

No	Name	Biological source	Family	Chemical constituents	Pharmacological uses	Nature of chemical compound	Structure of main constituents present in latex	Refs.
1.	Yellow oleander	*Thevetia peruviana* L.	Apocynaceae	Rutin, quercetin, naringenin, hesperidin, chlorogenic, syringenic, ferulic acid	Anticancer, anticoagulant, antioxidant, antimicrobial	Flavonoid	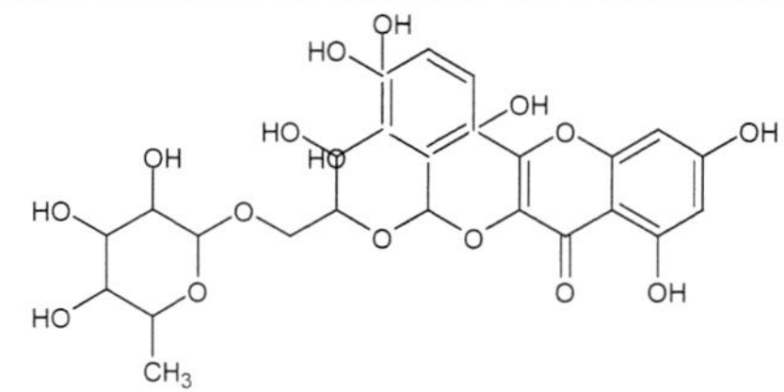 Rutin	[191]
2.	Soma	*Sarcostemma acidum* (Roxb.) Voigt.	Asclepiadaceae	Sacidumlignans A-D, sacidumols A, pinoresinol, perforatic acid	Anticancer, antiasthmatic, ulcer protective, antimutagenic, anti-inflammatory	Lignans	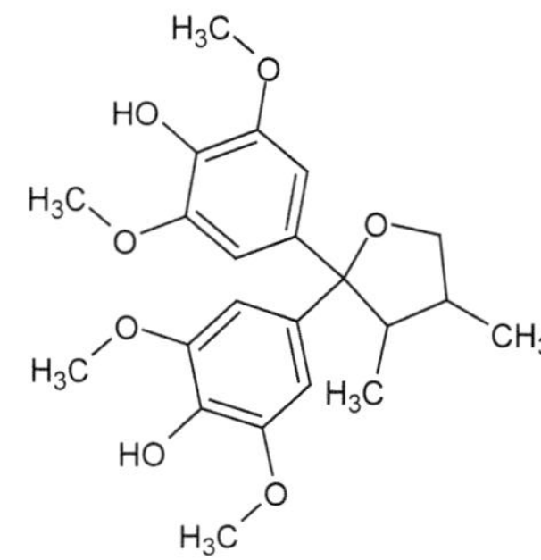 Sacidumlignans D	[192–194]

No	Name	Biological source	Family	Chemical constituents	Pharma-cological uses	Nature of chemical compound	Structure of main constituents present in latex	Refs.
3.	–	*Euphorbia coccinea* Roth	Euphor-biaceae	Gossypetin, euphogossypins, gossypilignan, pinoresinol, naringenin, quercitrin	Antimicrobial	Flavanoids	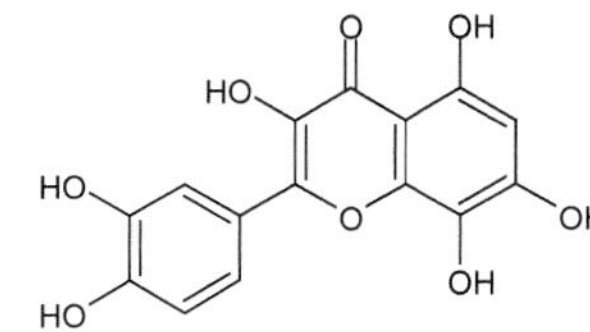 Gossypetin	[195, 196]
4.	Karandang, Kerenda	*Carissa carandus* L. Mant.	Apocynaceae	Epicatechingallate, Flavonoids, terpenoids and phenolic acids	Antioxidant	Catechins	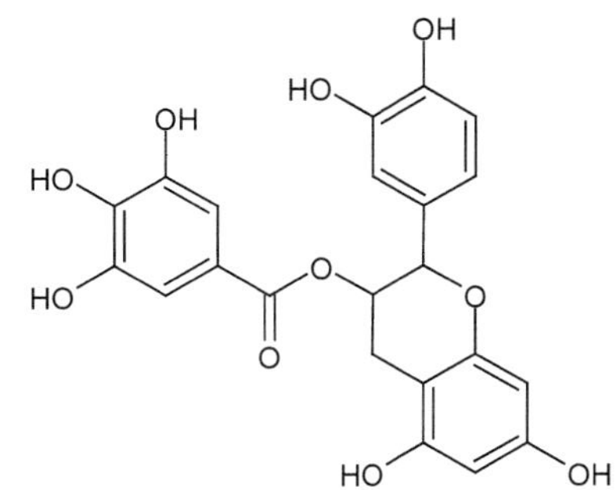 Epicatechingallate	[197–199]

(*Continued*)

Table 3.4 (*Continued*)

No	Name	Biological source	Family	Chemical constituents	Pharma-cological uses	Nature of chemical compound	Structure of main constituents present in latex	Refs.
5.	Golden Trumpet Vine	*Allamanda cathartica* L.	Apocynaceae	Glabiridin, heneicosane, heneicosanoic acid, β-ocineme, β-farnesene protocatechuic acid	Antimicrobial, wound healing, anti-depressant	Isoflavane	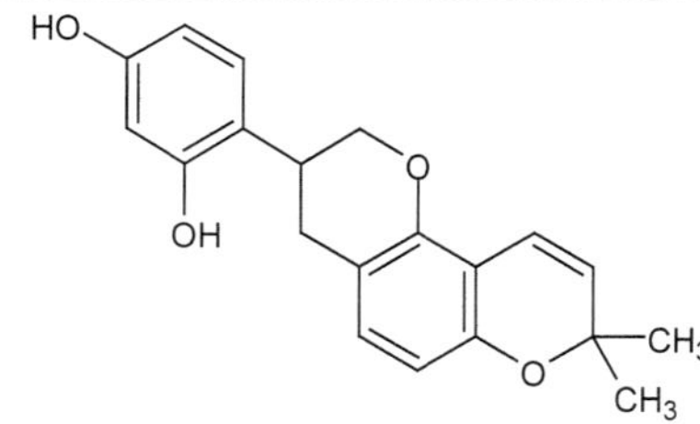 Glabridin	[200, 201]
6.	Aak or Madar	*Calotropis procera*	Apocynaceae	Isoquercitrin, quercetagetin, quercetin, isorhamnetin	Antimicrobial, anticancer, anti-genotoxicity, antifilarial	Flavonoid	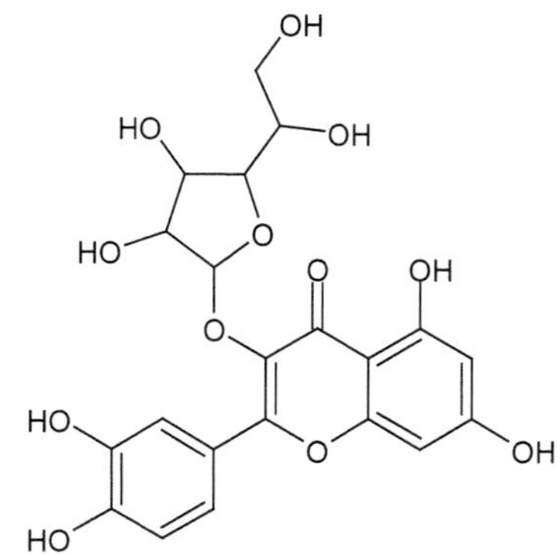 Isoquercitrin	[101, 202–206]

No	Name	Biological source	Family	Chemical constituents	Pharma-cological uses	Nature of chemical compound	Structure of main constituents present in latex	Refs.
7.	Dudhi bel	*Holostemma Ada*-kodien Schult.	Asclepia-daceae	Dihydroxycinnamic acid methyl ester, lup-20(29)-ene-3β, 28-diol, 3β-hydroxy-12-ursen-28-ic acid, 3,4-7-hydroxy-6-methoxycoumarin, 3,4-dihydroxycinnamic acid	Antimicrobial	Phenyl-propanoids	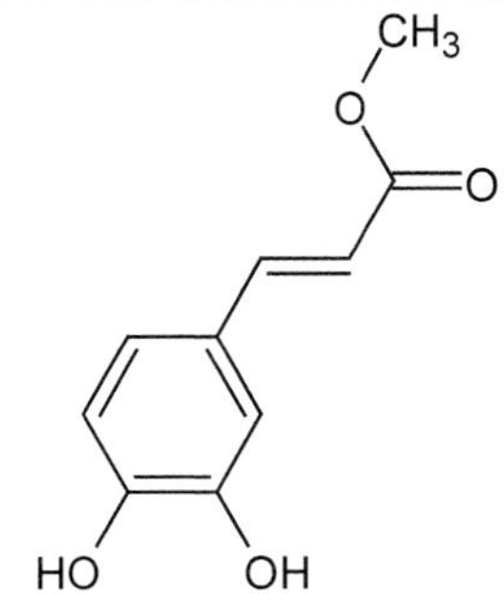 3,4-Dihydroxycinnamic acid methyl ester	[207]
8.	Aak family	*Pentatropis spiralis* (Forsk.)	Asclepia-daceae	Astragalin, Flavonoids, terpenoids and phenolic acids	Antioxidant	Flavonols	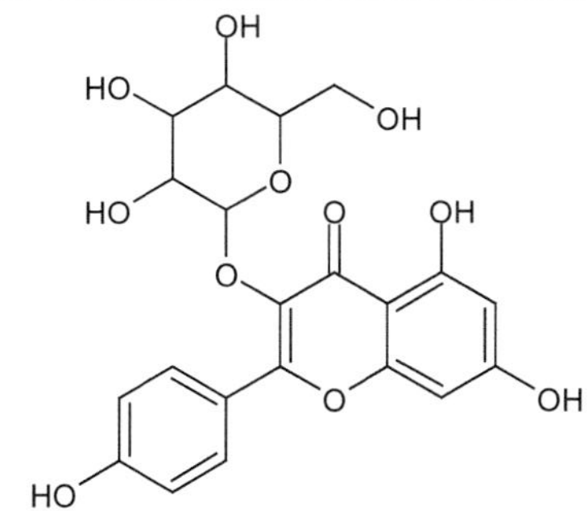 Astragalin	[208, 209]

(*Continued*)

Table 3.4 (*Continued*)

No	Name	Biological source	Family	Chemical constituents	Pharma-cological uses	Nature of chemical compound	Structure of main constituents present in latex	Refs.
9.	Trellis vine	Pergularia daemia	Asclepia-daceae	Formononetin, qurecetin, chrysoeriol, taxifolin, naringenin	Hepato-protective, antioxidant, antiepilepto-genic, neuro-protective	Isoflavones	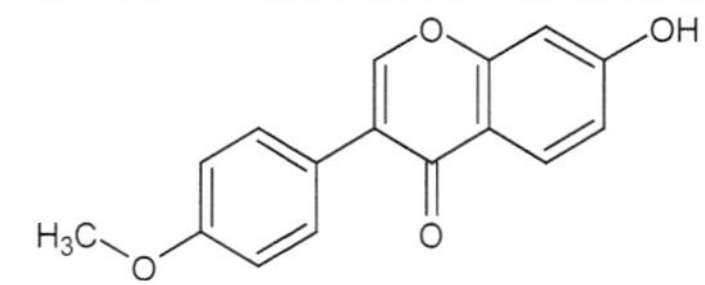 Formononetin	[210–215]
10.	Rubber tree	*Hevea brasiliensis*	Euphor-biaceae	Flavonoids, terpenoids and phenolic acids	Drug delivery systems, antioxidant	Phenol	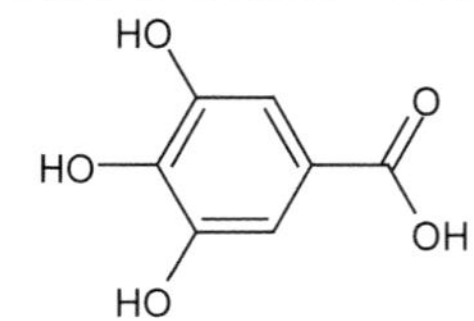 Gallic acid	[216, 217]
11.	Jackfruit	*Artocarpus heterophyllus L.*	Moraceae	Cycloartomunin, cyclomorusin, cudraflavone, cyclocommunin, artonins, artocarpanone	Antimicrobial, antifungal, fibrinolytic	Flavones and Flavonols	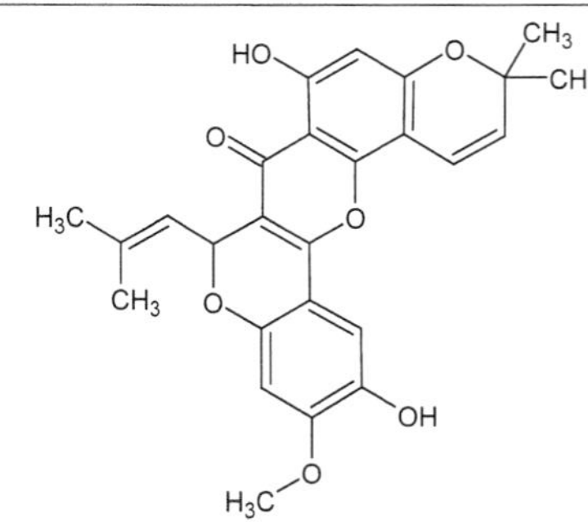 Cycloartomunin	[218–221]

No	Name	Biological source	Family	Chemical constituents	Pharma-cological uses	Nature of chemical compound	Structure of main constituents present in latex	Refs.
12.	Coral plant	*Jatropha multifida* L.	Euphor-biaceae	Scoparone, scopoletin, glaberidei, pinoresinol, medioresinol, syringaresinol, matairesinol, secoisolariciresinol, multifolone, (4E)-jatrogrossi-dentadione acetate, jatromultones A-I	Antimicrobial, anti-inflammatory, antioxidant	Coumarins	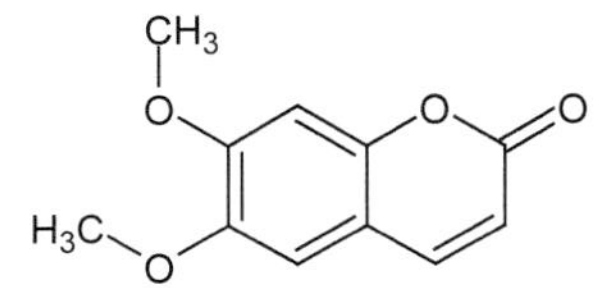 Scoparone	[222–227]
13.	Peregrina	*Jatropha integerrima* Jacq.	Euphor-biaceae	Fraxetin, vitexin, scopeltin, stigmasterol	Antimicrobial, antioxidant	Coumarins	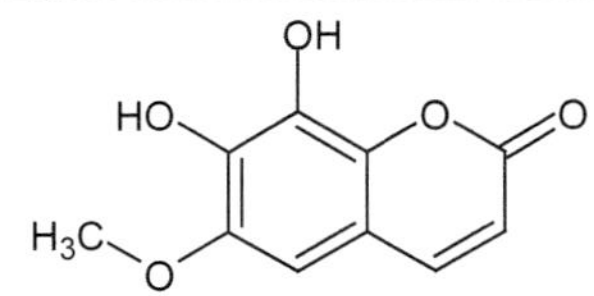 Fraxetin	[228, 229]

3.4.5 Proteases

Proteases are proteolytic enzymes present in the biological system. The latex of Caricaceae, Moraceae, Apocynaceae plant families contains cysteine and serine protease [230–232]. The mango, *Mangifera indica*, belongs to family Anacardiaceae, produces resinous fluid that resembles latex and includes both cysteine and serine proteases [233]. Earlier studies showed that the considerable toxic effects of papaya and wild fig (*Ficus virgata*) latex on the Eri silkworm and the cabbage worm was due to the presence of cysteine protease [234]. These findings showed that cysteine proteases possess a significant function in plant resistance and are present in the latex of fig (ficin) and papaya (papain). The protease activity in the latex of papaya is almost 200 times higher in latex than in leaves [234].

Cysteine proteases degrade membrane of insect's midgut [235]. Characteristics of proteases from latex bearing plants are presented in Table 3.5.

3.4.6 Protease Inhibitors

Protease inhibitors (PIs) are the agents that inhibit the breakdown of proteins and are considered as anti-nutritive agents. *Ficus carica* and *Carica papaya* latex contains inhibitors of trypsin (serine protease) [240, 241]. The laticifers of *H. brasiliensis* also express the trypsin inhibitor gene [242]. In addition, the *Cucurbita maxima* plant (pumpkin) belonging to family cucurbitaceae has a range of protease inhibitors such as trypsin and chymotrypsin [243, 244]. In many plants lacking latex protease inhibitors play recognized protective functions against herbivores and fungus by inhibiting proteolysis and protein consumption [245]. When leaves are injured, trypsin blockers, are highly stimulated in latex [246].

3.4.7 Proteins

Plant latex is considered as significant source of proteins and enzymes like proteases, oxidases, chitinases, lectins and their inhibitors.

Table 3.5 Characteristics of proteases from latex-bearing plants

No	Name	Biological source	Family	Chemical constituents	Pharmacological uses	Nature of chemical compound	Structure of main constituents present in latex	Reference
1.	Papaya	*Carica papaya* L.	Caricaceae	Papain, chymopapain, endopeptidase, papain III and IV, peptidase A and B, lysozymes	Anticancer, molluscicidal	Cysteine proteases	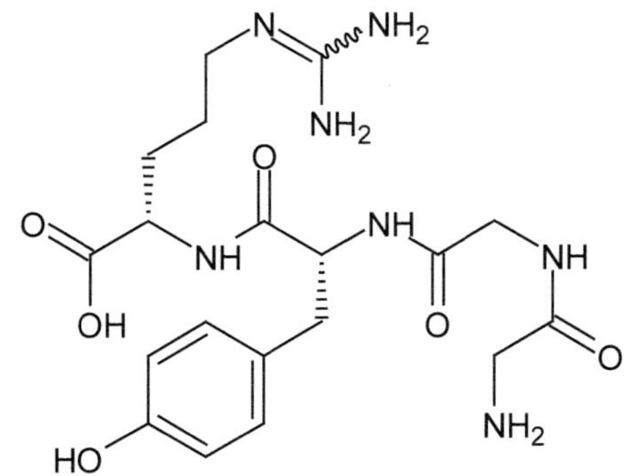 Papain	[236–239]

These proteins carry out several physiological processes in signaling pathways and antioxidant defense [246, 247]. Latex contains a diversity of proteins for defense against natural plant's predator [248]. It was interesting to learn that several of the proteins have direct toxicity for herbivores [249, 250].

Cystatins from *Chelidonium majus* L. latex are protective in nature and possess antimicrobial properties [251, 252]. Furthermore, it is reported that a variety of latex proteins are connected to a variety of inside molecular pathways as well as changing in the outside environment [253]. The antiviral effects of latex proteins are also reported [254].

Latex proteins are also acting against bugs including hemipteran, lepidopteran and co-leopteran [255–259] and are involved in latex coagulation process [251].

3.4.8 Lectins and Chitin-Binding Proteins that Resemble Hevein

Proteins that bind to carbohydrates and have an affinity for particular sugar moieties are known as lectins. Many of them are known for their insecticidal properties [259]. Latex from the families Euphorbiaceae, Moraceae, Apocynaceae and Cucurbitaceae has been discovered to contain a variety of lectins. The important latex protein hevein derived from *H. brasiliensis* is crucial for binding and adhesion of rubber fragments. Rubber coagulates when exposed to air due to hevein attachment to specific receptor protein and cross-linked rubber particles. The phloem sap of cucurbits has also been examined for its coagulation processes. This coagulation process can glue beetle mouthparts and prevent grazing [260]. Numerous caterpillars especially the cabbage worm *Mamestra brassicae*, are extremely poisonous to the chitin-binding proteins from Morus latex [261]. Proteins from *Calotropis procera* latex such as chitin-binding, chitinolytic and cysteine protease were found to have insecticidal effect [262]. Wheat germ lectin is a poisonous chitin-binding protein with hevein-like domains prevents the development of the peritrophic membrane in the stomach of insects [263].

3.4.9 Chitinases

Chitinases are the enzymes involved in the breakdown of chitin. They are distributed among the member of Caricaceae, Moraceae and Euphorbiaceae families. The *F. carica* and *C. papaya* latex expresses more chitinases that transmits signals and stimulate plants' response against herbivores [241]. It is reported that chitinases of insect origins have harmful effects on other insects. Also, chitinases may play a protective role in latex bearing plants as antifungal agents [264, 265]. The poisonous action and protective activities of plant chitinases are not well known [266].

3.4.10 Oxidases

Common plant oxidases include polyphenol oxidase and peroxidase which are found in plant families like Euphorbiaceae, Moraceae and Anacardiaceae [233, 267]. Such enzymes are the part of plant defense system and oxidize hydroxyphenolic compounds into their carbonyl derivatives and lower the nutritional value of proteins [245]. Proteomic studies of the phloem sap secreted by *Cucurbita sativa* and *Cucurbita maxima* found the presence of lipoxygenases (LOX) [243]. LOXs are assumed to be defense proteins and are frequently triggered by damage or jasmonic acid [245].

3.4.11 Others

As per earlier reports plant defense system involve few latex proteins. These substances include lipase in Caricaceae, Euphorbiaceae and Apocynaceae families and phosphatase in Euphorbiaceae family [268, 269]. In the Caricaceae family glutaminylcyclase and in Fabaceae family arabinogalactan proteins are reported [270, 271]. Linamarase is a protein found in the latex and leaves of the Cassava plant. The enzymatic activity is more in latex than leaves [272]. Characteristics of other important compounds from latex-bearing plants are given in Table 3.6.

Table 3.6 Characteristics of other important compounds from latex-bearing plants

No	Name	Biological source	Family	Chemical constituents	Pharmacological uses	Nature of chemical compound	Structure of main constituents present in latex	Refs.
1.	Upright myrtle spurge	*Euphorbia biglandulosa* Desf.,	Euphorbiaceae	Cerebrocides, atractyloside	Fish poison, tumor-promoting effects	Glycolipids	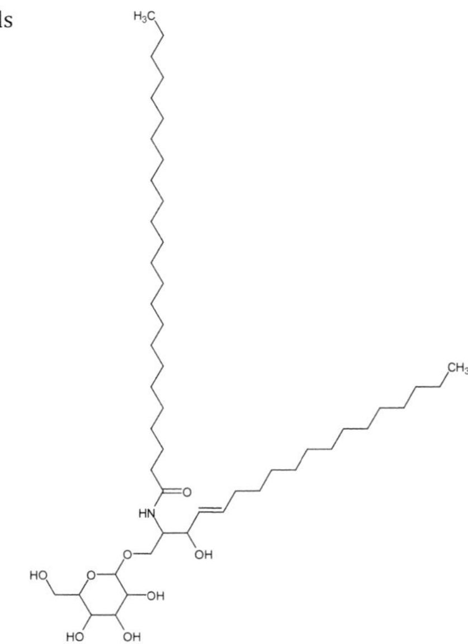 Cerebrocides	[273, 274]

No	Name	Biological source	Family	Chemical constituents	Pharmacological uses	Nature of chemical compound	Structure of main constituents present in latex	Refs.
2.	Tikiout	*Euphorbia officinarum* L.	Euphorbiaceae	Sterols	Antileishmania, antitrypanosomal, cytotoxic, antimicrobial, antiinflammatory, dermal and ocular disorders	Cholestane steroids	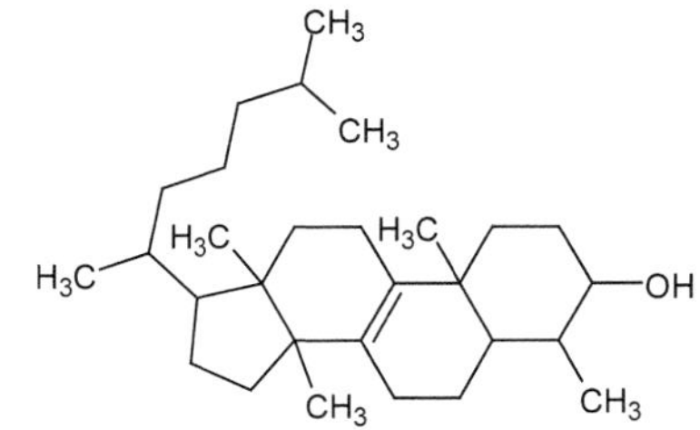 31-Norlanostenol	[275–277]
3.	White mulberry	Morus sp.	Moraceae	1,4-dideoxy-1,4-imino-D-arabinitol, 1,4-dideoxy-1,4-imino-D-ribitol	Antidiabetic	Pyrrolidine	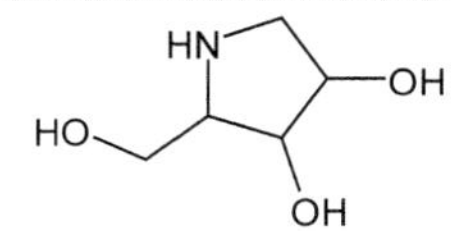 1,4-dideoxy-1,4-imino-D-ribitol	[278, 279]
4.	Rubber vine	*Cryptostegia grandiflora* Roxb.	Apocynaceae	Fatty acids	Milk clotting	Fatty acids	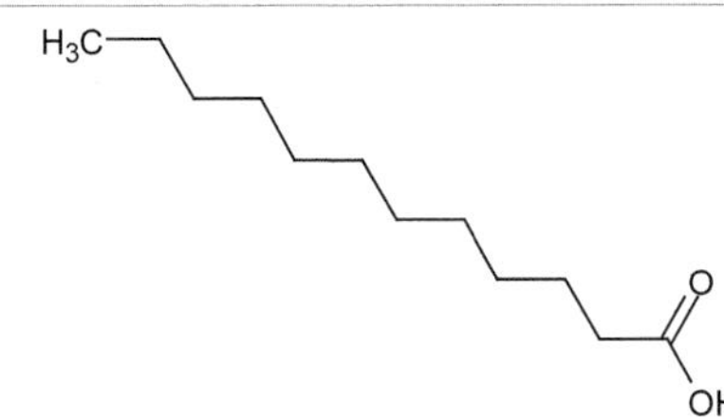 Lauric acid	[280]

(*Continued*)

Table 3.6 (*Continued*)

No	Name	Biological source	Family	Chemical constituents	Pharmacological uses	Nature of chemical compound	Structure of main constituents present in latex	Refs.
5.	Finbriata	*Caralluma adscendens* (Roxb.) R. Br.	Apocynaceae	3,11,17,21-Tetra-hydroxypregnan-20-one	Antioxidant		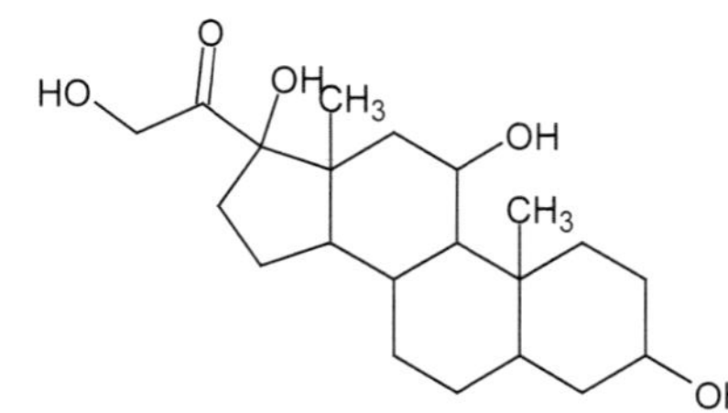 3,11,17,21-Tetrahydroxypregnan-20-one	[281–283]
6.	Lantern flower	*Ceropegia bulbosa* Roxb	Apocynaceae	Cerpegin, 5-allyl cerpegin, cis-vaccenic acid	Antimicrobial, anticancer, mosquitocidal	Fatty acid	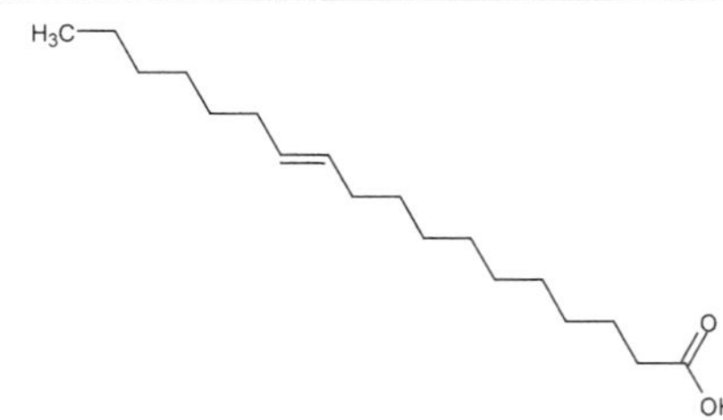 cis-Vaccenic acid	[284, 285]

No	Name	Biological source	Family	Chemical constituents	Pharmacological uses	Nature of chemical compound	Structure of main constituents present in latex	Refs.
7.	Cotton milk plant	*Dregea volubilis (L.f.)* Benth. Ex Hook	Apocynaceae	Volubilosides, 17β-marsdenin, stavaroside, hoyacarnoside G	Anti-tumor, anti-inflammatory, larvicidal, antioxidant	Saponins	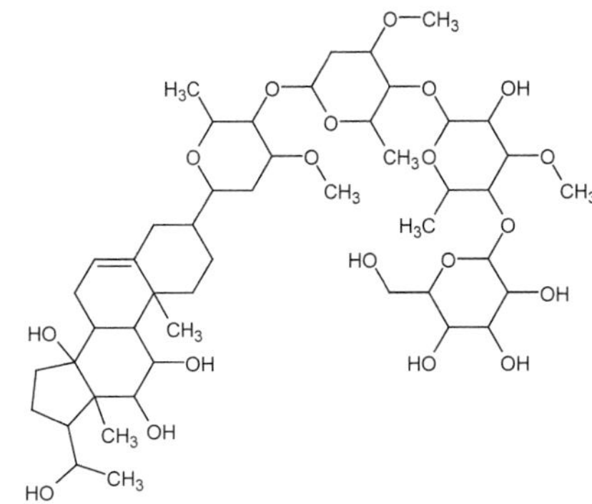 Volubiloside A	[286–292]
8.	Spotted euphorbia	*Euphorbia chamaesyce* L.	Euphorbiaceae	3β-OH-4α, 14α-dimethyl-5α-ergosta-8,24(28)-dien-11-one	Anti-infective, antimicrobial, antibiofilm, antioxidant	Phyto-sterols	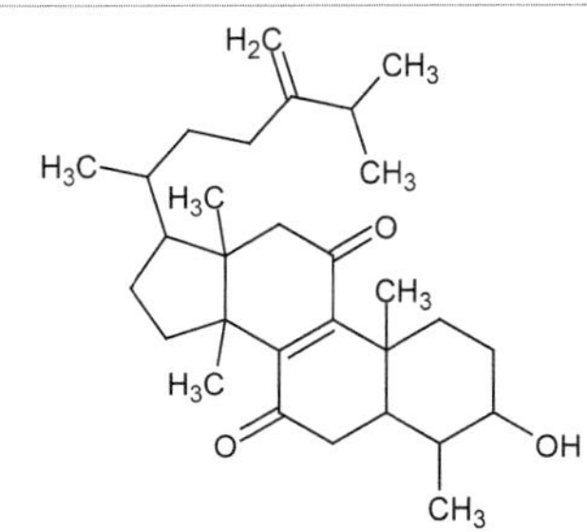 3β-OH-4α, 14α-dimethyl-5α-ergosta-8,24(28)-dien-11-one	[293–296]

(*Continued*)

Table 3.6 (*Continued*)

No	Name	Biological source	Family	Chemical constituents	Pharmacological uses	Nature of chemical compound	Structure of main constituents present in latex	Refs.
9.	Sijwu	*Euphorbia splendens* Boj. Ex. Hook.	Euphorbiaceae	Afzelin, millaminas, quercetin, myricetin	Antimicrobial, molluscicidal, ovicidal	Flavonol glycoside	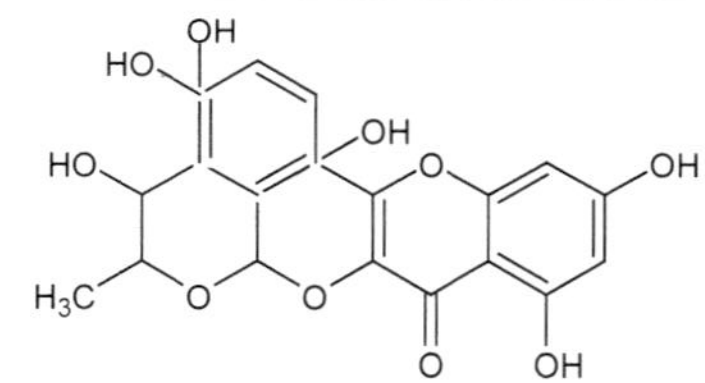 Afzelin	[297–299]
10.	Rakta Snuhi	*Euphorbia caducifolia* Haines	Euphorbiaceae	Quorumolides A-C, 3β-hydroxy-4α, 14α-dimethyl-5α-ergosta-8,24(28)-dien-11-one	Antimicrobial, anticancer, wound healing, anticancer	Steroidal	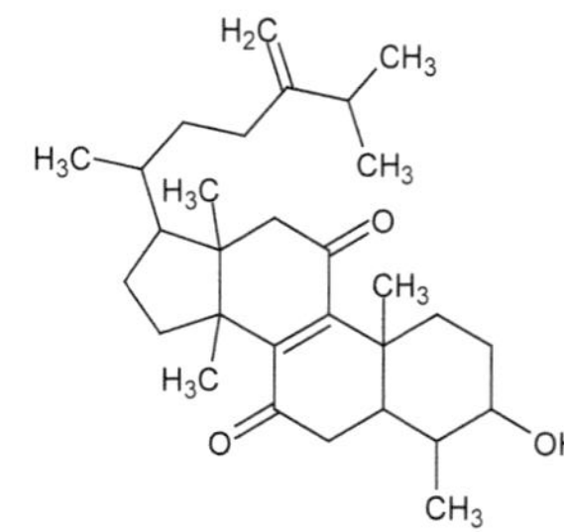 3β-OH-4α,14-dimethyl-5α-ergosta-8,24(28)-diene-7,11-dione	[300–303]

No	Name	Biological source	Family	Chemical constituents	Pharmacological uses	Nature of chemical compound	Structure of main constituents present in latex	Refs.
11.	Indian Oleander	*Nerium indicum* Mill.Gard. Dict	Apocynaceae	Oleandrin, uzarigenin, cardenolide N-1	Antioxidant, nematicidal	Cardiac glycosides	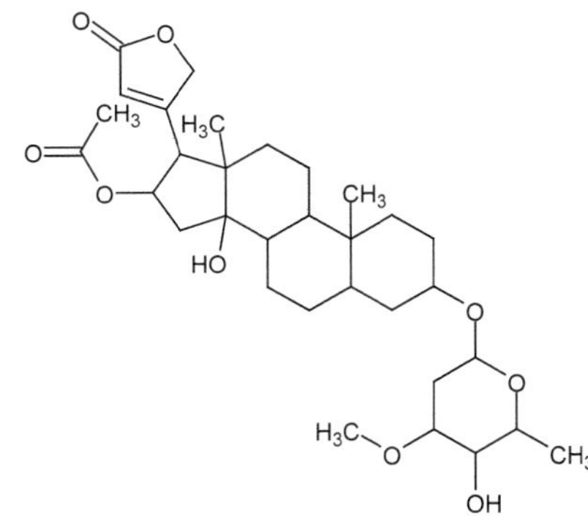 Oleandrin	[304–307]
12.	West Indian jasmine, pagoda tree	*Plumeria alba* L.	Apocynaceae	phenyl acetaldehyde, linalool, n-nonanal, neryl acetone	Antiarthritic, fibrinolytic, thrombin and plasmin like activities	Aldehydes	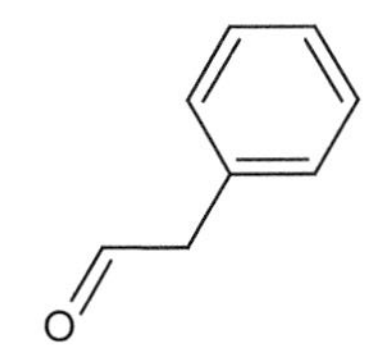 phenyl acetaldehyde	[308–310]

(*Continued*)

Table 3.6 (*Continued*)

No	Name	Biological source	Family	Chemical constituents	Pharmacological uses	Nature of chemical compound	Structure of main constituents present in latex	Refs.
13.	Champo, Khad-Champo	*Plumeria rubra* L.	Apocynaceae	Linalool, n-nonanal, neryl acetone	Antidiabetic, cardiovascular illness, antileprotic, antiinflammatory	Acyclic mono-terpenes	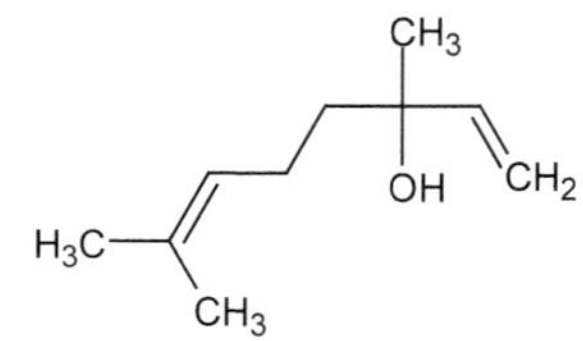 Linalool	[311–314]
14.	Karwanda	*Carissa congesta* Wt.Icon.t	Apocynaceae	Quercetin 3-xyloside, rutin, Vicenin, quercetin-3-O-xyloside/ arabinoside	Immuno-modulatory	Flavonol glycosides	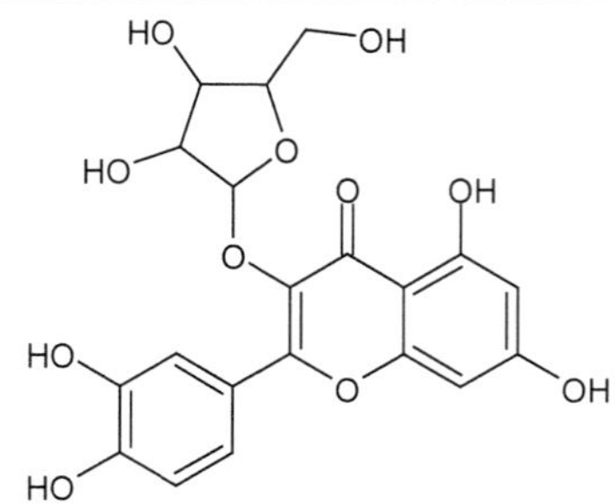 Quercetin 3-xyloside	[315–318]

No	Name	Biological source	Family	Chemical constituents	Pharmacological uses	Nature of chemical compound	Structure of main constituents present in latex	Refs.
15.	Wild Karanda	*Carissa opaca* Stapf ex Haines	Apocynaceae	Quinic acid, protocatechuic acid, gluconic acid, chlorogenic acid, apigenin-8-C-glucoside, rutin	Antioxidant, anti-proliferative, antimicrobial, antiasthmatic, antiviral, cardiotonic	Cyclic polyols	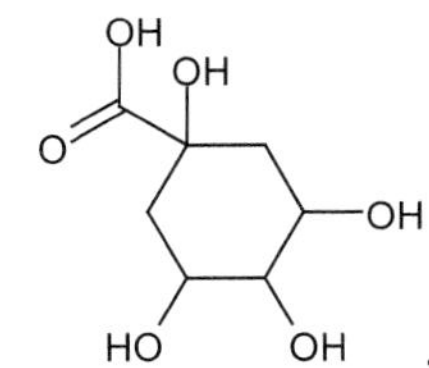 Quinic acid	[319–325]
16.	Lochnera-pusilla, Vinca pusilla	*Catha-ranthus pusillus* (Murr.) G. Don	Apocynaceae	N-benzoyl-L-phenylalaninol, vindoline, leurosine, ursolic acid	Antipyretic, anticancer, antidiabetic	Amino alcohol	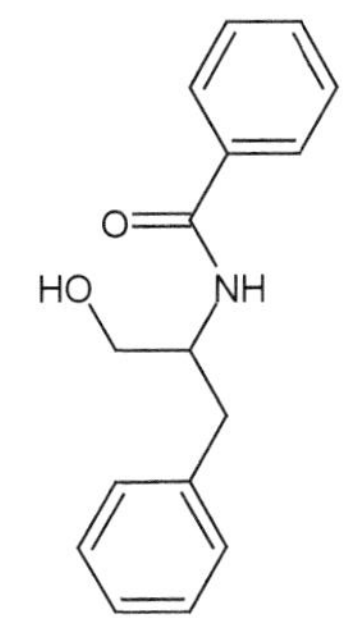 N-Benzoyl-L-phenylalaninol	[326–329]

(*Continued*)

Table 3.6 (*Continued*)

No	Name	Biological source	Family	Chemical constituents	Pharmacological uses	Nature of chemical compound	Structure of main constituents present in latex	Refs.
17.	Yellow oleander	*Thevetia peruviana* (Pers.) K.Schum.	Apocynaceae	Thevetin C, phenolic acids	Spermicidal, anticancer, abortifacient	Cardiac glycosides	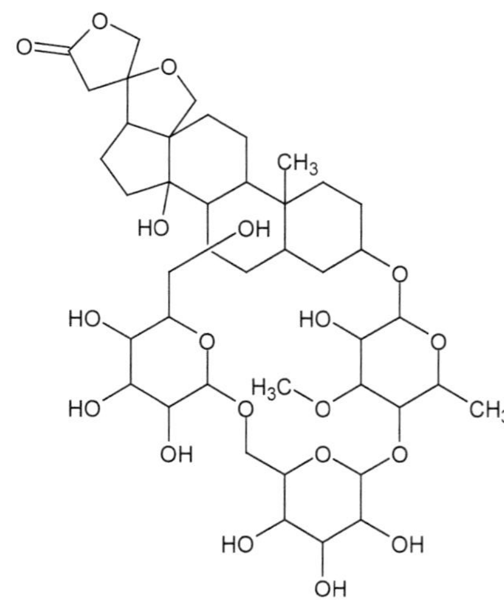 Thevetin C	[330–335]

No	Name	Biological source	Family	Chemical constituents	Pharmacological uses	Nature of chemical compound	Structure of main constituents present in latex	Refs.
18.	Bread Flower	*Vallaris solanacea* (Roth) O. Ktze.	Apocynaceae	Mono-O-acetyl-vallaroside, Mono-O-acetyl-solanoside	Thrombin like activity, reduce bleeding	Iridoid glycosides	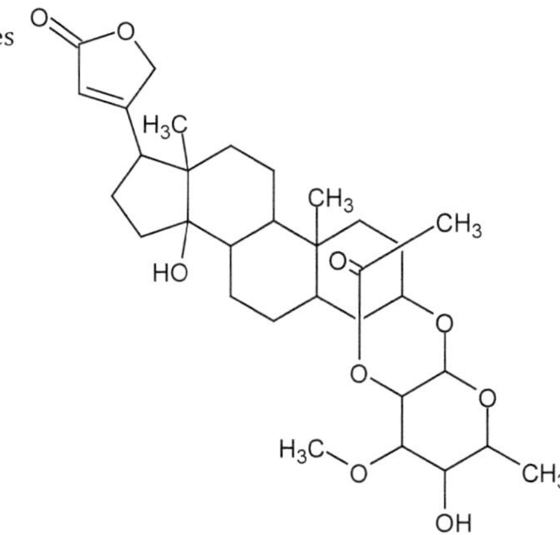 Mono-O-acetylvallaroside	[336, 337]

(*Continued*)

Table 3.6 (*Continued*)

No	Name	Biological source	Family	Chemical constituents	Pharmacological uses	Nature of chemical compound	Structure of main constituents present in latex	Refs.
19.	Banyan, banyan fig and Indian banyan	*Ficus benghalensis L.*	Moraceae	Leucopelargonidin-3-O-β-d-glucopyranoside, terpenoids, germacrene, γ-cadinene, δ-cadinene	Wound healing, neuroprotective, antioxidant, antidiabetic	Flavonoid glycosides	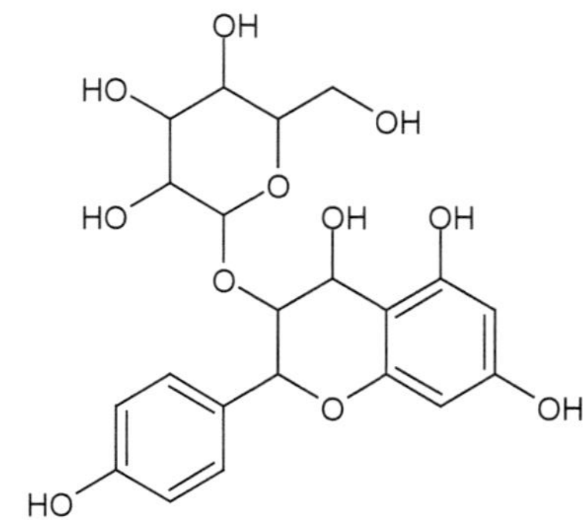 Leucopelargonidin-3-O-β-d-glucopyranoside	[338–343]

No	Name	Biological source	Family	Chemical constituents	Pharmacological uses	Nature of chemical compound	Structure of main constituents present in latex	Refs.
20.	Weeping fig	*Ficus benjamina L.*	Moraceae	α-Pinene, abietadiene, cis-α-bisabolene, germacrene-D-4-ol, germacrene-D-4-ol, 1,10-di-epi-cubenol	Antimycobacterial, antifungal	Mono-terpene hydro-carbon	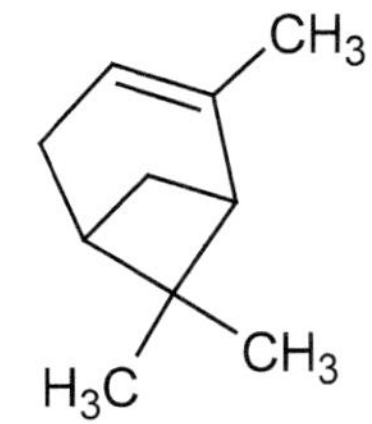 α-pinene	[344–347]
21.	Common fig, common fig tree	*Ficus carica L. Sp.*	Moraceae	Methylbutanal, methyl salicylate, limonene, menthol, cubenene, copaene, *β*-elemene	Anticancer, antibacterial, antifungal, antipyretic, antioxidant	Aldehydes	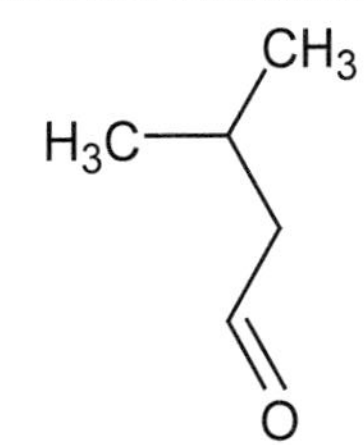 3-Methylbutanal	[348–350]

(*Continued*)

3.5 Chemical Evaluation of Components

It includes chemical assays, quantitative and qualitative chemical evaluation and instrumental methods of analysis of plant latex. It covers separation, purification and identification of active components of latex.

3.5.1 Qualitative Analysis

Qualitative chemical analysis includes identification tests for numerous phytoconstituents present in latex, and fingerprinting analysis by various chromatographic techniques. This will help in identifying chemical nature, and chemical composition of latex phytoconstituents. This can be done as per the method reported earlier [351–353].

3.5.1.1 Preliminary phytochemical analysis

It will give idea about the nature of phytoconstituents present in latex extract. The protocols followed is listed in research articles or as per WHO standards [93–96]. The latex-bearing plant also produces many additional chemicals and similar secondary metabolites that have physiological effects in addition to carbohydrates, proteins and lipids. The presence of phytoconstituents like alkaloids, tannins and flavonoids from *Calotropis procera* is determined [209].

3.5.1.2 Phytochemical fingerprinting analysis

There are many reports on chromatographic and spectroscopic methods which are primarily used to address the identity and quantity of botanicals, phytoconstituents present in latex-containing plants. These methods are typically successful and can produce a wealth of information about intricate samples. The process is subjective and in some cases the limited information provided by conventional fingerprints may not be adequate [341]. The identification, legitimacy and consistency of chromatographic fingerprints from one lot to another might be determined.

3.5.1.3 Thin-layer chromatography

Thin-layer chromatography fingerprinting of *Calotropis gigantea* extracts is reported. The solvent system used is Toluene: Chloroform in the ratio 3:7 [354]. Anisaldehyde-sulfuric acid spray reagents are employed to detect the bands on the TLC plates for spot detection or visualization [98]. The chromatogram is given in Fig. 3.2.

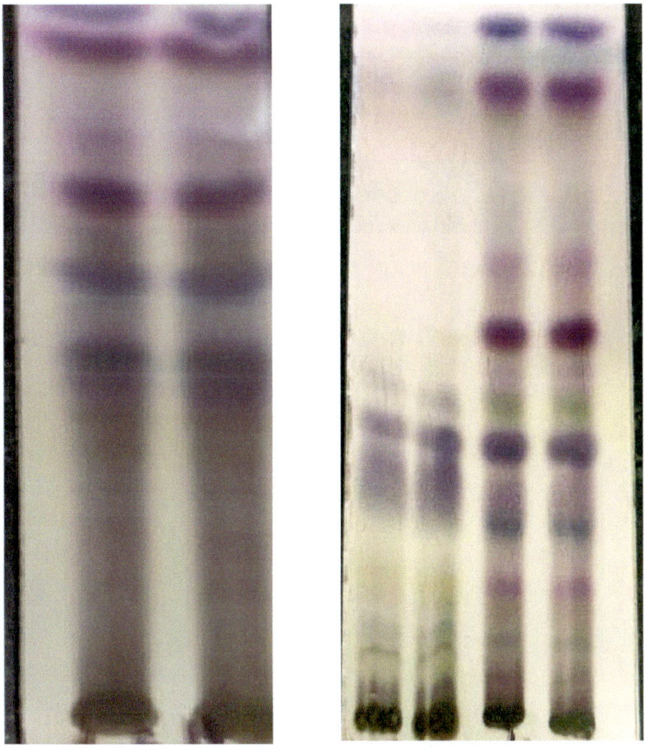

Figure 3.2 Thin-layer chromatography fingerprinting of *Calotropis gigantea* extracts.

3.5.1.4 High-performance thin-layer chromatography

By using HPTLC, chromatographic fingerprinting profiles of several latex containing samples have been studied [99, 100]. Figure 3.3 shows high-performance TLC fingerprinting (a) and chromatogram

(b) of *Ficus carica* leaves extract. The solvent system used is Ethanol: Chloroform (9:1) [355].

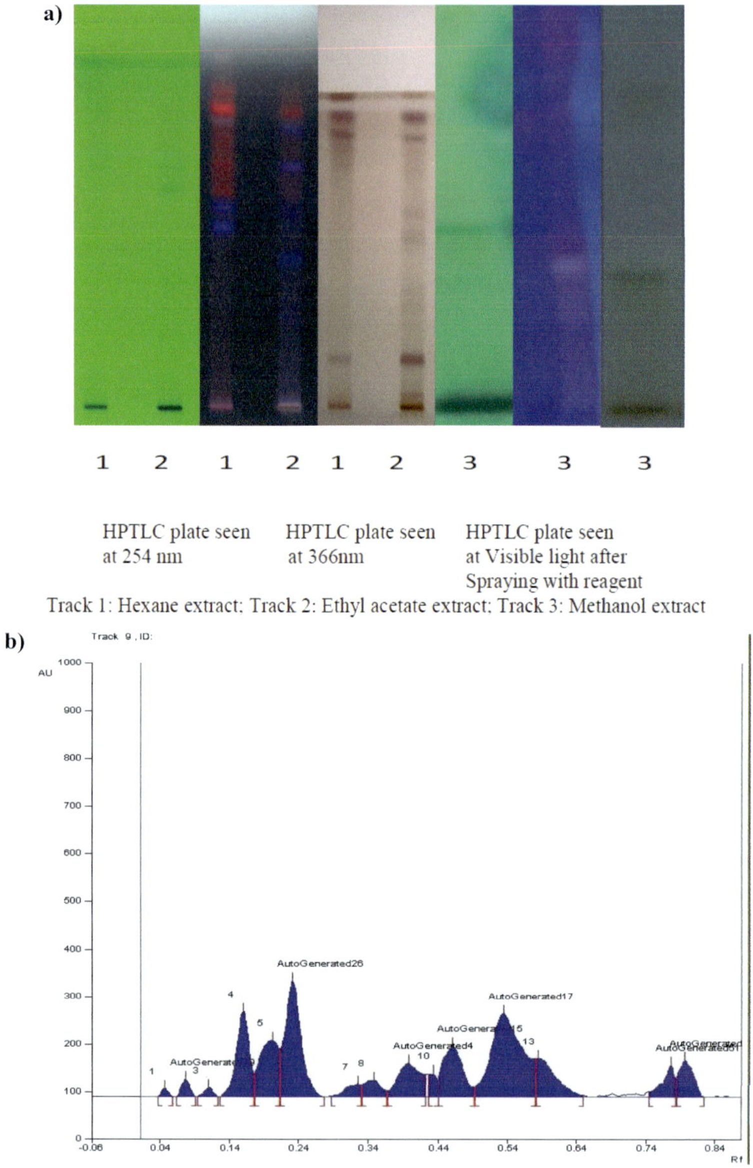

Figure 3.3 High-performance thin-layer chromatography fingerprinting (a) and chromatogram (b) of *Ficus carica* leaves extract.

3.5.1.5 High-resolution liquid chromatography and mass spectroscopy analysis (HRLCMS/QTOF/orbit rap)

Latex-containing extracts are analyzed for the identification of secondary metabolites through HR-LCMS. The process is carried out with mass spectroscopy mainly for the class of compounds like terpenoids, flavonoids, phenols and alkaloids. LC-QTOF-MS/MS analysis data of phytoconstituents from the crude extracts of *C. procera* extracts is reported earlier. The result indicated various phenols, phenolic acid and flavonoids [356].

3.5.1.6 Gas chromatography-mass spectroscopy analysis (GC-MS)

The analysis of volatile phytoconstituents is carried out by GC-MS analysis [357]. Based on the peak area of chromatogram, the relative amount of phytoconstituents present in latex extract is given as a percentage. The GC-MS chromatogram reveals the relative concentration of substances. Every peak height varies according to the chemicals that are now present in it. The NIST and mass spectrometers can identify the composition and structure of a chemical as well as its molecular weight, formula, retention time and percentage of area. The NIST and WILEY 8 databases allow for the identification of the structure and nature of the compound as well as its molecular weight, formula, retention time and area percentages. The analysis of terpenes in *Cannabis sativa* extract and *Jatropha curcas* latex is determined by GC-MS [358, 359].

3.5.2 Quantitative Phytochemical Evaluation

The quantitative phytochemical analysis of bioactive marker compounds in the latex sample can be done by various chromatographic and spectroscopic techniques. Gallic acid, quercetin and atropine are used as the standard chemicals to calculate the levels of total phenolics, total flavonoid and total alkaloidal content in the latex samples.

3.5.2.1 Quantitative estimation of total phenolic content (TPC)

Total phenolic content is determined as per Folin–Ciocalteu method [100] and is expressed as gallic acid equivalents (GAE) in mg GAE/g extract [100]. The total phenolic content of *Calotropis procera* and *Tabernaemontana divaricata* samples is determined by using gallic acid as the standard [360, 361].

3.5.2.2 Quantitative estimation of total flavonoid content (TFC)

Total flavonoid content is calculated with respect to quercetin and the result is expressed in micrograms (µg) of quercetin dehydrate equivalent (QE) per ml of the extract [100]. The total flavonoid content of *Calotropis procera* samples is determined by this method [209].

3.5.2.3 Quantitative estimation of total alkaloid content (TAC)

Total alkaloidal content of latex-bearing plant *Tabernaemontana divaricata* is determined using calibration curve of atropine and the concentration of alkaloid content is expressed as mg/gm [100, 361]. Analytical methods of phytoconstituents obtained from latex-bearing plants are given in Table 3.7.

3.6 Conclusions

In this chapter, several physicochemical properties including organoleptic properties (such as taste, odor, and appearance), solubility, melting point, density, pH, moisture content, ash value and extractive value of plant latex have been covered. Also, the chapter covers detailed chemical composition, structure and classification of various latex-bearing phytoconstituents along with their mechanism and pharmacological uses. It covers modern analytical methods such as LC-MS, GC-MS, FTIR, GC-MS-MS, and LC-MS-MS for quantitative and qualitative estimation of phytoconstituents in latex containing medicinal plants.

Table 3.7 Analytical methods of phytoconstituents from latex-bearing plants

No	Biological source	Phytoconstituents identified	Analytical methods	Refs
1	*Chelidonium majus*	Isoquinoline alkaloid	MS, HPLC	[362]
2	*Tabernaemontana citrifolia*	Indole alkaloids	GC/MS, HPLC	[363]
3	*Tabernaemontana divaricata*	Iboga alkaloids	HPLC-DAD–MS/MS, HPTLC	[364]
4	*Jatropha podagrica*	Alkaloid	HPLC-DAD–MS/MS, HPTLC	[364]
5	*Catharanthus roseus*	Indole alkaloids	TLC, FT-IR, HPLC, GC-MS	[365]
6	*Tylophora indica*	Indole alkaloids	TLC, FT-IR, HPLC, GC-MS	[365]
7	*Alstonia scholaris*	Phenanthro-indolizidine alkaloids	HPTLC, LC-MS, UV-VIS Spectroscopy	[366]
8	*Calotropis procera*	Cardenolides	HPLC, GC-MS, UV-VIS Spectroscopy	[367]
9	*Jatropha glandulifera*	Triterpenoid	HPLC, GC-MS	[368]
10	*Ficus amplissima*	Pentacyclic triterpenoids	LC-MS/MS, HPLC	[369]
11	*Ficus arnottiana*	Pentacyclic triterpenoids	HPLC, UV, LC-MS	[370
12	*Euphorbia indica*	Triterpenes	RP-HPLC, LC-MS/MS	[371]
13	*Euphorbia heyneana*	Diterpenes	HPLC, GC-MS	[372]
14	*Ficus carica*	Triterpenes	HPTLC, HPLC, LC-MS	[373]
15	*Euphorbia tirucalli*	Diterpenoid	LC-MS/MS, UV-VIS Spectroscopy	[374]
16	*Leptadenia reticulata*	Triterpenes	LC-MS/MS, GC, NMR	[375]
17	*Euphorbia hirta*	Diterpenes	HPLC, ELSD	[376]
18	*Thevetia peruviana*	Flavonoid	HPLC, GC-MS/MS	[377]
19	*Euphorbia coccinea*	Flavonoids	UV-VIS Spectroscopy, IR, LC-MS/MS, NMR	[378]
20	*Allamanda cathartica*	Isoflavone	UV-VIS Spectroscopy, FTIR, HPLC	[379]
21	*Pentatropis spiralis*	Flavonols	HPLC-Q-TOF-MS, UPLC-Q-TOF-MS	[380]
22	*Hevea brasiliensis*	Phenol	HPLC, GC-MS, UHPLC-MS/MS	[381, 382]

No	Biological source	Phytoconstituents identified	Analytical methods	Refs
23	*Jatropha multifida*	Flavonoids	HPLC-MS/MS	[383]
24	*Carica papaya*	Cysteine proteases	SDS-PAGE, UV-VIS Spectroscopy, HPLC	[337, 384]
25	*Euphorbia rigida*	Glycolipids	HPLC-MS/MS	[385]
26	*Tithymalus officinarum*	Cholestane steroids	HPLC, UV-VIS Spectroscopy, LC-MS	[386]
27	*Cryptostegia grandiflora*	Fatty acids	GC-FID, HPLC-ELSD	[387]
28	*Ceropegia edulis*	Fatty acid	UHPLC–ESI–MS/MS	[388]
29	*Dregea volubilis*	Saponins	HPLC, UV-VIS Spectroscopy, UHPLC-MS	[389]
30	*Euphorbia splendens*	Flavonol glycoside	LC-MS, UV-VIS Spectroscopy	[390]
31	*Nerium indicum*	Cardiac glycosides	HPLC, LC-MS/MS	[307, 391]
32	*Plumeria rubra*	Acyclic monoterpenes	HPTLC-MS/MS	[392]
33	*Carissa opaca*	Cyclic polyols	HPLC-UV, LC-MS/MS, NMR, UV-VIS Spectroscopy	[393]
34	*Thevetia peruviana*	Cardiac glycosides	HPLC-UV, LC-MS/MS, GC-MS, HPLC-DAD	[394]
35	*Ficus benjamina*	Monoterpene hydrocarbon	HPLC, GC-MS/MS	[395]
36	*Ficus carica*	Aldehydes	GC-MS/MS, UV-VIS Spectroscopy, HPLC	[396, 397]

Abbreviations: MS, Mass Spectrometry; HPLC, High-Performance Liquid Chromatography; LC-MS, Liquid Chromatography-Mass Spectrometry; HPLC-DAD, High-Performance Liquid Chromatography-Diode Array Detector; UHPLC-ESI-MS/MS, Ultra-High-Performance Liquid Chromatography-Electrospray Ionizer-Mass Spectrometry/Mass Spectrometry; GC-MS/MS, Gas Chromatography-Mass Spectrometry/Mass Spectrometry; GC-FID, Gas Chromatography-Fluid Ionization Detector; SDS-PAGE, Sodium Dodecyl Sulfate–Polyacrylamide Gel Electrophoresis; ELSD, Evaporative Light Scattering Detector; UHPLC-MS, Ultra-High-Performance Liquid Chromatography-Mass Spectrometry; UV-VIS, UV-Visible spectroscopy; TLC, Thin-Layer Chromatography; HPTLC, High-Performance Thin-Layer Chromatography; FTIR, Fourier Transform Infrared Spectroscopy; RP-HPLC, Reverse Phase High-Performance Liquid Chromatography; LCMS, Liquid Chromatography-Mass Spectrometry; HPLC-Q-TOF-MS, High-Performance Liquid Chromatography-Quadrupole-Time-of-Flight-Mass Spectrometry.

References

1. Samrot AV, Mun CY, Qi NX, Saigeetha S, Wilson S, Rajesh VV. Plant latex: Phytochemistry, medicinal properties and application – A review. *J Pharm Neg Results.* 2022; 13(7): 5351–75.
2. Kalita D, Saikia CN. Chemical constituents and energy content of some latex bearing plants. *Bioresour Technol.* 2004; 92(3): 219–27.
3. Kitajima S, Aoki W, Shibata D, Nakajima D, Sakurai N, Yazaki K, Munakata R, Taira T, Kobayashi M, Aburaya S, Savadogo EH. Comparative multi-omics analysis reveals diverse latex-based defense strategies against pests among latex-producing organs of the fig tree (Ficus carica). *Planta.* 2018; 247(6): 1423–38.
4. Matheson C. Characterisation of the physico-chemical and optical properties of Australian native plant exudates for a simple method to determine archaeological resins and gums. *J Archaeol Sci Reports.* 2022; 42: 103360.
5. Feller RL, Curran ME. Changes in solubility and removability of varnish resins with age. *J Am Inst Conserv.* 1975; 15: 17–48.
6. Vanzo E, Marchessault RH, Stannett V. The solubility and swelling of latex particles. *J Colloid Sci.* 1965; 20(1): 62–71.
7. Maria HJ, Lyczko N, Nzihou A, Mathew C, George SC, Joseph K, Thomas S. Transport of organic solvents through natural rubber/nitrile rubber/organically modified montmorillonite nano-composites. *J Mater Sci.* 2013; 48(15): 5373–86.
8. Zhong JP, Li CP, Li SD, Kong LX, Yang L, Liao SQ, She XD. Study on the properties of natural rubber during maturation. *J Polym Mater.* 2009; 26(3): 351–60.
9. Wagner I, Lackner M. Extraction and analysis of natural rubber from the latex of Ficus carica, Artocarpus heterophyllus and polymer analysis of Durio zibethinus. *Austin J Biotechnol Bioeng.* 2021; 8(2): 1112.
10. Edomwonyi-Otu L. Distortion of velocity profiles of water flow with heavy molecular weight polymers. *Defect Diffus Forum.* 2019; 392: 228–38.
11. Mazou M, Djossou AJ, Tchobo FP, Villeneuve P, Soumanou MM. Catalytic properties of lipase from Ficus trichopoda and Euphorbia unispina latex: Study of their typoselectivity. *J Appl Biosci.* 2017; 110: 10790–801.
12. Yahaya NZ, Abbas Z, Ismail MA, Mohd Ali B. Determination of moisture content of hevea rubber latex using a microstrip patch antenna. *Prog Electromagn Res Symp.* 2012 1290–93.

13. Kalaskar MG, Tatiya AU, Lamale JJ, Surana SJ. Morpho-anatomical studies of Euphorbia heterophylla Linn. leaves. *Biosci Discov.* 2017; 8(2): 146–52.

14. Dosumu A, Edomwonyi-Otu L, Yusuf N, Abubakar A. Guar gum as flow improver in single-phase water and liquid–liquid flows. *Arab J Sci Eng.* 2020; 45: 7267–73.

15. Edomwonyi Out LC. Natural rubber latex and gum arabic: A comparison of physico-chemical properties. *Mod Approaches Mater Sci.* 2020; 3(1): 329–36.

16. Nascimento RM do, Ramos SMM, Bechtold IH, Hernandes AC. Wettability study on natural rubber surfaces for applications as biomembranes. *ACS Biomater Sci Eng.* 2018; 4(8): 2784–93.

17. Patil V, Pimprikar R, Patil V. Pharmacognostical studies and evaluation of anti-inflammatory activity of Ficus bengalensis linn. *J Young Pharm.* 2009; 1(1): 49–53.

18. Momin RK, Kadam VB. Determination of ash values of some medicinal plants of genus sesbania of Marathwada region in Maharashtra. *J Phytol.* 2011; 3(12): 31–34.

19. Krishanu S, Divyansh M. Preliminary physico-phytochemical and phyto-cognostical evaluation of the leaves of Calotropis gigantea (L.) R. BR. *Int J Pharm Chem Sci.* 2013; 2(4): 1832–8.

20. Macalood JS, Vicente HJ, Boniao RD, Gorospe JG, Roa EC. Chemical analysis of Carica papaya L. crude latex. *Am J Plant Sci.* 2013; 4(10): 1941–8.

21. Okore VC. African Tropical Plant Gums: Grossly unexploited carriers or adjuncts in drug delivery systems. In *Biopolymers in Drug Delivery: Recent Advances and Challenges*, 2009; 3: 7–26.

22. DuBois M, Gilles KA, Hamilton JK, Rebers PA, Smith F. Colorimetric method for determination of sugars and related substances. *Anal Chem.* 1956; 28(3): 350–6.

23. Zhang Z, Zhang G, Li D, Liu Z, Chen X. Chlorohydrin water-swellable rubber compatibilized by an amphiphilic graft copolymer. II. Effects of PVA-g-PBA and CPA on water-swelling behaviors. *J Appl Polym Sci.* 1999; 74: 2509–16.

24. Silva MJ, Claro PI, da Silva JC, Júnior EJ, de Souza Gonçalves P, Martins MA, Mattoso LH. Evaluation of the physicochemical properties of natural rubber from *Hevea brasiliensis* clones. *Ind Crops Prod.* 2021; 171: 113925.

25. Bateman LC, Sekhar BC. Significance of PRI in raw and vulcanized natural rubber. *Rubber Chem Technol.* 1966; 39: 1608–16.

26. Jones C, Nelson J. Multi-element analysis of cannabis and hemp using ICP-MS. *Appl Note Cannabis Hemp Test.* 2018; 1–8.
27. Cai HH, Li SD, Tian GR, Wang HB, Wang JH. Reinforcement of natural rubber latex film by ultrafine calcium carbonate. *J Appl Polym Sci.* 2002; 87(6): 982–5.
28. Linh V, Plank J. Evaluation of natural rubber latex as film forming additive in cementitious mortar. *Constr Build Mater.* 2018; 169: 93–9.
29. Agrawal AA, Lajeunesse MJ, Fishbein M. Evolution of latex and its constituent defensive chemistry in milkweeds (Asclepias): A phylogenetic test of plant defense escalation. *Entomol Exp Appl.* 2008; 128(1): 126–38.
30. Agrawal AA, Konno K. Latex: A model for understanding mechanisms, ecology, and evolution of plant defense against herbivory. *Annu Rev Ecol Evol Syst.* 2009; 40: 311–31.
31. Konno K. Plant latex and other exudates as plant defense systems: Roles of various defense chemicals and proteins contained therein. *Phytochemistry.* 2011; 72(13): 1510–30.
32. Ramos MV, Grangeiro TB, Freire EA, Sales MP, Souza DP, Araújo ES, Freitas CD. The defensive role of latex in plants: Detrimental effects on insects. *Arthropod Plant Interact.* 2010; 4(1): 57–67.
33. Huber M, Triebwasser-Freese D, Reichelt M, Heiling S, Paetz C, Chandran JN, Bartram S, Schneider B, Gershenzon J, Erb M. Identification, quantification, spatiotemporal distribution and genetic variation of major latex secondary metabolites in the common dandelion (Taraxacum officinale agg.). *Phytochemistry.* 2015; 115: 89–98.
34. Zaman K, Adibah M, Azzreena MA. Plant toxins: Alkaloids and their toxicities. *GSC Biol Pharm Sci.* 2019; 06(02): 21–9.
35. Bird DA, Franceschi VR, Facchini PJ. A tale of three cell types: Alkaloid biosynthesis is localized to sieve elements in opium poppy. *Plant Cell.* 2003; 15(11): 2626–35.
36. Musidlak O, Warowicka A, Broniarczyk J, Adamczyk D, Goździcka-Józefiak A, Nawrot R. The activity of Chelidonium majus L. latex and its components on HPV reveal insights into the antiviral molecular mechanism. *Int J Mol Sci.* 2022; 23(16): 9241–70.
37. Nawrot R, Warowicka A, Rudzki PJ, Musidlak O, Dolata KM, Musijowski J, Stolarczyk EU, Goździcka-Józefiak A. Combined protein and alkaloid research of Chelidonium majus latex reveals CmMLP1

accompanied by alkaloids with cytotoxic potential to human cervical carcinoma cells. *Int J Mol Sci.* 2021; 22(21): 11838–53.

38. Marie-Magdeleine C, Mahieu M, D'Alexis S, Philibert L, Archimede H. In vitro effects of Tabernaemontana citrifolia extracts on Haemonchus contortus. *Res Vet Sci.* 2010; 89(1): 88–92.
39. Chander MP, Kartick C, Gangadhar J, Vijayachari P. Ethnomedicine and healthcare practices among Nicobarese of Car Nicobar—an indigenous tribe of Andaman and Nicobar Islands. *J Ethnopharmacol.* 2014; 158 Pt A: 18–24.
40. Han L, Huang K, Chen C, Zhu W, Ma Y, Hao X, He H, Zhang Y. Taberdines L and M, two new alkaloids from Tabernaemontana divaricata. *Nat Prod Res.* 2022; 36(21): 5470–5.
41. Pratchayasakul W, Pongchaidecha A, Chattipakorn N, Chattipakorn S. Ethnobotany & ethnopharmacology of Tabernaemontana divaricata. *Indian J Med Res.* 2008; 127(4): 317–35.
42. Chattipakorn S, Pongpanparadorn A, Pratchayasakul W, Pongchaidacha A, Ingkaninan K, Chattipakorn N. Tabernaemontana divaricata extract inhibits neuronal acetylcholinesterase activity in rats. *J Ethnopharmacol.* 2007; 110(1): 61–8.
43. Li XM, Jiang XJ, Wei GZ, Ren LH, Wang LX, Cheng XL, Wang F. New iboga-type indole alkaloids from Tabernaemontana divaricata. *Nat Products Bioprospect.* 2019; 9(6): 425–9.
44. Chanchal R, Balasubramaniam A, Navin R, Nadeem S. Tabernaemontana divaricata leaves extract exacerbate burying behavior in mice. *Avicenna J Phytomed.* 2015; 5(4): 282–7.
45. Zhang Y, Bai X, Yuwen HS, Guo LL, Liu JW, Hao XJ. Alkaloids from Tabernaemontana divaricata combined with fluconazole to overcome fluconazole resistance in Candida albicans. *Bioorg Chem.* 2021; 107: 104515.
46. Deng Y, Bao MF, Shi BB, Wu J, Cai XH. Three new indole alkaloids from Tabernaemontana divaricata. *Nat Products Bioprospect.* 2018; 8(3): 183–8.
47. Yuwen H, Yuan Y, Hao X, He H, Zhang Y. Two new monoterpenoid indole alkaloids from Tabernaemontana divaricata. *Nat Prod Res.* 2019; 33(15): 2139–44.
48. Jain S, Jain A, Paliwal P, Solanki SS. Antifertility effect of chronically administered Tabernaemontana divaricata leaf extract on male rats. *Asian Pac J Trop Med.* 2012; 5(7): 547–51.

49. Muzammil S, Andleeb R, Hayat S, Ijaz MU, Ashraf A, Zafar N, Naz S, Shaheen M. Integrating in silico and in vitro approaches to screen the antidiabetic properties from Tabernaemontana divaricata (Jasmine) Flowers. *Evid Based Complement Alternat Med.* 2022; 2022: 1–17.

50. Kumar A, Selvakumar S. Antiproliferative efficacy of Tabernaemontana divaricata against HEP2 cell line and Vero cell line. *Pharmacogn Mag.* 2015; 11(Suppl 1): S46–52.

51. Chen L, Huang JL, Zhang L, Tian HY, Yin S. Jatrogricaine A: A new diterpenoid with a 5/6/6/4 carbon ring system from the stems of Jatropha podagrica. *Chin J Nat Med.* 2019; 17(4): 298–302.

52. Van den Berg AJ, Horsten SF, Kettenes-van den Bosch JJ, Beukelman CJ, Kroes BH, Leeflang BR, Labadie RP. Podacycline A and B, two cyclic peptides in the latex of Jatropha podagrica. *Phytochemistry.* 1996; 42(1): 129–33.

53. Odebiyi OO. Antibacterial property of tetramethylpyrazine from the stem of Jatropha podagrica. *Planta Med.* 1980; 38(2): 144–6.

54. Aiyelaagbe OO, Adesogan EK, Ekundayo O, Adeniyi BA. The antimicrobial activity of roots of Jatropha podagrica (Hook). *Phytother Res.* 2000; 14(1): 60–2.

55. Ojewole JA, Odebiyi OO. Mechanism of the hypotensive effect of tetramethylpyrazine, an amide alkaloid from the stem of Jatropha podagrica. *Planta Med.* 1981; 41(3): 281–7.

56. Rajashekara S, Reena D, Mainavi MV, Sandhya LS, Baro U. Biological isolation and characterization of Catharanthus roseus (L.) G. Don methanolic leaves extracts and their assessment for antimicrobial, cytotoxic, and apoptotic activities. *BMC Complement Med Ther.* 2022; 22(1): 328–45.

57. Mardani-Nejad S, Khavari-Nejad RA, Saadatmand S, Najafi F, Aberoomand Azar P. Potent antioxidant properties of rolB-transformed Catharanthus roseus (L.) G. Don. *Iran J Pharm Res.* 2016; 15(2): 537–50.

58. Tiong SH, Looi CY, Hazni H, Arya A, Paydar M, Wong WF, Cheah SC, Mustafa MR, Awang K. Antidiabetic and antioxidant properties of alkaloids from Catharanthus roseus (L.) G. Don. *Molecules.* 2013; 18(8): 9770–84.

59. Gururani R, Patel S, Yaduvanshi N, Dwivedi J, Paliwal S, Sharma S. Tylophora indica (Burm. f.) merr: An insight into phytochemistry and pharmacology. *J Ethnopharmacol.* 2020; 262: 113122.

60. Ganguly T, Badheka LP, Sainis KB. Immunomodulatory effect of Tylophora indica on Con A induced lymphoproliferation. *Phytomedicine*. 2001; 8(6): 431–7.

61. Rao KV, Wilson RA, Cummings B. Alkaloids of tylophora. New alkaloids of Tylophora indica (Burm) Merrill and Tylophora dalzellii Hook. f. *J Pharm Sci*. 1971; 60(11): 1725–6.

62. Mostafa EM, Musa A, Mohammed HA, Alzarea AI, Abdelgawad MA, Al-Sanea MM, Ismail A, Zafar A, Elmowafy M, Selim S, Khan RA. Phenanthroindolizidine alkaloids secondary metabolites diversity in medicinally viable plants of the genus Tylophora. *Plants*. 2023; 12(5): 1143–65.

63. Mostafa EM, Mohammed HA, Musa A, Abdelgawad MA, Al-Sanea MM, Almahmoud SA, Ghoneim MM, Gomaa HA, Rahman FE, Shalaby K, Selim S. In vitro anti-proliferative, and kinase inhibitory activity of phenanthroindolizidine alkaloids isolated from Tylophora indica. *Plants*. 2022; 11(10): 1295–13.

64. Gupta V, Guleri R, Gupta M, Kaur N, Kaur K, Kumar P, Anand M, Kaur G, Pati PK. Anti-neuroinflammatory potential of Tylophora indica (Burm. f) Merrill and development of an efficient in vitro propagation system for its clinical use. *PLoS One*. 2020; 15(3): e0230142–61.

65. Choodej S, Hanthanong S, Aree T, Pudhom K. Diterpenoids from the aerial parts of Euphorbia antiquorum and their efficacy on nitric oxide inhibition. *Phytochemistry*. 2020; 180: 112523.

66. An L, Liang Y, Yang X, Wang H, Zhang J, Tuerhong M, Li D, Wang C, Lee D, Xu J, Shuai L.. NO inhibitory diterpenoids as potential anti-inflammatory agents from Euphorbia antiquorum. *Bioorg Chem*. 2019; 92: 103237.

67. Qi WY, Zhao JX, Wei WJ, Gao K, Yue JM. Quorumolides A-C, three cembranoids from Euphorbia antiquorum. *J Org Chem*. 2018; 83(2): 1041–5.

68. Madhavan V, Murali A, Lalitha DS, Yoganarasimhan S. Studies on anti-hyperglycemic effect of Euphorbia antiquorum L. root in diabetic rats. *J Intercult Ethnopharmacol*. 2015; 4(4): 308–13.

69. Dong M, Chen XQ, Chen CH, Li RT. Terpenes from Euphorbia antiquorum and their in vitro anti-HIV activity. *Chem Biodivers*. 2018; 15(5): e1700560.

70. Liang Y, An L, Shi Z, Zhang X, Xie C, Tuerhong M, Song Z, Ohizumi Y, Lee D, Shuai L, Xu J. Bioactive diterpenoids from the stems of Euphorbia antiquorum. *J Nat Prod*. 2019; 82(6): 1634–44.

71. Tran TN, Sichaem J, Nguyen VK, Nguyen HH, Cao TT, Nguyen TP, Vo VG, Niamnont N, Nguyen NH, Duong TH. New diterpenoids from the stems of Euphorbia antiquorum growing in Vietnam. *Nat Prod Res.* 2022; 36(2): 523–30.

72. Tran CL, Dao TB, Tran TN, Mai DT, Tran TM, Tran NM, Dang VS, Vo TX, Duong TH, Sichaem J. Alpha-Glucosidase inhibitory diterpenes from Euphorbia antiquorum growing in Vietnam. *Molecules.* 2021; 26(8): 2257–63.

73. Goel N, Gupta VK, Garg A, Bhoumik A, Biswas R, Natarajan R, Majumder HK, Jaisankar P. Holanamine, a steroidal alkaloid from the bark of Holarrhena pubescens Wall. ex G. Don inhibits the growth of leishmania donovani by targeting DNA topoisomerase 1B. *ACS Infect Dis.* 2023; 9(1): 162–77.

74. Bhusal A, Jamarkattel N, Shrestha A, Lamsal NK, Shakya S, Rajbhandari S. Evaluation of antioxidative and antidiabetic activity of bark of holarrhena pubescens wall. *J Clin Diagn Res.* 2014; 8(9): HC05–8.

75. Cheenpracha S, Boapun P, Limtharakul Née Ritthiwigrom T, Laphookhieo S, Pyne SG. Antimalarial and cytotoxic activities of pregnene-type steroidal alkaloids from Holarrhena pubescens roots. *Nat Prod Res.* 2019; 33(6): 782–8.

76. Tuntiwachwuttikul P, Pootaeng-on Y, Phansa P, Limpachayaporn P, Charoenchai P, Taylor WC. Constituents of the leaves of Holarrhena pubescens. *Fitoterapia.* 2007; 78(3): 271–3.

77. Siddiqui BS, Usmani SB, Ali ST, Begum S, Rizwani GH. Further constituents from the bark of Holarrhena pubescens. *Phytochemistry.* 2001; 58(8): 1199–204.

78. Zahara K, Panda SK, Swain SS, Luyten W. Metabolic diversity and therapeutic potential of Holarrhena pubescens: An important ethnomedicinal plant. *Biomolecules.* 2020; 10(9): 1341–65.

79. Jain S, Choudhary GP, Jain DK. Pharmacological evaluation and antifertility activity of Jatropha gossypifolia in rats. *Biomed Res Int.* 2013; 8: 2013–18.

80. Adeosun OG, Oduola T, Fagbomedo FA, Oyedeji SO, Akanji BO, Adesina AA. Suitability of stem latex of jatropha gossypifolia as a protein precipitant for biochemical analysis. *Indian J Clin Biochem.* 2014; 29(2): 210–2.

81. Drafor G, Duah E, Ankamah NA, Kpene GE, Mante PK. Investigating the anticonvulsant properties of aqueous ethanolic extracts of the

leaves, roots, and fruits of Jatropha gossypifolia L. (Euphorbiaceae). *Adv Pharmacol Pharm Sci.* 2021; 2021: 1–7.

82. Granados S, Balcázar N, Guillén A, Echeverri F. Evaluation of the hypoglycemic effects of flavonoids and extracts from Jatropha gossypifolia L. *Molecules.* 2015; 20(4): 6181–93.
83. Das B, Rao SP, Srinivas KV. Isolation of isogadain from Jatropha gossypifolia. *Planta Med.* 1996; 62(1): 90.
84. Apu AS, Hossain F, Rizwan F, Bhuyan SH, Matin M, Jamaluddin ATM. Study of pharmacological activities of methanol extract of Jatropha gossypifolia fruits. *J Basic Clin Pharm.* 2012; 4(1): 20–4.
85. Ravindranath N, Venkataiah B, Ramesh C, Jayaprakash P, Das B. Jatrophenone, a novel macrocyclic bioactive diterpene from Jatropha gossypifolia. *Chem Pharm Bull.* 2003; 51(7): 870–1.
86. Khyade MS, Kasote DM, Vaikos NP. Alstonia scholaris (L.) R. Br. and Alstonia macrophylla Wall. ex G. Don: A comparative review on traditional uses, phytochemistry and pharmacology. *J Ethnopharmacol.* 2014; 153(1): 1–18.
87. Hu BY, Zhao YL, Ma DY, Xiang ML, Zhao LX, Luo X-D. Anti-hyperuricemic bioactivity of Alstonia scholaris and its bioactive triterpenoids in vivo and in vitro. *J Ethnopharmacol.* 2022; 290: 115049.
88. Baliga MS. Alstonia scholaris Linn R Br in the treatment and prevention of cancer: Past, present, and future. *Integr Cancer Ther.* 2010; 9(3): 261–9.
89. Zhao YL, Gou ZP, Shang JH, Li WY, Kuang Y, Li MY, Luo XD. Anti-microbial effects in vitro and in vivo of Alstonia scholaris. *Nat Products Bioprospect.* 2021; 11(1): 127–35.
90. Bagheri G, Ayatollahi SA, Ramírez-Alarcón K, Fernández M, Salehi B, Forman K, Martorell M, Moghadam MH, Sharifi-Rad J. Phytochemical screening of Alstonia scholaris leaf and bark extracts and their antimicrobial activities. *Cell Mol Biol.* 2020; 66(4): 270–9.
91. Sultana N, Qazi MS, Kamal M. New anti-inflammatory triterpene esters and glycosides from Alstonia scholaris. *Antiinflamm Antiallergy Agents Med Chem.* 2020; 19(4): 370–86.
92. Chattopadhyay D, Maiti K, Kundu AP, Chakraborty MS, Bhadra R, Mandal SC, Mandal AB. Antimicrobial activity of Alstonia macrophylla: A folklore of bay islands. *J Ethnopharmacol.* 2001; 77(1): 49–55.

93. Tan MCS, Carranza MSS, Linis VC, Malabed RS, Oyong GG. Antioxidant, cytotoxicity, and antiophidian potential of Alstonia macrophylla Bark. *ACS Omega*. 2019; 4(5): 9488–96.

94. Changwichit K, Khorana N, Suwanborirux K, Waranuch N, Limpeanchob N, Wisuitiprot W, Suphrom N, Ingkaninan K. Bisindole alkaloids and secoiridoids from Alstonia macrophylla Wall. ex G. Don. *Fitoterapia*. 2011; 82(6): 798–804.

95. Cheenpracha S, Ritthiwigrom T, Laphookhieo S. Alstoniaphyllines A-C, unusual nitrogenous derivatives from the bark of Alstonia macrophylla. *J Nat Prod*. 2013; 76(4): 723–6.

96. Arai H, Zaima K, Mitsuta E, Tamamoto H, Saito A, Hirasawa Y, Rahman A, Kusumawati I, Zaini NC, Morita H. Alstiphyllanines I-O, ajmaline type alkaloids from Alstonia macrophylla showing vasorelaxant activity. *Bioorg Med Chem*. 2012; 20(11): 3454–9.

97. Kam TS, Choo YM. New indole alkaloids from Alstonia macrophylla. *J Nat Prod*. 2004; 67(4): 547–52.

98. Lim SH, Low YY, Sinniah SK, Yong KT, Sim KS, Kam TS. Macroline, akuammiline, sarpagine, and ajmaline alkaloids from Alstonia macrophylla. *Phytochemistry*. 2014; 98: 204–15.

99. Hirasawa Y, Arai H, Zaima K, Oktarina R, Rahman A, Ekasari W, Widyawaruyanti A, Indrayanto G, Zaini NC, Morita H. Alstiphyllanines A-D, indole alkaloids from Alstonia macrophylla. *J Nat Prod*. 2009; 72(2): 304–7.

100. Abarca LF, Klinkhamer PG, Choi YH. Plant latex, from ecological interests to bioactive chemical resources. *Planta Med*. 2019; 85(11/12): 856–68.

101. Wadhwani BD. A review on phytochemical constituents and pharmacological potential of Calotropis procera. *RSC Adv*. 2021; 11(57): 35854–78.

102. Kumar VL, Basu N. Anti-inflammatory activity of the latex of Calotropis procera. *J Ethnopharmacol*. 1994; 44(2): 123–5.

103. Ramos MV, Freitas AP, Leitão RF, Costa DV, Cerqueira GS, Martins DS, Martins CS, Alencar NM, Freitas LB, Brito GA. Anti-inflammatory latex proteins of the medicinal plant Calotropis procera: A promising alternative for oral mucositis treatment. *Inflamm Res Off J Eur Histamine Res Soc*. 2020; 69(9): 951–66.

104. Dai HF, Gan YJ, Que DM, Wu J, Wen ZC, Mei WL. A new cytotoxic 19-nor-cardenolide from the latex of Antiaris toxicaria. *Molecules*. 2009; 14(9): 3694–9.

105. Dai HF, Gan YJ, Que DM, Wu J, Wen ZC, Mei WL. Two new cytotoxic cardenolides from the latex of Antiaris toxicaria. *J Asian Nat Prod Res.* 2009; 11(9): 832–7.

106. Alonso-Castro AJ, Arana-Argáez V, Yáñez-Barrientos E, Torres-Romero JC, Chable-Cetz RJ, Worbel K, de Jesús Euan-Canto A, Wrobel K, González-Ibarra A, Solorio-Alvarado CR, del Carmen Juárez-Vázquez M. Pharmacological activities of Asclepias curassavica L. (Apocynaceae) aerial parts. *J Ethnopharmacol.* 2021; 281: 114554.

107. Ai-Jia JI, Qing MA, Mu-Yan KO, Le-Yan LI, Xin-Lian CH, Zhong-Qiu LI, Jin-Jun WU, Zhang RR.. Two cardenolide glycosides from the seed fairs of Asclepias curassavica and their cytotoxic activities. *Chin J Nat Med.* 2022; 20(3): 202–9.

108. Rubiano-Buitrago P, Pradhan S, Paetz C, Rowland HM. New structures, spectrometric quantification, and inhibitory properties of cardenolides from Asclepias curassavica seeds. *Molecules.* 2022; 28(1): 105–18.

109. Kanchan T, Atreya A. Calotropis gigantea. *Wilderness Environ Med.* 2016; 27(2): 350–1.

110. He YL, Yang HY, Huang PZ, Feng WJ, Gao K. Cytotoxic cardenolides from Calotropis gigantea. *Phytochemistry.* 2021; 192: 112951.

111. Nguyen MTT, Nguyen KDH, Dang PH, Nguyen HX, Awale S, Nguyen NT. A new cytotoxic cardenolide from the roots of Calotropis gigantea. *Nat Prod Res.* 2021; 35(23): 5096–101.

112. Bhatia S, Al-Harrasi A, Kumar A, Behl T, Sehgal A, Singh S, Sharma N, Anwer MK, Kaushik D, Mittal V, Chigurupati S.. Anti-migraine activity of freeze-dried latex obtained from Calotropis gigantea Linn. *Environ Sci Pollut Res Int.* 2022; 29(18): 27460–78.

113. Sawong S, Pekthong D, Suknoppakit P, Winitchaikul T, Kaewkong W, Somran J, Intapa C, Parhira S, Srisawang P. Calotropis gigantea stem bark extracts inhibit liver cancer induced by diethylnitrosamine. *Sci Rep.* 2022; 12(1): 12151–80.

114. Winitchaikul T, Sawong S, Surangkul D, Srikummool M, Somran J, Pekthong D, Kamonlakorn K, Nangngam P, Parhira S, Srisawang P. Calotropis gigantea stem bark extract induced apoptosis related to ROS and ATP production in colon cancer cells. *PLoS One.* 2021; 16(8): e0254392-115.

115. Pintus F, Medda R, Rinaldi AC, Spanò D, Floris G. Euphorbia latex biochemistry: Complex interactions in a complex environment. *Plant Biosyst.* 2010; 144(2): 381–91.

116. Sessa RA, Bennett MH, Lewis MJ, Mansfield JW, Beale MH. Metabolite profiling of sesquiterpene lactones from Lactuca species. Major latex components are novel oxalate and sulfate conjugates of lactucin and its derivatives. *J Biol Chem.* 2000; 275(35): 26877–84.

117. Dussourd DE. Chemical stimulants of leaf-trenching by cabbage loopers: Natural products, neurotransmitters, insecticides, and drugs. *J Chem Ecol.* 2003; 29(9): 2023–47.

118. Mazid M, Khan TA, Mohammad F. Role of secondary metabolites in defense mechanisms of plants. *Biol Med.* 2011; 3(2): 232–49.

119. Trinel M, Jullian V, Le Lamer AC, Mhamdi I, Mejia K, Castillo D, Cabanillas BJ, Fabre N. Profiling of Hura crepitans L. latex by ultra - high - performance liquid chromatography/atmospheric pressure chemical ionisation linear ion trap Orbitrap mass spectrometry. *Phytochem Anal.* 2018; 29(6): 627–638.

120. Barbieri L, Falasca A, Franceschi C, Licastro F, Rossi CA, Stirpe F.. Purification and properties of two lectins from the latex of the euphorbiaceous plants *Hura crepitans* L. (sand-box tree) and Euphorbia. *Biochem J.* 1983; 215: 433–9.

121. Papageorgiou VP, Assimopoulou AN, Ballis AC. Alkannins and shikonins: A new class of wound healing agents. *Curr Med Chem.* 2008; 15(30): 3248–67.

122. Sharma SK, Singh H. A review on pharmacological significance of genus Jatropha (Euphorbiaceae). *Chin J Integr Med.* 2012; 18(11): 868–80.

123. Inbaneson SJ, Ravikumar S, Suganthi P. In vitro antiplasmodial effect of ethanolic extracts of coastal medicinal plants along Palk Strait against Plasmodium falciparum. *Asian Pac J Trop Biomed.* 2012; 2(5): 364–7.

124. Arunachalam K, Parimelazhagan T. Anti-inflammatory, wound healing and in vivo antioxidant properties of the leaves of Ficus amplissima Smith. *J Ethnopharmacol.* 2013; 145(1): 139–45.

125. Arunachalam K, Parimelazhagan T. Antidiabetic activity of Ficus amplissima Smith. bark extract in streptozotocin induced diabetic rats. *J Ethnopharmacol.* 2013; 147(2): 302–10.

126. Barbalho SM, Bueno PC, Delazari DS, Guiguer EL, Coqueiro DP, Araújo AC, de Souza MD, Farinazzi-Machado FM, Mendes CG, Groppo M. Antidiabetic and antilipidemic effects of Manilkara zapota. *J Med Food.* 2015; 18(3): 385–91.

127. Rajamanickam SK, Kasinathan S. Fatty acid ethyl ester from Manilkara zapota seed oil: A completely renewable biofuel for sustainable development. *Environ Sci Pollut Res Int.* 2021; 28(43): 61790–800.

128. Chunhakant S, Chaicharoenpong C. Antityrosinase, antioxidant, and cytotoxic activities of phytochemical constituents from Manilkara zapota L. bark. *Molecules.* 2019; 24(15): 2798–15.

129. Mourão Mulvaney LC, Xavier-Júnior FH, Rodrigues AMS, Stien D, Allegretti SM, Garcia VL. Antimicrobial and anthelmintic activities of the ethanolic extract, fractions and isolated compounds from Manilkara zapota L. P. Royen (Sapotaceae). *J Pharm Pharmacol.* 2021; 73(3): 377–87.

130. Ganguly A, Al Mahmud Z, Kumar Saha S, Abdur Rahman SM. Evaluation of antinociceptive and antidiarrhoeal properties of Manilkara zapota leaves in Swiss albino mice. *Pharm Biol.* 2016; 54(8): 1413–9.

131. Pansare AV, Shedge AA, Sonawale MC, Pansare SV, Mahakal AD, Khairkar SR, Chhatre SY, Kulal DK, Patil VR. Deciphering the sensing of α-amyrin acetate with hs-DNA: A multipronged biological probe. *RSC Adv.* 2022; 12(3): 1238–43.

132. Kalaskar MG, Surana SJ. Pharmacognostic and phytochemical studies on Ficus microcarpa L. fil. *Anc Sci Life.* 2012; 32(2): 107–11.

133. Rizk AM, Rimpler H. Constituents of egyptian euphorbiaceae IV investigation of Euphorbia indica. *Planta Med.* 1977; 32(2): 177–80.

134. Awaad AS, Alothman MR, Zain YM, Zain GM, Alqasoumi SI, Hassan DA. Comparative nutritional value and antimicrobial activities between three Euphorbia species growing in Saudi Arabia. *Saudi Pharm J.* 2017; 25(8): 1226–30.

135. Ahmad S, Perveen S, Arshad MA, Rehman T. Pharmacological and nutritive potential of Euphorbia granulata. *J Complement Integr Med.* 2018; 16(2): 20180149.

136. Eida AA, Ziegler M, Lafi FF, Michell CT, Voolstra CR, Hirt H, Saad MM. Desert plant bacteria reveal host influence and beneficial plant growth properties. *PLoS One.* 2018; 13(12): e0208223.

137. Hussein G, Miyashiro H, Nakamura N, Hattori M, Kawahata T, Otake T, Kakiuchi N, Shimotohno K. Inhibitory effects of Sudanese plant extracts on HIV-1 replication and HIV-1 protease. *Phytother Res.* 1999; 13(1): 31–6.

138. Wang Y, Zhou D, Meng Q, Xu L, Yao X, Ni X, Xie H, Wu G, Chen G, Hou Y, Li N. Anti-neuroinflammatory effects in vitro and in vivo,

and chemical profile of Jatropha curcas L. *Bioorg Chem.* 2022; 122: 105720.

139. Bastos EM, da Silva AB, Coelho PL, Borges JM, da Silva VD, da Cunha VH, Costa SL. Anti-inflammatory activity of Jatropha curcas L. in brain glial cells primary cultures. *J Ethnopharmacol.* 2021; 264: 113201–12.

140. Tinpun K, Nakpheng T, Padmavathi AR, Srichana T. In vitro studies of Jatropha curcas L. latex spray formulation for wound healing applications. *Turkish J Pharm Sci.* 2020; 17(3): 271–9.

141. Fujiki H, Suttajit M, Rawangkan A, Iida K, Limtrakul P, Umsumarng S, Suganuma M. Phorbol esters in seed oil of Jatropha curcas L. (saboodam in Thai) and their association with cancer prevention: From the initial investigation to the present topics. *J Cancer Res Clin Oncol.* 2017; 143(8): 1359–69.

142. Debnath M, Bisen PS. Jatropha curcas L., a multipurpose stress resistant plant with a potential for ethnomedicine and renewable energy. *Curr Pharm Biotechnol.* 2008; 9(4): 288–306.

143. Albino RC, Antoniassi R, de Faria-Machado AF, Ferraris FK, Amendoeira FC, Ramos DF, Silva PE, Leitao SG, Oliveira DR. Traditional detoxification of Jatropha curcas L. seeds. *J Ethnopharmacol.* 2019; 241: 111970.

144. Maravi DK, Mazumdar P, Alam S, Goud VV, Sahoo L. Jatropha (Jatropha curcas L.). *Methods Mol Biol.* 2015; 1224: 25–35.

145. Chen S, Tsang NY, Jiang H, Zhang H, Ge X, Wu Y, Shen Y, Xie W, Li J. Cytotoxic tigliane diterpenoids from Euphorbia neorubella. *Nat Prod Res.* 2022; 36(11): 2753–7.

146. Zhan ZJ, Li S, Chu W, Yin S. Euphorbia diterpenoids: Isolation, structure, bioactivity, biosynthesis, and synthesis (2013–2021). *Nat Prod Rep.* 2022; 39(11): 2132–74.

147. Amtaghri S, Akdad M, Slaoui M, Eddouks M. Traditional uses, pharmacological, and phytochemical studies of Euphorbia: A review. *Curr Top Med Chem.* 2022; 22(19): 1553–70.

148. Lazreg-Aref H, Mars M, Fekih A, Aouni M, Said K. Chemical composition and antibacterial activity of a hexane extract of Tunisian caprifig latex from the unripe fruit of Ficus carica. *Pharm Biol.* 2012; 50(4): 407–12.

149. Jeivad F, Yassa N, Ostad SN, Hassannejad Z, Gheshlaghi GH, Sabzevari O. Ficus Carica L. latex: Possible chemo-preventive, apoptotic activity and safety assessment. *Iran J Pharm Res IJPR.* 2020; 19(3): 231–40.

150. Soltana H, Pinon A, Limami Y, Zaid Y, Khalki L, Zaid N, Salah D, Sabitaliyevich UY, Simon A, Liagre B, Hammami M. Antitumoral activity of Ficus carica L. on colorectal cancer cell lines. *Cell Mol Biol.* 2019; 65(6): 6–11.

151. de Amorin A, Borba HR, Carauta JP, Lopes D, Kaplan MA. Anthelmintic activity of the latex of Ficus species. *J Ethnopharmacol.* 1999; 64(3): 255–8.

152. Shahinuzzaman M, Yaakob Z, Anuar FH, Akhtar P, Kadir NH, Hasan AM, Sobayel K, Nour M, Sindi H, Amin N, Sopian K. In vitro antioxidant activity of Ficus carica L. latex from 18 different cultivars. *Sci Rep.* 2020; 10(1): 10852–65.

153. Wang L, Yang J, Kong L-M, Deng J, Xiong Z, Huang J, Jianying L, Yijun Y, Yikao H, Xiao NL, Yan L, Yong Z, Sheng XH. Natural and semisynthetic tigliane diterpenoids with new carbon skeletons from Euphorbia dracunculoides as a WNT signaling pathway inhibitor. *Org Lett.* 2017; 19(14): 3911–4.

154. Prasad DN, Gode KD, Singh RS, Das PK. Preliminary phytochemical and pharmacological studies on Euphorbia dracunculoides, Lam. *Indian J Med Res.* 1967; 55(1): 73–9.

155. Batool R, Khan MR, Majid M. Euphorbia dracunculoides L. abrogates carbon tetrachloride induced liver and DNA damage in rats. *BMC Complement Altern Med.* 2017; 17(1): 1–11.

156. Yu CX, Wang RY, Qi FM, Su PJ, Yu YF, Li B, Zhao Y, Zhi DJ, Zhang ZX, Fei DQ. Eupulcherol A, a triterpenoid with a new carbon skeleton from Euphorbia pulcherrima, and its anti-Alzheimer's disease bioactivity. *Org Biomol Chem.* 2019; 18(1): 76–80.

157. Dai LF, Liang Q, Liu T, He MY, Zhao P, Xu WH. A new tigliane-type diterpene from Euphorbia dracunculoides Lam. *Nat Prod Res.* 2016; 30(14): 1639–45.

158. Majid M, Khan MR, Shah NA, Haq IU, Farooq MA, Ullah S, Sharif A, Zahra Z, Younis T, Sajid M. Studies on phytochemical, antioxidant, anti-inflammatory and analgesic activities of Euphorbia dracunculoides. *BMC Complement Altern Med.* 2015; 15: 1–5.

159. Bennett MH, Mansfield JW, Lewis MJ, Beale MH. Cloning and expression of sesquiterpene synthase genes from lettuce (Lactuca sativa L.). *Phytochemistry.* 2002; 60(3): 255–61.

160. Moulin-Traffort J, Giordani R, Régli P. Antifungal action of latex saps from Lactuca sativa L. and Asclepias curassavica L. *Mycoses.* 1990; 33(7–8): 383–92.

161. Kikuyama F, Suzuki S, Jibiki A, Yokoyama Y, Kawazoe H, Kitanaka S, Nakamura T. Ingenol mebutate inhibits the growth of pancreatic cancer cells in vitro via STING with an efficacy comparable to that of clinically used anticancer agents. *J Nat Med*. 2023; 77(2): 343–51.

162. Hua J, Liu Y, Xiao CJ, Jing SX, Luo SH, Li SH. Chemical profile and defensive function of the latex of Euphorbia peplus. *Phytochemistry*. 2017; 136: 56–64.

163. Binckley S, Zahra F. Euphorbia Tirucalli. In: StatPearls, Treasure Island (FL), StatPearls Publishing; 2023; 452: 1753.

164. Duong TH, Beniddir MA, Genta-Jouve G, Nguyen HH, Nguyen DP, Mac DH, Boustie J, Chavasiri W, Le Pogam P. Further terpenoids from Euphorbia tirucalli. *Fitoterapia*. 2019; 135: 44–51.

165. Brunetti RL, Da Paz DP, Mackowiak Da Fonseca II, Nagamine MK, Mori CM, Del Grande MP, Queiroz-Hazarbassanov N, Salles-Gomes CO, Dagli ML, Hernandez-Blazquez FJ. Inhibitory effects of Euphorbia tirucalli latex on murine B16/F10 melanoma cells and lung metastasis. *Mol Clin Oncol*. 2019; 11(5): 511–6.

166. Rodrigues ML, Gomes A de J, Funez MI, Marques MA da S, Lunardi CN. Euphorbia tirucalli latex loaded polymer nanoparticles: Synthesis, characterization, in vitro release and in vivo antinociceptive action. *PLoS One*. 2022; 17(11): e0274432-49.

167. Weng HZ, Tian Y, Zhang JS, Huang JL, Tang GH, Yin S. A new tigliane-type diterpenoid from Euphorbia tirucalli. *Nat Prod Res*. 2022; 36(20): 5380–6.

168. Franco-Salla GB, Prates J, Cardin LT, Dos Santos AR, Silva Jr WA, da Cunha BR, Tajara EH, Oliani SM, Rodrigues-Lisoni FC. Euphorbia tirucalli modulates gene expression in larynx squamous cell carcinoma. *BMC Complement Altern Med*. 2016; 16: 1–10.

169. Tiwari P, Ahmad K, Baig MH. *Gymnema sylvestre* for diabetes: From traditional herb to future's therapeutic. *Curr Pharm Des*. 2017; 23(11): 1667–76.

170. Pothuraju R, Sharma RK, Chagalamarri J, Jangra S, Kumar Kavadi P. A systematic review of *Gymnema sylvestre* in obesity and diabetes management. *J Sci Food Agric*. 2014; 94(5): 834–40.

171. Devangan S, Varghese B, Johny E, Gurram S, Adela R. The effect of *Gymnema sylvestre* supplementation on glycemic control in type 2 diabetes patients: A systematic review and meta-analysis. *Phytother Res*. 2021; 35(12): 6802–12.

172. Khan F, Sarker MM, Ming LC, Mohamed IN, Zhao C, Sheikh BY, Tsong HF, Rashid MA. Comprehensive review on phytochemicals,

pharmacological and clinical potentials of *Gymnema sylvestre. Front Pharmacol.* 2019; 10: 1223–41.

173. Di Fabio G, Romanucci V, Di Marino C, Pisanti A, Zarrelli A. *Gymnema sylvestre* R. Br., an Indian medicinal herb: Traditional uses, chemical composition, and biological activity. *Curr Pharm Biotechnol.* 2015; 16(6): 506–16.
174. Tian J, Wang YG, Ma J, Yang JB, Zhou L, Ji TF, Wang AG, Su YL. Hepatoprotective benzofurans and furanolignans from Gymnema tingens. *J Asian Nat Prod Res.* 2015; 17(3): 268–73.
175. Tian J, Ma QG, Yang JB, Wang AG, Ji TF, Wang YG, Su YL. Hepatoprotective phenolic glycosides from Gymnema tingens. *Planta Med.* 2013; 79(9): 761–7.
176. He J, Tang P, Liu M, Liao G, Lu R, Yang X. Triterpenoid saponins and C(21) steroidal glycosides from Gymnema tingens and their glucose uptake activities. *RSC Adv.* 2023; 13(11): 7503–13.
177. Dhalani JM, Nariya PB. A pharmacological review: Leptadenia reticulata (Wight & Arn.); Jivanti: The real life giving plant. *Folia Med (Plovdiv).* 2017; 59(4): 405–12.
178. Girishkumar V, Sreepriya M, Praveenkumar S, Bali G, Jagadeesh MS. Modulating effect of Leptadenia reticulata (Retz) Wight & Arn. against chromate (VI)-induced immunosuppression and oxidative stress on mouse splenic lymphocytes and bone marrow derived macrophages. *J Ethnopharmacol.* 2010; 131(2): 505–8.
179. Mohanty SK, Swamy MK, Sinniah UR, Anuradha M. Leptadenia reticulata (Retz.) Wight & Arn. (Jivanti): Botanical, agronomical, phytochemical, pharmacological, and biotechnological aspects. *Molecules.* 2017; 22(6): 1019–46.
180. Pal A, Sharma PP, Pandya TN, Acharya R, Patel BR, Shukla VJ, Ravishankar B. Phytochemical evaluation of dried aqueous extract of Jivanti [Leptadenia reticulata (Retz.) Wt. et Arn]. *Ayu.* 2012; 33(4): 557–60.
181. Kalidass C, Glory M, Borgio F, Manickam VS. Antibacterial activity of Leptadenia reticulata (Retz.) Wight. & Arn. (Asclepidaceae). *Anc Sci Life.* 2009; 28(4): 10–2.
182. Zhang L, Wang XL, Wang B, Zhang L-T, Gao H-M, Shen T, Hong XL, Dong MR, Xiao NW. Lignans from Euphorbia hirta L. *Nat Prod Res.* 2022; 36(1): 26–36.
183. Cayona R, Creencia E. Phytochemicals of Euphorbia hirta L. and their inhibitory potential against SARS-CoV-2 main protease. *Front Mol Biosci.* 2021; 8: 801401–16.

184. Iskandar B, Lukman A, Syaputra S, Al-Abrori UNH, Surboyo MDC, Lee CK. Formulation, characteristics and anti-bacterial effects of Euphorbia hirta L. mouthwash. *J Taibah Univ Med Sci.* 2022; 17(2): 271–82.

185. Ali MZ, Mehmood MH, Saleem M, Gilani AH. The use of Euphorbia hirta L. (Euphorbiaceae) in diarrhea and constipation involves calcium antagonism and cholinergic mechanisms. *BMC Complement Med Ther.* 2020; 20(1): 1–6.

186. Ramesh KV, Padmavathi K. Assessment of immunomodulatory activity of Euphorbia hirta L. *Indian J Pharm Sci.* 2010; 72(5): 621–5.

187. Rosenthal GA, Berenbaum MR. Herbivores: Their interactions with secondary plant metabolites ecological and evolutionary processes, 2nd ed. 2012; 2: 493.

188. Appel HM. Phenolics in ecological interactions: The importance of oxidation. *J Chem Ecol.* 1993; 19(7): 1521–52.

189. Kandungan P, Getah P. Determination of polyphenol contents in *Hevea brasiliensis* AND rubber-processing effluent. *Malaysian J Anal Sci.* 2018; 22(2): 185–96.

190. Snook, Data EM, Kays S. Characterization and quantification of hexadecyl-, octadecyl-, and eicosyl-esters of p-coumaric acids in the vine and root latex of sweetpotato [Ipomoea batatas (L.) Lam.]. *J Agric Food Chem.* 1994; 42: 2589–95.

191. Al-Rajhi AM, Yahya R, Abdelghany TM, Fareid MA, Mohamed AM, Amin BH, Masrahi AS. Anticancer, anticoagulant, antioxidant and antimicrobial activities of Thevetia peruviana latex with molecular docking of antimicrobial and anticancer activities. *Molecules.* 2022; 27(10): 3165–83.

192. Gan LS, Yang SP, Fan CQ, Yue JM. Lignans and their degraded derivatives from Sarcostemma acidum. *J Nat Prod.* 2005; 68(2): 221–5.

193. Rathore MS, Shekhawat NS. In vitro regeneration in Sarcostemma acidum (Roxb.)—an important medicinal plant of semi-arid ecosystem of Rajasthan, India. *Physiol Mol Biol Plants.* 2013; 19(2): 269–75.

194. Venma PK, Sharma A, Mathur A, Sharma P, Gupta RS, Joshi SC, Dixit VP. Effect of Sarcostemma acidum stem extract on spermatogenesis in male albino rats. *Asian J Androl.* 2002; 4(1): 43–7.

195. Ndip RN, Tarkang AE, Mbullah SM, Luma HN, Malongue A, Ndip LM, Nyongbela K, Wirmum C, Efange SM. In vitro anti-Helicobacter pylori

activity of extracts of selected medicinal plants from North West Cameroon. *J Ethnopharmacol.* 2007; 114(3): 452–7.

196. Hammadi R, Kúsz N, Dávid CZ, Mwangi PW, Berkecz R, Szemerédi N, Spengler G, Hohmann J, Vasas A. Polyoxypregnane ester derivatives and lignans from Euphorbia gossypina var. coccinea Pax. *Plants.* 2022; 11(10): 1299–14.

197. Patil RP, Pai SR, Pawar NV, Shimpale VB, Patil RM, Nimbalkar MS. Chemical characterization, mineral analysis, and antioxidant potential of two underutilized berries (Carissa carandus and Eleagnus conferta) from the Western Ghats of India. *Crit Rev Food Sci Nutr.* 2012; 52(4): 312–20.

198. Charteejee ML, Roy AR. Pharmacological action of Carissa carandus root (N.O. apocynaceae): Preliminary observations. *Bull Calcutta Sch Trop Med.* 1965; 13: 14–16.

199. Prakash D, Upadhyay G, Pushpangadan P, Gupta C. Antioxidant and free radical scavenging activities of some fruits. *J Complement Integr Med.* 2011; 8: 1–6.

200. Matignon L, Lo MM, Monpierre M, Correia MV, Valencia DP, Palmeira-Mello MV, Sylvestre MN, Pruneau L, Sylvestre M, Domenech A, Benfodda Z. Phytochemical and biological study of trophic interaction between Pseudosphinx tetrio L. larvae and Allamanda Cathartica L. *Plants.* 2023; 12(3): 520–37.

201. Bonomini TJ, Holzmann I, Thiesen LC, Fratoni E, Muller AF, Lucinda-Silva RM, Yunes RA, Malheiros A, Gonçalves AE, Dalmagro AP, da Silva KA. Neuropharmacological and acute toxicological evaluation of ethanolic extract of Allamanda cathartica L. flowers and plumieride. *Regul Toxicol Pharmacol.* 2017; 91: 9–19.

202. Melo IM, Sarte MF, Tavares SJS, Lustosa MS, Oliveira JS, Alencar NMN, Ramos MV, Lima V. Calotropis procera latex protein reduces inflammation and bone loss in ligature-induced periodontitis in male rats. *Arch Oral Biol.* 2023; 147: 105613.

203. Rabelo AC, Miglino MA, Arbizu S, Carreira AC, Carneiro FJ, Layosa MA, Noratto G. Calotropis procera induced caspase-dependent apoptosis and impaired Akt/mTOR signaling in 4T1 breast cancer cells. *Anticancer Agents Med Chem.* 2022; 22(18): 3136–47.

204. Rabelo AC, Borghesi J, Carreira AC, Hayashi RG, Bessa F, Barreto RD, da Costa RP, Cantanhede Filho AJ, Carneiro FJ, Miglino MA. Calotropis procera (Aiton) Dryand (Apocynaceae) as an anti-cancer agent against canine mammary tumor and osteosarcoma cells. *Res Vet Sci.* 2021; 138: 79–89.

205. Mohan A, Shaji S, Padmanabhan S, Naisam S, Sreekumar N. The potentials of Calotropis procera against filarial elephantiasis: An insilico approach. *J Parasit Dis.* 2022; 46(2): 1–11.

206. Saddiq AA, Tag HM, Doleib NM, Salman AS, Hagagy N. Antimicrobial, antigenotoxicity, and characterization of Calotropis procera and its rhizosphere-inhabiting actinobacteria: In vitro and in vivo studies. *Molecules.* 2022; 27(10): 3123–42.

207. Sini S, Malathy NS. Antimicrobial properties of roots of medicinal plants. *Anc Sci Life.* 2005; 25(2): 62–65.

208. Lateef M, Farheen S, Naqvi B, Iqbal L, Siddiqui K. Native wild plants in Karachi, Pakistan: Rich source of antioxidant raw material. *Pak J Pharm Sci.* 2020; 33(3): 1131–38.

209. Al-Dalahmeh Y, Al-Bataineh N, Al-Balawi SS, Lahham JN, Al-Momani IF, Al-Sheraideh MS, Mayyas AS, Abu Orabi ST, Al-Qudah MA. LC-MS/MS screening, total phenolic, flavonoid and antioxidant contents of crude extracts from three asclepiadaceae species growing in Jordan. *Molecules.* 2022; 27(3): 859–70.

210. Vaithiyanathan V, Mirunalini S. Assessment of antioxidant potential and acute toxicity studies of whole plant extract of Pergularia Daemia (Forsk). *Toxicol Int.* 2015; 22(1): 54–60.

211. Ananth DA, Rameshkumar A, Jeyadevi R, Aseervatham GSB, Sripriya J, Bose PC, Sivasudha T. Amelioratory effect of flavonoids rich Pergularia daemia extract against CFA induced arthritic rats. *Biomed Pharmacother.* 2016; 80: 244–52.

212. Kavaye Kandeda A, Okomolo Moto FC, Mbomo Ayissi RE, Omam Omam JP, Ojong L, Ngo Bum E. Pergularia daemia hydro-ethanolic extract protects against pentylenetetrazole kindling-induced seizures, oxidative stress, and neuroinflammation in mice. *J Ethnopharmacol.* 2021; 279: 114338.

213. Sureshkumar SV, Mishra SH. Hepatoprotective effect of extracts from Pergularia daemia Forsk. *J Ethnopharmacol.* 2006; 107(2): 164–8.

214. Kandeda AK, Taiwe GS, Moto FC, Ngoupaye GT, Nkantchoua GC, Njapdounke JS, Omam JP, Pale S, Kouemou N, Ngo Bum E. Antiepileptogenic and neuroprotective effects of Pergularia daemia on pilocarpine model of epilepsy. *Front Pharmacol.* 2017; 8: 440–54.

215. Bhaskar VH, Balakrishnan N. Protective effects of Pergularia daemia roots against paracetamol and carbon tetrachloride-induced hepatotoxicity in rats. *Pharm Biol.* 2010; 48(11): 1265–72.

216. Guerra NB, Sant'Ana Pegorin G, Boratto MH, de Barros NR, de Oliveira Graeff CF, Herculano RD. Biomedical applications of natural rubber

latex from the rubber tree *Hevea brasiliensis. Mater Sci Eng C Mater Biol Appl.* 2021; 126: 112126–44.

217. Jacob JL, d'Auzac J, Prevôt JC. The composition of natural latex from *Hevea brasiliensis. Clin Rev Allergy.* 1993; 11(3): 325–37.

218. Wei BL, Weng JR, Chiu PH, Hung CF, Wang JP, Lin CN. Antiinflammatory flavonoids from Artocarpus heterophyllus and Artocarpus communis. *J Agric Food Chem.* 2005; 53(10): 3867–71.

219. Siritapetawee J, Thammasirirak S, Samosornsuk W. Antimicrobial activity of a 48 kDa protease (AMP48) from Artocarpus heterophyllus latex. *Eur Rev Med Pharmacol Sci.* 2012; 16(1): 132–7.

220. Sibi G, Wadhavan R, Singh S, Shukla A, Dhananjaya K, Ravikumar KR, Mallesha H. Plant latex: A promising antifungal agent for post harvest disease control. *Pakistan J Biol Sci PJBS.* 2013; 16(23): 1737–43.

221. Siritapetawee J, Thumanu K, Sojikul P, Thammasirirak S. A novel serine protease with human fibrino(geno)lytic activities from Artocarpus heterophyllus latex. *Biochim Biophys Acta.* 2012; 1824(7): 907–12.

222. Anani K, Adjrah Y, Ameyapoh Y, Karou SD, Agbonon A, De Souza C, Gbeassor M. Antimicrobial, anti-inflammatory and antioxidant activities of Jatropha multifida L. (Euphorbiaceae). *Pharmacognosy Res.* 2016; 8(2): 142–46.

223. Woo SY, Wong CP, Win NN, Lae KZ, Woo B, Elsabbagh SA, Liu QQ, Ngwe H, Morita H. Anti-melanin deposition activity and active constituents of Jatropha multifida stems. *J Nat Med.* 2019; 73(4): 805–13.

224. Das B, Ravikanth B, Reddy KR, Thirupathi P, Raju TV, Sridhar B. Diterpenoids from Jatropha multifida. *Phytochemistry.* 2008; 69(14): 2639–41.

225. Zhang JS, Zhang Y, Li S, Ahmed A, Tang GH, Yin S. Cytotoxic macrocyclic diterpenoids from Jatropha multifida. *Bioorg Chem.* 2018; 80: 511–8.

226. Vieira DS, Oliveira FT de, Suarez JAG, Silva DP da, Bernardo THL, Bastos ML de A. Biological activities: Anti-infectious, antioxidant and healing of the vegetable species Jatropha multifida. *Rev Bras Enferm.* 2021; 74(2): e20200451.

227. Rampadarath S, Puchooa D, Ranghoo-Sanmukhiya VM. Antimicrobial, phytochemical and larvicidal properties of Jatropha multifida Linn. *Asian Pac J Trop Med.* 2014; 7S1: S380–3.

228. Suriyakala G, Sathiyaraj S, Devanesan S, AlSalhi MS, Rajasekar A, Maruthamuthu MK, Babujanarthanam R. Phytosynthesis of silver nanoparticles from Jatropha integerrima Jacq. flower extract and their possible applications as antibacterial and antioxidant agent. *Saudi J Biol Sci.* 2022; 29(2): 680–8.

229. Mahrous EA, Elosaily AH, Salama AAA, Salama AM, El-Zalabani SM. Oral and topical anti-inflammatory activity of Jatropha integerrima leaves extract in relation to its metabolite profile. *Plants.* 2022; 11(2): 218–36.

230. Arribére MC, Cortadi AA, Gattuso MA, Bettiol MP, Priolo NS, Caffini NO. Comparison of Asclepiadaceae latex proteases and characterization of Morrenia brachystephana Griseb. Cysteine peptidases. *Phytochem Anal.* 1998; 9(6): 267–73.

231. Arima K, Uchikoba T, Yonezawa H, Shimada M, Kaneda M. Cucumisin-like protease from the latex of Euphorbia supina. *Phytochemistry.* 2000; 53(6): 639–44.

232. Tomar R, Kumar R, Jagannadham MV. A stable serine protease, wrightin, from the latex of the plant Wrightia tinctoria (Roxb.) R. Br.: Purification and biochemical properties. *J Agric Food Chem.* 2008; 56(4): 1479–87.

233. John KS, Bhat SG, Prasada Rao UJS. Biochemical characterization of sap (latex) of a few Indian mango varieties. *Phytochemistry.* 2003; 62(1): 13–19.

234. Konno K, Hirayama C, Nakamura M, Tateishi K, Tamura Y, Hattori M, Kohno K.. Papain protects papaya trees from herbivorous insects: Role of cysteine proteases in latex. *Plant J.* 2004; 37(3): 370–8.

235. Pechan T, Cohen A, Williams WP, Luthe DS. Insect feeding mobilizes a unique plant defense protease that disrupts the peritrophic matrix of caterpillars. *Proc Natl Acad Sci.* 2002; 99(20): 13319–23.

236. Haber RA, Garcia RD, Hernandez JN, Jamieson S, Mondal A, Bishayee A. Papaya (Carica papaya L.) for cancer prevention: Progress and promise. *Crit Rev Food Sci Nutr.* 2022; 63(30): 10499–519.

237. Jaiswal P, Singh DK. Molluscicidal activity of Carica papaya and Areca catechu against the freshwater snail Lymnaea acuminata. *Vet Parasitol.* 2008; 152(3–4): 264–70.

238. Khodarahmi GA, Ghasemi N, Hassanzadeh F, Safaie M. Cytotoxic effects of different extracts and latex of Ficus carica L. on HeLa cell line. *Iran J Pharm Res IJPR.* 2011; 10(2): 273–7.

239. Baohong L, Zhongyuan L, Ying T, Beibei Y, Wenting N, Yiming Y, Qinghua C, Qingjun Z. Latex derived from Ficus carica L. inhibited the growth of NSCLC by regulating the caspase/gasdermin/AKT signaling pathway. *Food Funct.* 2023; 14(4): 2239–48.

240. Kim JS, Kim YO, Ryu HJ, Kwak YS, Lee JY, Kang H. Isolation of stress-related genes of rubber particles and latex in fig tree (Ficus carica) and their expressions by abiotic stress or plant hormone treatments. Plant Cell Physiol. 2003; 44(4): 412–14.

241. Azarkan M, Wintjens R, Looze Y, Baeyens-Volant D. Detection of three wound-induced proteins in papaya latex. *Phytochemistry.* 2004; 65(5): 525–34.

242. Han KH, Shin DH, Yang J, Kim IJ, Oh SK, Chow KS. Genes expressed in the latex of *Hevea brasiliensis. Tree Physiol.* 2000; 20(8): 503–10.

243. Walz C, Giavalisco P, Schad M, Juenger M, Klose J, Kehr J. Proteomics of curcurbit phloem exudate reveals a network of defence proteins. *Phytochemistry.* 2004; 65(12): 1795–804.

244. Kehr J. Phloem sap proteins: Their identities and potential roles in the interaction between plants and phloem-feeding insects. *J Exp Bot.* 2006; 57(4): 767–74.

245. Zhu-Salzman K, Luthe DS, Felton GW. Arthropod-inducible proteins: Broad spectrum defenses against multiple herbivores. *Plant Physiol.* 2008; 146(3): 852–8.

246. Rani P, Yasur J, Usha Rani P, Jyothsna Y. Biochemical and enzymatic changes in rice as a mechanism of defense. *Acta Physiol Plant.* 2010; 32: 695–701.

247. Gulsen O, Eickhoff T, Heng-Moss T, Shearman R, Baxendale F, Sarath G, Lee D. Characterization of peroxidase changes in resistant and susceptible warm-season turfgrasses challenged by Blissus occiduus. *Arthropod Plant Interact.* 2010; 4(1): 45–55.

248. Gidrol X, Chrestin H, Tan HL, Kush A. Hevein, a lectin-like protein from *Hevea brasiliensis* (rubber tree) is involved in the coagulation of latex. *J Biol Chem.* 1994 Mar; 269(12): 9278–83.

249. Wititsuwannakul R, Rukseree K, Kanokwiroon K, Wititsuwannakul D. A rubber particle protein specific for Hevea latex lectin binding involved in latex coagulation. *Phytochemistry.* 2008; 69(5): 1111–18.

250. Wang Y, Yang L, Chen X, Ye T, Zhong B, Liu R, Wu Y, Chan Z. Major latex protein-like protein 43 (MLP43) functions as a positive regulator during abscisic acid responses and confers drought tolerance in Arabidopsis thaliana. *J Exp Bot.* 2016; 67(1): 421–34.

251. Zielińska S, Jezierska-Domaradzka A, Wójciak-Kosior M, Sowa I, Junka A, Matkowski AM. Greater celandine's ups and downs-21 centuries of medicinal uses of Chelidonium majus from the viewpoint of today's pharmacology. *Front Pharmacol.* 2018; 9: 299–328.

252. Decker G, Wanner G, Zenk MH, Lottspeich F. Characterization of proteins in latex of the opium poppy (Papaver somniferum) using two-dimensional gel electrophoresis and microsequencing. *Electrophoresis.* 2000; 21(16): 3500–516.

253. Cho WK, Jo Y, Chu H, Park SH, Kim KH. Integration of latex protein sequence data provides comprehensive functional overview of latex proteins. *Mol Biol Rep.* 2014; 41(3): 1469–81.

254. Nawrot R. Defense-related proteins from Chelidonium majus L. as important components of its latex. *Curr Protein Pept Sci.* 2017; 18(8): 864–80.

255. Vandenborre G, Smagghe G, Van Damme EJM. Plant lectins as defense proteins against phytophagous insects. *Phytochemistry.* 2011; 72(13): 1538–50.

256. Chakraborti D, Sarkar A, Mondal HA, Das S. Tissue specific expression of potent insecticidal, Allium sativum leaf agglutinin (ASAL) in important pulse crop, chickpea (Cicer arietinum L.) to resist the phloem feeding Aphis craccivora. *Transgenic Res.* 2009; 18(4): 529–44.

257. Macedo MLR, das Graças Machado Freire M, da Silva MBR, Coelho LCBB. Insecticidal action of Bauhinia monandra leaf lectin (BmoLL) against Anagasta kuehniella (Lepidoptera: Pyralidae), Zabrotes subfasciatus and Callosobruchus maculatus (Coleoptera: Bruchidae). Comparative Biochemistry and Physiology Part A. *Mol Integr Physiol.* 2007; 146(4): 486–98.

258. Saha P, Majumder P, Dutta I, Ray T, Roy SC, Das S. Transgenic rice expressing Allium sativum leaf lectin with enhanced resistance against sap-sucking insect pests. *Planta.* 2006; 223(6):1329–43.

259. Damme EJM Van, Peumans WJ, Barre A, Rougé P. Plant lectins: A composite of several distinct families of structurally and evolutionary related proteins with diverse biological roles. *CRC Crit Rev Plant Sci.* 1998; 17(6): 575–692.

260. Mccloud ES, Tallamy DW, Halaweish FT. Squash beetle trenching behaviour: Avoidance of cucurbitacin induction or mucilaginous plant sap? *Ecol Entomol.* 1995; 20: 51–59.

261. Wasano N, Konno K, Nakamura M, Hirayama C, Hattori M, Tateishi K. A unique latex protein, MLX56, defends mulberry trees from insects. *Phytochemistry*. 2009; 70(7): 880–88.

262. Ramos MV, Freitas CD, Stanisçuaski F, Macedo LL, Sales MP, Sousa DP, Carlini CR. Performance of distinct crop pests reared on diets enriched with latex proteins from Calotropis procera: Role of laticifer proteins in plant defense. *Plant Sci*. 2007; 173: 349–57.

263. Hopkins TL, Harper MS. Lepidopteran peritrophic membranes and effects of dietary wheat germ agglutinin on their formation and structure. *Arch Insect Biochem Physiol*. 2001; 47(2): 100–109.

264. Kramer KJ, Muthukrishnan S. Insect chitinases: Molecular biology and potential use as biopesticides. *Insect Biochem Mol Biol*. 1997; 27(11): 887–900.

265. Kabir KE, Sugimoto H, Tado H, Endo K, Yamanaka A, Tanaka S, Koga D. Effect of Bombyx mori chitinase against Japanese pine sawyer (*Monochamus alternatus*) adults as a biopesticide. *Biosci Biotechnol Biochem*. 2006; 70(1): 219–29.

266. Lawrence SD, Novak NG. Expression of poplar chitinase in tomato leads to inhibition of development in colorado potato beetle. *Biotechnol Lett*. 2006; 28(8): 593–99.

267. Wititsuwannakul D, Chareonthiphakorn N, Pace M, Wititsuwannakul R. Polyphenol oxidases from latex of *Hevea brasiliensis*: Purification and characterization. *Phytochemistry*. 2002; 61(2): 115–21.

268. Fiorillo F, Palocci C, Soro S, Pasqua G. Latex lipase of *Euphorbia characias* L.: An aspecific acylhydrolase with several isoforms. *Plant Sci*. 2007; 172: 722–27.

269. Gandhi NN, Mukherjee KD. Specificity of papaya lipase in esterification with respect to the chemical structure of substrates. *J Agric Food Chem*. 2000; 48(2): 566–70.

270. Zerhouni S, Amrani A, Nijs M, Smolders N, Azarkan M, Vincentelli J, Looze Y. Purification and characterization of papaya glutamine cyclotransferase, a plant enzyme highly resistant to chemical, acid and thermal denaturation. *Biochim Biophys Acta*. 1998; 1387(1–2): 275–90.

271. Goodrum LJ, Patel A, Leykam JF, Kieliszewski MJ. Gum arabic glycoprotein contains glycomodules of both extensin and arabinogalactan-glycoproteins. *Phytochemistry*. 2000; 54(1): 99–106.

272. Nambisan B. Cassava latex as a source of linamarase for determination of linamarin. *J Agric Food Chem*. 1999; 47(2): 372–73.

273. Noack EA, Crea AE, Falsone G. Inhibition of mitochondrial oxidative phosporylation by 4-deoxyphorbol triester, a poisonous constituent of the latex sap of Euphorbia biglandulosa Desf. *Toxicon*. 1980 1; 18(2): 165–74.

274. Falsone G, Cateni F, Katusian F. New Cerebrosides from Euphorbia characias L. 1993; 59(S1), A617.

275. Mazoir N, Benharref A, Bailén M, Reina M. Antileishmanial and antitrypanosomal activity of triterpene derivatives from latex of two Euphorbia Species. *Z Naturforsch C J Biosci*. 2011; 66(7–8): 360–66.

276. Chamkhi I, Hnini M, Aurag J. Conventional medicinal uses, phytoconstituents, and biological activities of *Euphorbia officinarum* L.: A Systematic Review. *Adv Pharmacol Pharm Sci*. 2022; 2022: 9971085.

277. Bailen M, Khamlichi MD, Benharref A, Martinez-Diaz RA, Gonzalez-Coloma A. New bioactive semisynthetic derivatives of 31-norlanostenol and obtusifoliol from *Euphorbia officinarum*. *Nat Prod Commun*. 2016; 11(6): 733–38.

278. Konno K, Ono H, Nakamura M, Tateishi K, Hirayama C, Tamura Y, Hattori M, Koyama A, Kohno K. Mulberry latex rich in antidiabetic sugar-mimic alkaloids forces dieting on caterpillars. *Proc Natl Acad Sci*. 2006; 103(5): 1337–41.

279. Hirayama C, Konno K, Wasano N, Nakamura M. Differential effects of sugar-mimic alkaloids in mulberry latex on sugar metabolism and disaccharidases of Eri and domesticated silkworms: Enzymatic adaptation of Bombyx mori to mulberry defense. *Insect Biochem Mol Biol*. 2007; 37(12): 1348–58.

280. Freitas CD, Leite HB, Oliveira JP, Amaral JL, Egito AS, Vairo-Cavalli S, Lobo MD, Monteiro-Moreira AC, Ramos MV. Insights into milk-clotting activity of latex peptidases from Calotropis procera and Cryptostegia grandiflora. *Food Res Int*. 2016; 87: 50–59.

281. Maheshu V, Priyadarsini DT, Sasikumar JM. Antioxidant capacity and amino acid analysis of Caralluma adscendens (Roxb.) Haw var. fimbriata (wall.) Grav. & Mayur. aerial parts. *J Food Sci Technol*. 2014; 51(10): 2415–24.

282. Kunert O, Rao VG, Babu GS, Sujatha P, Sivagamy M, Anuradha S, Rao BV, Kumar BR, Alex RM, Schühly W, Kühnelt D.. Pregnane glycosides from Caralluma adscendens var. fimbriata. *Chem Biodivers*. 2008; 5(2): 239–50.

283. Jyothi MV, Bhargav E, Babu CN, Reddy KS, Kumar BP, Rao RP. Apoptosis-mediated cytotoxic effect of Caralluma adscendens var.

attenuata on Colon (HT29) and Hepatic (HepG2) Cancer Cell Lines. *J Pharm Bioallied Sci.* 2019; 11(1): 38–42.

284. Gharat SA, Shinde BA, Mule RD, Punekar SA, Dholakia BB, Jayaramaiah RH, Ramaswamy G, Giri AP. High-throughput metabolomic and transcriptomic analyses vet the potential route of cerpegin biosynthesis in two varieties of Ceropegia bulbosa Roxb. *Planta.* 2019; 251(1): 1–13.

285. Cittrarasu V, Kaliannan D, Dharman K, Maluventhen V, Easwaran M, Liu WC, Balasubramanian B, Arumugam M. Green synthesis of selenium nanoparticles mediated from Ceropegia bulbosa Roxb extract and its cytotoxicity, antimicrobial, mosquitocidal and photocatalytic activities. *Sci Rep.* 2021; 11(1): 1032–47.

286. Biswas M, Haldar PK, Ghosh AK. Antioxidant and free-radical-scavenging effects of fruits of Dregea volubilis. *J Nat Sci Biol Med.* 2010; 1(1): 29–34.

287. Gong F, Shen Y, Zhang C, Xu J, Wu X, Hua Z, Xu Q. Dregea volubilis ameliorates concanavalin A-induced liver injury by facilitating apoptosis of activated T cells. *Exp Biol Med (Maywood).* 2008; 233(9): 1124–32.

288. Thuy NT, Phuong PT, Hien NT, Trang DT, Huan NV, Anh PT, Tai BH, Nhiem NX, Hung NT, Kiem PV. Pregnane glycosides from the leaves of Dregea volubilis and their α-glucosidase and α-amylase inhibitory activities. *Nat Prod Res.* 2021; 35(21): 3931–38.

289. Hossain E, Rawani A, Chandra G, Mandal SC, Gupta JK. Larvicidal activity of Dregea volubilis and Bombax malabaricum leaf extracts against the filarial vector Culex quinquefasciatus. *Asian Pac J Trop Med.* 2011; 4(6): 436–41.

290. Hossain E, Sarkar D, Maiti A, Chatterjee M, Mandal SC, Gupta JK. Anti-inflammatory effect of a methanolic extract of leaves of Dregea volubilis. *J Ethnopharmacol.* 2010; 132(2): 525–28.

291. Hossain E, Chakroborty S, Milan A, Chattopadhyay P, Mandal SC, Gupta JK. In vitro and in vivo antitumor activity of a methanol extract of Dregea volubilis leaves with its antioxidant effect. *Pharm Biol.* 2012; 50(3): 338–43.

292. Sahu N, Panda N, Mandal N, Banerjee S, Koike K, Nikaido T. Polyoxypregnane glycosides from the flowers of Dregea volubilis. *Phytochemistry.* 2002; 61(4): 383–88.

293. Rugaie OA, Mohammed HA, Alsamani S, Messaoudi S, Aroua LM, Khan RA, Almahmoud SA, Altaleb AD, Alsharidah M, Aldubaib M,

Al-Regaiey KA. Antimicrobial, antibiofilm, and antioxidant potentials of four halophytic plants, Euphorbia chamaesyce, Bassia arabica, Fagonia mollis, and Haloxylon salicornicum, growing in Qassim region of Saudi Arabia: Phytochemical profile and in vitro and in silico. *Antibiot.* 2023; 12(3): 501–16.

294. Küçükbay FZ, Kuyumcu E. Determination of elements by atomic absorption spectrometry in medicinal plants employed to alleviate common cold symptoms. *Guang Pu Xue Yu Guang Pu Fen Xi.* 2014; 34(9): 2548–56.

295. Tanaka R, Kasubuchi K, Kita S, Tokuda H, Nishino H, Matsunaga S. Bioactive steroids from the whole herb of Euphorbia chamaesyce. *J Nat Prod.* 2000; 63(1): 99–103.

296. Tanaka R, Wada S, Yamada T, Yamori T. Potent antitumor activity of 3,4-seco-8betaH-Ferna-4(23),9(11)-dien-3-oic acid (EC-2) and 3,4-seco-Oleana-4(23),18-dien-3-oic acid (EC-4), evaluated by an in vitro human cancer cell line panel. *Planta Med.* 2006; 72(14): 1347–49.

297. de Vasconcellos MC, Schall VT. Latex of "coroa de cristo" (Euphorbia splendens): An effective molluscicide. *Mem Inst Oswaldo Cruz.* 1986; 81(4): 475–76.

298. Freitas JC, Presgrave OA, Fingola FF, Menezes MA, Vasconcellos MC, Schall VT, Paumgartten FJ.. Toxicological study of the molluscicidal latex of Euphorbia splendens: Irritant action on skin and eye. *Mem Inst Oswaldo Cruz.* 1991; 86 Suppl 2: 87–88.

299. Schall VT, Vasconcellos MC, Valent GU, Sato MI, Furlan E V, Sanchez PS. Evaluation of the genotoxic activity and acute toxicity of Euphorbia splendens latex, a molluscicide for the control of schistosomiasis. *Braz J Med Biol Res.* 1991; 24(6): 573–82.

300. Goyal M, Nagori BP, Sasmal D. Wound healing activity of latex of Euphorbia caducifolia. *J Ethnopharmacol.* 2012; 144(3): 786–90.

301. Bano S, Siddiqui BS, Farooq AD, Begum S, Siddiqui F, Kashif M, Azhar M. In vitro growth inhibition and cytotoxicity of Euphorbia caducifolia against four human cancer cell lines and its phytochemical characterisation. *Nat Prod Res.* 2017; 31(24): 2936–40.

302. Gupta S, Acharya R. Management of Padadari (cracked feet) with Rakta Snuhi (Euphorbia caducifolia Haines.) based formulation: An open-labeled clinical study. *Ayu.* 2018; 39(4): 208–12.

303. Kumar L, Deshmukh RK, Gaikwad KK. Antimicrobial packaging film from cactus (Cylindropuntia fulgida) mucilage and gelatine. *Int J Biol Macromol.* 2022; 215: 596–605.

304. Wang XB, Li GH, Zheng LJ, Ji KY, Lü H, Liu FF, Dang LZ, Mo MH, Zhang KQ. *Chem Biodivers.* 2009; 6(3): 431–36.

305. Mohale DS, Tripathi AS, Shrirao AV, Jawarkar AG, Chandewar AV. Evaluation of antioxidant effect of Nerium indicum in anxious rats. *Indian J Pharmacol.* 2016; 48(4): 430–33.

306. Ghate NB, Chaudhuri D, Panja S, Mandal N. Nerium indicum leaf alleviates iron-induced oxidative stress and hepatic injury in mice. *Pharm Biol.* 2015; 53(7): 1066–74.

307. Arai MA, Akamine R, Hayashi N, Koyano T, Kowithayakorn T, Ishibashi M. The notch inhibitors isolated from Nerium indicum. *J Nat Prod.* 2018; 81(5): 1235–40.

308. Kusuma CG, Gubbiveeranna V, Sumachirayu CK, Bhavana S, Ravikumar H, Nagaraju S. Thrombin- and plasmin-like and platelet-aggregation-inducing activities of Plumeria alba L. latex: Action of cysteine protease. *J Ethnopharmacol.* 2021; 273: 114000.

309. Lawal OA, Ogunwande IA, Opoku AR. Constituents of essential oils from the leaf and flower of Plumeria alba grown in Nigeria. *Nat Prod Commun.* 2014; 9(11): 1613–14.

310. Choudhary M, Kumar V, Gupta P, Singh S. Investigation of antiarthritic potential of Plumeria alba L. leaves in acute and chronic models of arthritis. *Biomed Res Int.* 2014; 2014: 1–13.

311. Bihani T. Plumeria rubra L. A review on its ethnopharmacological, morphological, phytochemical, pharmacological and toxicological studies. *J Ethnopharmacol.* 2021; 264: 113291.

312. Khan IA, Hussain M, Syed SK, Saadullah M, Alqahtani AM, Alqahtani T, Aldahish AA, Asiri S, Zeng LH. Pharmacological justification for the medicinal use of Plumeria rubra Linn. in cardiovascular disorders. *Molecules.* 2021; 27(1): 251–67.

313. Zhang SN, Song HZ, Ma RJ, Liang CQ, Wang HS, Tan QG. Potential anti-diabetic isoprenoids and a long-chain δ-lactone from frangipani (Plumeria rubra). *Fitoterapia.* 2020; 146: 104684.

314. Wang DL, Liu YY, Tian D, Yu LY, Gui LJ. Characterization of the complete chloroplast genome of Plumeria rubra cv. Acutifolia (Apocynaceae). *Mitochondrial DNA Part B.* 2020; 5(1): 927–28.

315. Purohit P, Palamthodi S, Lele SS. Effect of karwanda (Carissa congesta Wight) and sugar addition on physicochemical characteristics of ash gourd (Benincasa hispida) and bottle gourd (Langenaria siceraria) based beverages. *J Food Sci Technol.* 2019; 56(2): 1037–45.

316. Doshi GM, Nalawade VV, Mukadam AS, Chaskar PK, Zine SP, Somani RR, Une HD. Elucidation of flavonoids from Carissa congesta, Polyalthia longifolia, and Benincasa hispida plant extracts by hyphenated technique of liquid chromatography-mass spectroscopy. *Pharmacognosy Res.* 2016; 8(4): 281–86.

317. Doshi GM, Nalawade VV, Mukadam AS, Chaskar PK, Zine SP, Somani RR, Une HD. Structural elucidation of chemical constituents from Benincasa hispida seeds and Carissa congesta roots by gas chromatography: Mass spectroscopy. *Pharmacognosy Res.* 2015; 7(3): 282–93.

318. Une HD, Doshi GM. Carissa congesta Wight and Benincasa hispida (Thunb.) Cogn. as budding immunomodulatory agents. *Indian J Exp Biol.* 2016; 54(10): 650–58.

319. Bashir K, Naz S, Farooq U, Wahid F, Shah AJ, McCauley EP, Crews P, Khan T. Assessing the ethnobotanical potential of Carissa opaca berries by merging outcomes from metabolomics profiling, enzyme assays, and in silico docking studies. *Food Chem.* 2021; 363: 130259.

320. Sahreen S, Khan MR, Khan RA, Alkreathy HM. Protective effects of Carissa opaca fruits against CCl4-induced oxidative kidney lipid peroxidation and trauma in rat. *Food Nutr Res.* 2015; 59: 28438–48.

321. Sahreen S, Khan MR, Khan RA. Evaluation of antioxidant profile of various solvent extracts of Carissa opaca leaves: An edible plant. *Chem Cent J.* 2017; 11(1): 1–7.

322. Alqahtani YS, Amir H, Ahmad T, Mahnashi MH, Alyami BA, Alqarni AO, Akram M, Khan HU. Cardiotonic potential of Carissa opaca Stapf ex Haines: A study on isolated rabbit heart. *Pak J Pharm Sci.* 2021; 34(1): 367–72.

323. Sahreen S, Khan MR, Khan RA. Effects of Carissa opaca fruits extracts on oxidative pulmonary damages and fibrosis in rats. *BMC Complement Altern Med.* 2014 14: 1–9.

324. Sahreen S, Khan MR, Khan RA, Shah NA. Estimation of flavoniods, antimicrobial, antitumor and anticancer activity of Carissa opaca fruits. *BMC Complement Altern Med.* 2013; 13: 1–7.

325. Malik W, Ahmed D, Izhar S. Tyrosinase inhibitory activities of Carissa opaca Stapf ex haines roots extracts and their phytochemical analysis. *Pharmacogn Mag.* 2017; 13(Suppl 3): S544–48.

326. Tin-Wa M, Farnsworth NR, Fong HH, Trojanek J. Catharanthus alkaloids. XXV. Isolation of leurosine and ursolic acid from C. pusillus. *Lloydia.* 1970; 33(2): 261–63.

327. Battersby AR, Kapil RS. N-Benzoyl-L-phenylalaninol from Catharanthus pusillus (Murr.) G. Don = Vinca pusilla Murr. *Tetrahedron Lett.* 1965; (39): 3529–31.

328. Zárate R, Dirks C, van der Heijden R, Verpoorte R. Terpenoid indole alkaloid profile changes in Catharanthus pusillus during development. *Plant Sci.* 2001; 160(5): 971–77.

329. Tin-Wa M, Fong HH, Blomster RN, Farnsworth NR. Catharanthus alkaloids. XXI. Isolation of lochnerinine, one of the cytotoxic principles of C. pusillus. *J Pharm Sci.* 1968; 57(12): 2167–69.

330. Cole JB, Corcoran JN. Yellow Oleander (Thevetia peruviana), a source of toxic cardiac glycosides, may be substituted for candlenuts (Aleurities moluccana) when taken as a weight-loss supplement. In *Cardiology in the Young,* vol. 30, England; 2020. pp. 1755–56.

331. Samanta J, Bhattacharya S, Rana AC. Antifertility activity of Thevetia peruviana (Pers.) K. Schum leaf in female Sprague-Dawley rat. *Indian J Pharmacol.* 2016; 48(6): 669–74.

332. Kohls S, Scholz-Böttcher BM, Teske J, Zark P, Rullkötter J. Cardiac glycosides from Yellow Oleander (Thevetia peruviana) seeds. *Phytochemistry.* 2012; 75: 114–27.

333. Ramos-Silva A, Tavares-Carreón F, Figueroa M, De la Torre-Zavala S, Gastelum-Arellanez A, Rodríguez-García A, Galán-Wong LJ, Avilés-Arnaut H. Anticancer potential of Thevetia peruviana fruit methanolic extract. *BMC Complement Altern Med.* 2017; 17(1): 1–11.

334. El Gizawy HA, El Zanaty SA, El Ghaly EM, Seif-Eldein NA. Thevetia peruviana leaves, HPLC profile, isolation, and in vitro cytotoxicity. *RSC Adv.* 2023; 13(18): 12072–79.

335. Mondal P, Maity R, Mallick C. In vitro spermicidal effect of Thevetia peruviana leaves on human spermatozoa. *Andrologia.* 2022; 54(2): e14323.

336. Kaufmann H, Wehrli W, Reichstein T. The glycosides of the seeds of Vallaris solanacea (Roth) O. K. Isolation. Glycosides and aglycones, 263. *Helv Chim Acta.* 1965; 48: 65–82.

337. Urs AP, Rudresha GV, Manjuprasanna VN, Suvilesh KN, Gowda MM, Yariswamy M, Hiremath V, Ramakrishnan C, Savitha MN, Jayachandra K, Sharanappa P. Plant latex thrombin-like cysteine proteases alleviates bleeding by bypassing factor VIII in murine model. *J Cell Biochem.* 2019; 120(8): 12843–58.

338. Khanal P, Patil BM. Integration of in silico, in vitro and ex vivo pharmacology to decode the anti-diabetic action of Ficus benghalensis L. bark. *J Diabetes Metab Disord.* 2020; 19(2): 1325–37.

339. Murugesu S, Selamat J, Perumal V. Phytochemistry, pharmacological properties, and recent applications of Ficus benghalensis and Ficus religiosa. *Plants.* 2021; 10(12): 2749–78.

340. Iltaf J, Noreen S, Rehman MF, Ghumman SA, Batool F, Mehdi M, Hasan S, Ijaz B, Akram MS, Butt H. Ficus benghalensis as potential inhibitor of 5α-reductase for hair growth promotion: In vitro, in silico, and in vivo evaluation. *Front Pharmacol.* 2021; 12: 774583–594.

341. Ramasamy A, Anandakumar K, Kathiresan K. In vitro anti-oxidant potential and acetylcholinesterase inhibitory effect of Ficus benghalensis aerial root extract. *Afr Health Sci.* 2022; 22(4): 291–99.

342. Shim KH, Sharma N, An SSA. Mechanistic insights into the neuroprotective potential of sacred ficus trees. *Nutrients.* 2022; 14(22): 4731–90.

343. Garg VK, Paliwal SK. Wound-healing activity of ethanolic and aqueous extracts of Ficus benghalensis. *J Adv Pharm Technol Res.* 2011; 2(2): 110–14.

344. Mota HR, Oliveira JT, Martins TF, Vasconcelos IM, Costa HP, Neres DP, Silva FD, Souza PF. Chitinase from the latex of Ficus benjamina L. Displays antifungal activity by inducing ROS generation and structural damage to the fungal cell wall and plasma membrane. *Protein Pept Lett.* 2022; 29(10): 869–81.

345. da Cruz RC, Agertt V, Boligon AA, Janovik V, Anraku de Campos MM, Guillaume D, Athayde ML. In vitro antimycobacterial activity and HPLC-DAD screening of phenolics from Ficus benjamina L. and Ficus luschnathiana (Miq.) Miq. leaves. *Nat Prod Res.* 2012; 26(23): 2251–54.

346. Salehi B, Prakash Mishra A, Nigam M, Karazhan N, Shukla I, Kiełtyka-Dadasiewicz A, Sawicka B, Głowacka A, Abu-Darwish MS, Hussein Tarawneh A, Gadetskaya AV. Ficus plants: State of the art from a phytochemical, pharmacological, and toxicological perspective. *Phytother Res.* 2021; 35(3): 1187–217.

347. Ogunwande IA, Jimoh R, Ajetunmobi AA, Avoseh NO, Flamini G. Essential oil composition of Ficus benjamina (Moraceae) and Irvingia barteri (Irvingiaceae). *Nat Prod Commun.* 2012; 7(12): 1673–75.

348. Hajam TA, Saleem H. Phytochemistry, biological activities, industrial and traditional uses of fig (Ficus carica): A review. *Chem Biol Interact.* 2022; 368: 110237.

349. Morovati MR, Ghanbari-Movahed M, Barton EM, Farzaei MH, Bishayee A. A systematic review on potential anticancer activities of Ficus carica L. with focus on cellular and molecular mechanisms. *Phytomedicine*. 2022; 105: 154333.

350. Mawa S, Husain K, Jantan I. Ficus carica L. (Moraceae): Phytochemistry, traditional uses and biological activities. *Evid Based Complement Alternat Med*. 2013; 2013: 1–9.

351. Ishnava KB, Chauhan JB, Garg AA, Thakkar AM. Antibacterial and phytochemical studies on Calotropis gigantia (L.) R. Br. latex against selected cariogenic bacteria. *Saudi J Biol Sci*. 2012; 19(1): 87–91.

352. Rutuba C, Sharma P, Modi N. Preliminary phytochemical screening, quantitative estimation of total phenols, total flavonoids and anti-oxidant activity of leaves of Plumeria pudica Jacq. *Indian J Nat Sci*. 202;12(67): 32926–35.

353. Ozarde YS, Choudhari VP. Preliminary qualitative phytochemical analysis and acute oral toxicity study of latex of an ethnomedicinal plant Euphorbia fusiformis buch.-Ham. Ex D. Don. *Res J Pharm Technol*. 2022; 15: 1123–27.

354. Madaan R, Kumar D, Kumar S. High performance thin layer chromatography method development for estimation of triterpenoids in Calotropis Gigantea Roots. *Indian J Pharm Sci*. 2022; 84: 1593–1596.

355. Sharma M, Abid R, Sajgotra M. Phytochemical screening and thin layer chromatography of Ficus carica leaves extract. *J Pharm Biosci*. 2017; 5(1): 18–23.

356. Sauvage FL, Saint-Marcoux F, Duretz B, Deporte D, Lachatre G, Marquet P. Screening of drugs and toxic compounds with liquid chromatography-linear ion trap tandem mass spectrometry. *Clin Chem*. 2006; 52(9): 1735–42.

357. Jegadeeswari P, Nishanthini A, Muthukumarasamy S. GC-MS Analysis of bioactive components of Aristolochia Krysagathra (Aristolochiaceae). *J Curr Chem Pharm Sci*. 2012; 2(4): 226–32.

358. Ibrahim EA, Wang M, Radwan MM, Wanas AS, Majumdar CG, Avula B, Wang YH, Khan IA, Chandra S, Lata H, Hadad GM. Analysis of terpenes in Cannabis sativa L. using GC/MS: Method development, validation, and application. *Planta Med*. 2019; 85(5): 431–38.

359. Solesi OA, Adesina FC, Adebayo-tayo BC, Abiodun AS. Gas chromatography/mass spectrometry (GC-MS) analysis of Jatropha

curcas latex and its antimicrobial activity on clinical isolates. *World J Adv Res Rev.* 2020; 8(1): 12–8.

360. Usman A, Onore RO, Oforghor OA, Mohammed J, Usman NL, Mohammed J. Total phenolic and flavonoid contents, antioxidant activity and phytochemical screening of Calotropis procera stem bark extracts. *Commun Phys Sci.* 2020; 5(3): 233–9.

361. Khatoon A, Jain S, Shukla K. Phytochemical investigation and determination of total phenols, flavonoid, saponins and alkaloids concentration in root extract of Tabernaemontana. *J Pharm Negat Results.* 2022; 13(10): 4822–26.

362. Chen YR, Wen K, Her G. Analysis of coptisine and palmatine in adulterated Chinese medicine by capillary electrophoresis-electrospray ion trap mass spectrometry. *J Chromatogr A.* 2000; 866: 273–80.

363. Bouso JC, Fornís I, Vilamala MV, Loenen BD, Sainz-Cort A, Jiménez-Garrido DF, Santos RG, Hallak JE, Alcázar-Córcoles MÁ, Jenks CW. An analytical study of iboga alkaloids contained in Tabernanthe iboga-derived products offered by ibogaine treatment providers. *Rev Psiquiatr Clin.* 2020; 47(2): 51–54.

364. Tang DQ, Zheng XX, Chen X, Yang DZ, Du Q. Quantitative and qualitative analysis of common peaks in chemical fingerprint of Yuanhu Zhitong tablet by HPLC-DAD-MS/MS. *J Pharm Anal.* 2014; 4(2): 96–106.

365. Nataraj R, Jagadish MS, Varadaraju KR, Bindu J. Isolation, purification and characterization of vindoline from Catharanthus roseus. *Int J Sci Res Arch.* 2023; 8(1): 310–19.

366. Kaur H, Anand M, Goyal D. Chromatographic determination of tylophorine content in in vitro raised plants of Tylophora indica-an endangered medicinal plant. *Life Sci Informatics Publ.* 2016; 2(3): 161–71.

367. Groeneveld HW, Steijl H, Van Den Berg B, Elings JC. Rapid, quantitative HPLC analysis of Asclepias fruticosa L. and Danaus plexippus L. cardenolides. *J Chem Ecol.* 1990; 16(12): 3373–82.

368. Sridevi G. Phytochemical analysis of pergularia daemia for its bioactive components through gas chromatographic mass spectrometry (GCMS). *IOSR J Pharm.* 2014; 04(05): 41–46.

369. Cheng X, Shin YG, Levine BS, Smith AC, Tomaszewski JE, Van Breemen RB. Quantitative analysis of betulinic acid in mouse, rat and dog plasma using electrospray liquid chromatography/mass spectrometry. *Rapid Commun Mass Spectrom.* 2003; 17(18): 2089–92.

370. Zacchigna M, Cateni F, Faudale M, Sosa S, Loggia R Della. Rapid HPLC analysis for quantitative determination of the two isomeric triterpenic acids, oleanolic acid and ursolic acid, in plantago major. *Sci Pharm*. 2009; 77(1): 79–86.

371. Sharma K, Zafar R. Simultaneous estimation of taraxerol and taraxasterol in root callus cultures of Taraxacum officinale Weber. *Int J Pharmacogn Phytochem Res*. 2014; 6(3): 540–46.

372. Zhu HY, Meng XY, Bao YL, Yu CL, Wu Y, Li YX. Simultaneous determination of lathyrane diterpenoids by NPLC-DAD in the Seeds of Euphorbia lathyris and "Zi-Jin-Ding" Pastille. *Chromatographia*. 2010; 71(7): 617–22.

373. Manjula K, Rajendran K, Eevera T, Kumaran S. Quantitative estimation of lupeol and stigmasterol in costus igneus by high-performance thin-layer chromatography. *J Liq Chromatogr Relat Technol*. 2013; 36(2): 197–212.

374. Zhang M, Otsuki K, Kikuchi T, Bai ZS, Zhou D, Huang L, Chen CH, Morris-Natschke SL, Lee KH, Li N, Koike K. LC-MS identification, isolation, and structural elucidation of anti-HIV tigliane diterpenoids from Wikstroemia lamatsoensis. *J Nat Prod*. 2021; 84(8): 2366–73.

375. Ji X, Lin S, Chen Y, Liu J, Yun X, Wang T, Qin J, Luo C, Wang K, Zhao Z, Zhan R. Identification of α-amyrin 28-carboxylase and glycosyltransferase from Ilex asprella and production of ursolic acid 28-O-β-D-Glucopyranoside in engineered yeast. *Front Plant Sci*. 2020; 11: 1–10.

376. Chen H, Wang Z, Yang L. Analysis of euphornin in Euphorbia helioscopia L. and its cytotoxicity to mice lung adenocarcinoma cells (LA795). *Nat Prod Res*. 2012; 26(22): 2112–16.

377. Musthaba SM, Athar MT, YTK, Baboota S, Ali J, Ahmad S. Fast analysis and validation of rutin in anti-psoriatic ayurvedic formulation by HPLC. *J Liq Chromatogr Relat Technol*. 2011; 34(6): 446–55.

378. González J, Cuéllar A, Gaysinski M, Massi L. Chemical characterization by TLC, UV, IR, HPLC-MS/MS and NMR of gossypetin-3-O-glucoside from Talipariti elatum (Sw.) *Malvaceae*. 2019; 24: 100–16.

379. Kumari A, Kumar J. Phyto-chemical screening of root extracts of Glycyrrhiza glabra by spectroscopic methods (UV-VIS Spectro-photometer, FTIR and HPLC). *Int J Pharm Sci Drug Res*. 2019; 11(06): 376–81.

380. Xu W, Fu ZQ, Lin J, Huang XC, Chen D, Yu HM, Huang ZH, Fan SM. Qualitative and quantitative analysis of major constituents in

Tetrastigma hemsleyanum by HPLC-Q-TOF-MS and UPLC-QqQ-MS. *Zhongguo Zhong Yao Za Zhi*. 2014; 39(22): 4365–72.

381. Newsome AG, Li Y, van Breemen RB. Improved quantification of free and ester-bound gallic acid in foods and beverages by UHPLC-MS/MS. *J Agric Food Chem*. 2016; 64(6): 1326–34.

382. Tor ER, Francis TM, Holstege DM, Galey FD. GC/MS Determination of pyrogallol and gallic acid in biological matrices as diagnostic indicators of oak exposure. *J Agric Food Chem*. 1996; 44(5): 1275–79.

383. Dah-Nouvlessounon D, Chokki M, Agossou EA, Houédanou JB, Nounagnon M, Sina H, Vulturar R, Heghes SC, Cozma A, Mavoungou JF, Fodor A. Polyphenol analysis via LC-MS-ESI and potent antioxidant, anti-inflammatory, and antimicrobial activities of jatropha multifida l. extracts used in benin pharmacopoeia. *Life*. 2023; 13(9):1898–918.

384. Muharram MM, Abdel-Kader MS. Utilization of gel electrophoreses for the quantitative estimation of digestive enzyme papain. *Saudi Pharm J*. 2017; 25(3): 359–64.

385. Bielawski J, Pierce JS, Snider J, Rembiesa B, Szulc ZM, Bielawska A. Sphingolipid analysis by high performance liquid chromatography-tandem mass spectrometry (HPLC-MS/MS). *Adv Exp Med Biol*. 2010; 688: 46–59.

386. Nes CR, Singha UK, Liu J, Ganapathy K, Villalta F, Waterman MR, Lepesheva GI, Chaudhuri M, Nes WD. Novel sterol metabolic network of Trypanosoma brucei procyclic and bloodstream forms. *Biochem J*. 2012; 443(1): 267–77.

387. Ponphaiboon J, Limmatvapirat S, Chaidedgumjorn A, Limmatvapirat C. Optimization and comparison of GC-FID and HPLC-ELSD methods for determination of lauric acid, mono-, di-, and trilaurins in modified coconut oil. *J Chromatogr B Anal Technol Biomed Life Sci*. 2018; 1099: 110–16.

388. Röhrig W, Waibel R, Perlwitz C, Pischetsrieder M, Hoch T. Identification of the oleic acid ethanolamide (OEA) isomer cis-vaccenic acid ethanolamide (VEA) as a highly abundant 18: 1 fatty acid ethanolamide in blood plasma from rats and humans. *Anal Bioanal Chem*. 2016; 408(22): 6141–51.

389. Bhaskar Das, Arnab DE, Piu Das, Amalesh Nanda, Amalesh Samanta. Pharmacognostic studies on flowers of Dregea volubilis: Evaluation for authentication and standardization. *Asian J Pharm Clin Res*. 2019; 12(5): 79–89.

390. Papp I, Apati P, Andrasek V, Blazovics A, Balazs A, Kursinszki L, Kite GC, Houghton PJ, Kery A. LC-MS analysis of antioxidant plant phenoloids. *Chromatographia*. 2004; 60(1): S93–100.

391. Ma C, Li X, Zheng C, Zhou B, Xu C, Xia T. Comparison of characteristic components in tea-leaves fermented by Aspergillus pallidofulvus PT-3, Aspergillus sesamicola PT-4 and Penicillium manginii PT-5 using LC-MS metabolomics and HPLC analysis. *Food Chem*. 2021; 350: 129228.

392. Goswami AK, Gogoi N, Sharma HK. Validated high-performance thin-layer chromatography-mass spectrometry method and stability study of linalool in the volatile oil of the rhizomes of Homalomena aromatica Schott. *J Chromatogr Sci*. 2023; 61(3): 269–78.

393. Blumberg S, Frank O, Hofmann T. Quantitative studies on the influence of the bean roasting parameters and hot water percolation on the concentrations of bitter compounds in coffee Brew. *J Agric Food Chem*. 2010; 58(6): 3720–28.

394. Khan AN, Ali Aldowairy NN, Saad Alorfi HS, Aslam M, Bawazir WA, Hameed A, Soomro MT. Excellent antimicrobial, antioxidant, and catalytic activities of medicinal plant aqueous leaf extract derived silver nanoparticles. *Processes*. 2022; 10(10): 1949–73.

395. Ghimire BK, Yoo JH, Yu CY, Kim SH, Chung IM. Profiling volatile and phenolic compound composition and characterization of the morphological and biological activities of Perilla frutescence Britton var. Japonica accessions. *Acta Physiol Plant*. 2019; 41(7): 1–16.

396. Hertel MO, Scheuren H, Sommer K, Glas K. Limiting separation factors and limiting activity coefficients for hexanal, 2-methylbutanal, 3-methylbutanal, and dimethylsulfide in water at (98.1 to 99.0)°C. *J Chem Eng Data*. 2007; 52(1): 148–50.

397. Meng HY, Piccand M, Fuchsmann P, Dubois S, Baumeyer A, Tena Stern M, von Ah U. Formation of 3-methylbutanal and 3-methylbutan-1-ol recognized as malty during fermentation in Swiss raclette-type cheese, reconstituted milk, and De Man, Rogosa, and Sharpe Broth. *J Agric Food Chem*. 2021; 69(2): 717–29.

Chapter 4

Processing of Herbal Latex

Laleh Khodaie,[a,b] Saurabh Bhatia,[c,d] Ahmed Al-Harrasi,[c] and Ajay Sharma[e]

[a]*Department of Traditional Pharmacy, Faculty of Traditional Medicine, Tabriz University of Medical Sciences, Tabriz, Iran*

[b]*Department of Pharmacognosy, Faculty of Pharmacy, Tabriz University of Medical Sciences, Tabriz, Iran*

[c]*Natural & Medical Sciences Research Center, University of Nizwa, Birkat Al Mauz, Nizwa, Oman*

[d]*School of Health Sciences, University of Petroleum and Energy Studies, Dehradun, Uttarakhand, India*

[e]*Department of Pharmacognosy & Phytochemistry, School of Pharmaceutical Sciences,*
Delhi Pharmaceutical Sciences and Research University, Mehrauli-Badarpur Road, New Delhi, India

khodaiel@gmail.com

The growth in the application and popularity of herbal materials has surged their economic value and global trade. Regulatory requirements should be met to process and manufacture herbal products to ensure their safety and efficacy. Also, the right time of harvesting to obtain the optimum amount of biologically active phytochemicals is highly substantial. This chapter encompasses primary post-harvest processing treatments required after collecting herbal products. First, for the betterment of the drying

Introduction to Plant-Based Latex: Phytochemistry, Ethnopharmacology, Biological, and Pharmacological Activities
Edited by Laleh Khodaie, Ajay Sharma, Saurabh Bhatia, and Ahmed Al-Harrasi

ISBN 978-981-5129-12-0 (Hardcover), 978-1-003-57071-4 (eBook)
www.jennystanford.com

process, pre-drying is applied. This approach includes diverse thermal and non-thermal processes. It prepares the herbal material for drying since dried herbs and herbal materials can be preserved for a long duration. Drying could improve the quality of dried herbs and their products by preserving their biologically active compounds. It also prevents the decomposition of phytochemicals by some enzymes and microbes. Each raw material requires a special method of drying. Various types of herb-drying approaches have been developed in the last decades. Novel drying methods were introduced to improve the efficiency and quality of herbal materials and products to preserve heat-sensitive active phytochemicals. Moreover, the storage of medicinal plants is a part of the post-harvest chain after dehydration, which affects the stability, quantity, and quality of the phyto-constituents. Stability testing requires assessing physicochemical stability, biological activity, and toxicity which form chemical and biological shelf life. Stability testing recommended by regulatory guidelines is designed to accelerate product chemical decomposition and physical alteration. These tests aimed to impose exaggerated stress conditions on herbal products. Packaging is another important part of the post-harvesting process. Various packaging materials are used to provide a barrier for herbs and herbal products. Glass and metal packaging are approximately superior barriers against environmental and chemical factors. Further investigations are required to introduce new methods for improving post-harvest process.

4.1 Introduction

Utilizing plant-based is as old as mankind's existence. Recently, people have shown great interest in using medicinal plants and their preparations. Thus, there is a steep increase in their importance and value economically. As an alternative medication, herbal products and preparations were improved to be produced on an industrial scale. In this regard, quality assurance of herbal products and preparations is demanding. WHO (World Health Organization) has suggested several technical guidelines for different phases of plant-based processing to guarantee the quality of raw materials and plant-based products. Processing medicinal plants is defined as some approaches which could be applied in different stages of herbal product preparation, from cultivation and collection of raw material to post-harvest processing and manufacturing of the finished product. Latex, as an herbal product, is collected from different plant parts, trunks, branches, fruits, and roots. Thus, this chapter first encompasses some general rules in herbal organ harvesting. Afterward, post-harvest approaches will be discussed.

4.2 Collecting Process of Herbal Organs Producing Latex

Harvesting of each herb should be carried out at a suitable time. One good approach to allocating the right harvesting time is monitoring their biologically active phytochemicals. The growth and development stage and season highly affect the amount of latex, phytochemicals, and their medicinal effects. Thus, the preparation of efficacious herbal products depends on the age and physiological development of the herb. However, for herbs with unknown active phyto-constituents, general rules of harvesting could be applied. The bark should be harvested in spring, summer, and fall. Regarding the trunk, peels should be removed horizontally to prevent sap circulation and plant death. Underground parts ought to be harvested in winter when the proportion of biologically active constituents is at the highest. When a plant starts flowering, it is an appropriate time to harvest the leaves. Tips of the shoot should be harvested before the formation of seeds. Fruits are better to be harvested before complete maturing. Pharmacopeias and reference manuscripts can provide comprehensive information regarding each herb's suitable harvesting time. While harvesting process requires precautions not to mix any foreign matter, toxic herbs, or weeds with the main herb. Generally, efficient harvesting of medicinal plants requires some conditions; dew, rain, and humidity should be avoided during collection. If it necessitates harvesting plant material in wet conditions, it should be transferred into an indoor place to be dried. Drying of herbal material prevents excess moisture, microbial contamination, and fermentation. Furthermore, any adhering soil should be removed from harvested underground herbal organs to lessen the microbial contamination of collected herbal materials. After collection, the primary processing of harvested herbal material will be initiated. Preliminary processing includes disposal of contaminants and unwanted material, washing, sorting out, trimming, and cutting [1].

Transportation of raw herbal materials should be performed in dry and clean conditions. Well-aerated, clean, and dry containers or baskets should be used for transportation. If the herbal processing site is distant from the harvesting site, the herbal material should be air-dried or sun-dried. If various plant latex-producing organs are collected, they should be separately collected and transported to the processing site. Compacting and packing of raw herbal material should be avoided, as mechanical damage could decrease the quality of raw herbal material and latex. The containers of herbal material collection should be free from foreign matter or contamination. Containers not in use should be kept in dry areas and preserved from pests, insects, rodents, and livestock. Moreover, mechanical tools used for harvesting, such as cutting devices or harvesters, should be kept clean since they are in direct contact with herbal material [2].

Hevea brasiliensis, or rubber tree, is one of the economically substantial latex-producing herbs. Latex from rubber trees is harvested from rubber trees using a tapping system. This process begins with cutting the tree bark and flow of latex into a cup attached to the cut's bottom. The latex collected from many trees is accumulated in large tanks. For collecting and transporting *Hevea* latex, the processing mentioned above conditions should be performed to avoid contamination [3].

The aim of post-harvest is to preserve herbs' organoleptic, chemical, physical, and pharmacological properties. It is performed to avoid microbial contamination, loss of biologically active phytochemicals, and low-quality herbal raw material. Typically, the processing should be performed carefully to prevent loss of phyto-constituents, degradation of metabolic pathways, hydrolysis of phytochemicals, enzymatic, heat, and light decomposition, essential oil evaporation, and bacteria or fungi contamination. The initial post-harvest processing includes cleaning herbal material from grains and manually separating damaged and stained parts. It is much better to wash some herbal parts with running water to clean them from pathogens. The raw material should be stored in a cool place to avoid loss of moisture and deterioration. After initial storage of herbal raw materials, they undergo further processing, defined as pre-drying and drying [4].

4.3 Pre-Drying Approaches

The pre-drying approach is a post-harvest and physical treatment applied to herbal products for the betterment of the drying process. One of the crucial requirements of modern society is healthy, high quality, and safe herbal material with higher biologically active ingredients besides better organoleptic qualities. Several conventional and novel drying approaches have several pros and cons. To overcome the drawbacks of each method, tandem and parallel drying methods are applied. The tandem drying approach is an application of diverse approaches in a row. In other words, pre-drying methods are applied before the final dehydration technique. While in parallel mode, various systems are used simultaneously [5].

To improve the drying process, pre-drying processes are applied (tandem drying) as a main approach to herbal material preservation method, pre-drying processes are applied (tandem drying). Diverse pre-drying approaches derived attention. The pre-drying process aims to improve the quality of dried herbs and reduce the duration of drying and energy consumption. Pretreatment approaches fall into two groups, chemical and physical approaches. Physical modification of samples, such as cutting or sizing into pieces, is carried out to improve moisture diffusion. It provides more exposure of the sample's surface to the drying environment. Freezing is another physical pretreatment in which disruption of cellular structure is facilitated. Frozen intracellular water molecules assist in disrupting the cell wall, leading to faster mass transfer [6]. Various pre-drying methods are exhibited in Fig. 4.1.

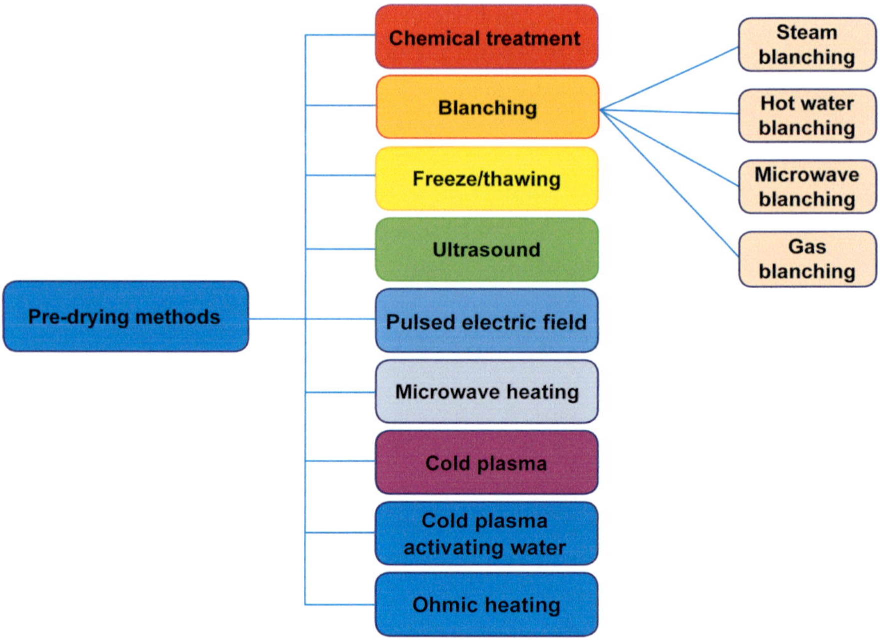

Figure 4.1 Different pre-drying approaches.

4.3.1 Chemical Pre-Drying Treatment

Several chemical compounds were introduced to dissociate cell walls and facilitate water diffusion. Chemicals accompanied by physical pretreatments could efficiently reduce drying time. Several chemicals used in the pre-drying treatment are as follows: citric acid, ethyl oleate solution, ethyl oleate and NaOH, ethyl oleate and K_2CO_3, fermented whey, potassium meta bi sulfite, KOH, K_2CO_3 with olive oil, $MgCO_3$, and NaOH [7].

4.3.2 Blanching

Conventional thermal technologies are steam and hot water blanching. Blanching is one of the pretreatment techniques applied to a wide array of herbs and herbal materials. In this process, herbal material is quickly heated to a specific degree and maintained at that temperature in chambers. Blanching can cause changes at the macroscopic level by changing the texture of the sample. Also, microscopic modifications occur; for instance, pectin dissolution, separation of cells, and cell-wall polymer gelation. In the middle lamella, separation of cells, polysaccharide dissolution, and softening of herbal tissue occur. Similarly, thermal conditions trigger the cleavage of methyl ester groups. This approach is beneficial due to the inactivation of enzymes causing biochemical changes and texture modification. Furthermore, it removes trapped air and oxygen from intracellular spaces. Thus, it prevents the color degradation of herbal material. It also reduces the duration of drying and leaves a high-quality product. In industry, steam and hot water blanching are the commonly used pretreatments by heat. There are other blanching types, such as microwave and hot gas blanching [8].

Steam blanching is performed in steam blanchers, where a conveyer transports herbal material into a chamber containing water steam. The chamber temperature is measured and controlled by headspace (near 100°C or 212°F). Steam flow rate should also be under control. This process is less time-consuming than water blanching due to the higher transfer of condensed steam heat. Despite being a good process, heat transfer through the herbal product is not efficiently performed. For the betterment of this process, forced convection blanchers were designed containing nested chambers where a fan recirculates the steam. This process offers high throughput, productivity, and efficiency. According to previous studies, 15s of blanching with steam multiplied the drying rate by ten times that of untreated herbal material. It is energy efficient and steam-blanched herbal materials show higher retention rates of phytochemicals and color [9].

Hot water blanching is another pretreatment technology that applies 70–100°C hot water. Special tools, namely rotary drums, are needed to convey and immerse the herbal material and product. Generally, it is a long-term process; however, the water's temperature determines the duration of this process. Based on water temperature, water bleaching is categorized into two categories: long-time low-temperature (LTLT) and short-time high-temperature (STHT). This approach triggers the decline of the product quality and loss of active ingredients due to the long duration of the process and high temperature.

To improve mechanical strength, some solutes, such as calcium chloride, are used for blanching to preserve the texture of herbal products. Calcium ions bring about tissue hardening by bridging pectin chains, where cross-links are formed between free carboxyl groups of chains. On the other hand, calcium chloride gives rise to an undesirable taste; thus, its usage is restricted to lower concentrations. Another solute for the retention of the textural quality of the material is sugar solutions. They are considered heat carriers that preserve cell wall integrity. Particularly, mono and disaccharide solutions protect cell walls by preserving fluidity and preventing protein denaturation. In this regard, disaccharides, namely trehalose and maltose, more than sucrose, have attracted attention to be used in the manufacturing section. Especially, maltose can efficiently preserve the structural integrity of cell walls and retain volatile components, pigments, and ascorbic acid. Trehalose, a non-reducing disaccharide consisting of two molecules of α-glucose, is also used as a bio-protectant solute to improve the blanching process. As a source of energy, several species synthesize it in stressful situations. This solute can protect chemical structures; by stabilizing and preservation of proteins of the membrane from denaturation. Therefore, it can protect the microstructure and texture of herbal samples during the blanching process. Notably, the combination of trehalose and maltose aqueous solution has shown a protective effect on the microstructure and texture of herbal samples, acting as heat carriers in the thermal pretreatment of herbal products [9].

Another approach for the pre-drying process is **microwave blanching**. It consists of electromagnetic waves generated by a magnetron or microwave generator. The frequency of electromagnetic waves ranges from 300 MHz to 300 GHz. Heating by microwave generally depends on some criteria; electric factors, namely dielectric permittivity, and ionic conductivity, microwave frequency, moisture, salt, and temperature of herbal material. Concerning the mechanism of action of this approach, the generated electromagnetic field generates water molecule polarization in samples and leads to polarization. Since ionized atoms and polarized molecules swing back and forth in that field, collision occurs between them, which produces heat and energy to free and evaporate bound water molecules of the sample. A demand for further improvement in the pretreatment process led to the development of this approach. Notably, loss of flavor and volatile ingredients, modification of texture and color, and reduction of nutrients are the major concerns of conventional thermal pretreatment methods. Thus, optimizing the quality of herbal products and materials led to the development of dry procedures [10].

Previous studies indicated that in contrast to water and steam blanching, heat-sensitive phytochemicals are preserved by this pretreatment technology. Due to having a very short duration of processing compared to the mentioned conventional approaches, degradation of phyto-constituents is less likely to occur. Like the methods above, enzyme inactivation, expelling of intracellular trapped air, and infection reduction could occur in microwave processing. However, considering the shortcomings of this approach, limited penetration of microwaves and varied depth of penetration into herbal material are limitations of this method. Consequently, this could trigger varied energy distributed within the product. The main obstacle to this processing is the heating and penetration uniformity in products which is difficult to measure and control. Some other criteria affect the pattern of heating: heating speed, power, frequency, products' mass, rotation of herbal material inside the oven, moisture, conductivity, density, and dielectric constant. Physical characteristics of herbal material are also substantial, such as shape, size, quantity, and thickness. The other limitation of this method is the generation of hot spots due to continuous heating of the product and uneven distribution of temperature. Thus, microwave-processed products should be assessed regarding absorbed and distributed temperature and power. Therefore, implementing this approach on an industrial scale is an uphill task, and it may take time to be practical. Microwave-processed herbal materials are much more nutritious than conventional pre-drying processing, such as water and steam blanching at the same temperature.

A combination of processes is also available. For instance, the mentioned approach could be combined with water and steam blanching to reduce heating duration further. It can also be used before the osmotic drying method to increase diffusivity and mass transfer to surge the drying rate effectively. However, the equipment is high-priced and has yet to be applied on an industrial scale [7].

Gas blanching is the combination of combustion gas and steam. The main benefit of this processing is improving humidity and preventing dehydration. Moreover, in this method, less waste is produced, and more active ingredients of products are retained. On the other hand, there is a loss of product weight. This method requires further studies and has yet to be practiced in the industry [8].

4.3.3 Freeze/Thawing Processing

This process is a non-thermo-physical approach in which herbal material is briskly frozen at −85°C and thawed at 6–10°C consecutively. This process is repeated 2–3 times. Following the decrease in temperature, the water inside the cells freezes and expands, leading to excess water penetration into cell walls. Consequently, overflow of free water and mass transfer occur. Although this method could reduce the drying duration, its limitations should not be underestimated. Its drawbacks are severe loss of bioactive ingredients, high cost, and incapability to inactivate some reaction enzymes [10].

Furthermore, it can stress and alter the structure of some proteins and macromolecules of latex. As a result, alteration and denaturation of macromolecular structures occur. Researchers opt for various freezing and thawing rates to increase macromolecule stability. These varying rates can influence the stability of macromolecules. Yet, there is no consistent and standard rate for the freezing and thawing of herbal material. The freezing and thawing rates are dependent on the equipment. Both are grouped into rapid, slow, and intermediate. To achieve a rapid freezing rate, the herbal material is soaked in baths filled with subzero solvents such as liquid nitrogen or alcohol. Several approaches should be evaluated to incorporate the optimum condition of freezing and thawing. Fast freezing rates trigger less formation of ice crystals. However, in this technique, proteins are concentrated and adsorbed at the surface and interface of ice crystals and ice liquid, respectively. As a consequence, fragmentary unfolding and aggregation of proteins take place. Thus, structural damage could give rise to weakened biological activity. One of the strategies to decrease protein denaturation is using stabilizers, such as cryo-protectants and surfactants. However, in slow freezing rates, cryo-concentration occurs. That is to say, proteins of herbal material are concentrated and excluded from the interface of ice-liquid. This could prompt the separation of phases, the shift of pH, protein unfolding, denaturation, and aggregation. To conclude, obtaining the best strategy for freezing/thawing latex containing several proteins requires diverse trials and assessments [11].

4.3.4 Ultrasound Processing

Ultrasound pre-drying is a non-thermo-physical pre-drying technique. It applies to heat-sensitive herbs, typically containing flavonoids and phenolic compounds. To explain the mechanism of action of this approach, due to ultrasonic radiations, small vaporous cavities are formed within the liquid, named cavitation. These cavities are propagated in the medium leading to the formation of the sponge effect. Consequently, mass transfer occurs, and heat-sensitive substances are protected. According to previous studies, this method can surge porosity, shrink density, and concentration of phytoconstituents [11].

For optimization of pre-drying processing, the incorporation of several techniques could be undertaken. In other words, both single and combination of approaches could be applied. To illustrate, some creatively designed patterns have been established, such as ultrasound-assisted thawing and ultrasound-freeze-thaw, to improve the drying process of herbs. All of the mentioned methods should be investigated in terms of altering the characteristics of latex drying, namely drying kinetics, energy consumption, and the material's microstructure. Furthermore, the selected method should be explored to see how it alters the quality, such as color, hardness, and flavor, and the content of biologically active ingredients of herbal products, namely, flavonoids, phenolics, and pharmacological properties. These investigations will assist in proposing the best pre-drying technique [12].

4.3.5 Pulsed Electric Field (PEF)

This method has been introduced as pre-drying processing to enhance herbal material dewatering. As a non-thermal method, herbal tissue is held at electric fields of moderate strength (E = 200–1000 V/cm) for a very short while (10^{-2}–10^{-4}s). This process brings about the plant cell walls' electroporation, disintegration, and damage. Thus, moisture diffusion and mass transfer occur; consequently, drying could be facilitated. If this treatment is associated with mild heat, the efficiency of conventional dehydration methods (air drying) would be further improved. PEF processing is typically selected for heat-sensitive herbal material. This process is highly energy-consuming. Also, some criteria affect the efficiency of PEF, namely, impulse energy, pulse number, and electric field strength [5]. However, this pretreatment method of dewatering occurs at lower temperatures (for instance, 20°C), which could be promising for heat-sensitive herbs [13]. By PEF, the duration of the dehydration process could be minimized due to the instant distribution of energy in the material. Furthermore, in some cases where the preservation of color, flavor, and phytochemical contents of herbal material is crucially important, PEF seems as an upstream method [14].

4.3.6 Microwave Heating

Different methods are used to increase the drying rate. Moisture diffusivity depends on the physical structure and surface resistance of herbal mass. The more porous the herbal material is, the better moisture transport occurs. Microwave heating (dielectric energy) is a pretreatment method to remove the surface resistance of herbal material for moisture diffusion. Since conventional drying processes, such as heat transfer, are time-consuming and rate-limiting, microwave heating as a pre-dehydration approach could be beneficial. As far as the mechanism is concerned, microwave heating brings about the dielectric heating of herbal material. Given that the dielectric constant of water is higher than other components, water molecules could better absorb microwave energy. Thus, the energy of microwaves could surge heating and drying rates.

However, although dielectric energy is time-saving, due to some technical difficulties in its operation and costly equipment, it has limited industrial and commercial applications [15].

4.3.7 Cold Plasma Pretreatment

Cold plasma pretreatment, a non-thermal, novel, promising, and innovative technology applies to thermally sensitive herbs and herbal materials to preserve their biologically active phytochemicals. It is commonly applied for disinfection in the food industry as it does not require extreme conditions and does not alter the original ingredients and physical characteristics of food [16].

At the same time, it has attracted attention in recent years owing to the insertion of physicochemical changes in samples. Generally, plasmas are grouped into thermal, applying high temperature, and non-thermal, applying a low temperature. This method applies UV light, electricity, and carrier gas, namely nitrogen, oxygen, and helium. At low-temperature plasma, ionized gases consist of reactive and energetic ions, mainly radicals of oxygen and hydroxyl, electrons, and even neutral species radiated to the herbal material. Moreover, other types of gases (Ar, N_2, N_2O, and He) and their combinations ($He+N_2$, $He+O_2$, Ar/O_2, N_2+N_2O, N_2+O_2) are applicable in non-thermal plasma. The reactions are attributed to the type of plasma, voltage, gas, duration of treatment, and humidity; Two systems are used for practicing this method; plasma jet system and gliding arc cold plasma discharge; both aimed to convert gaseous medium to a highly reacting state of plasma [16]. The duration of this process could vary from 3 to 120 s depending on the herbal material and the type of the generated (the type of ions and electrons) cold plasma. Various systems will be developed in subsequent years operating at either atmospheric or low-pressure chambers. Hence, as a pre-drying method, it can efficiently accelerate the drying process. Changing the topography of the material surface generates intracellular spaces and cavities, bringing about lower density and facilitating mass transfer and dehydration process [16].

Also, a lower amount of activation energy is required, offering an efficient way of power management and higher drying rates or improving drying kinetics. Due to the mentioned reasons, flavonoids, phenolic compounds and antioxidant activity of herbal material are retained. In this method, microbes are efficiently disinfected and enzymatic reactions are deactivated, improving quality and vouching for the safety and decontamination of herbal products. On the other hand, previous studies indicated that if cold plasma is applied directly to the material, it could trigger loss of color, biodegradation of some phytochemicals, and alteration of the topography of the material surface. However, on the whole, this technology requires further research regarding perceiving chemical processes and how it affects phyto-constituents of herbal products and latex [17].

4.3.8 Cold Plasma Activating Water (CPAW)

Since cold plasma could influence herbal material, CPAW was introduced as an alternative pre-drying processing. The mechanism of action of CPAW resembles that of CP; however, if plasma discharge is generated over the surface of pure water, CPAW is created. While CPAW is created, several chemical reactions occur. As sequential consequences, the pH level decreases, and reactive oxygen and nitrogen species are generated. Furthermore, a low pH level of CPAW could bring about potential anti-microbial and oxidation-reduction effects. Thus, this pretreatment process increases the sample's shelf-life and quality. Since CPAW is formed by treating water, it is an eco-friendly, resource-efficient, and economical approach. Hence, this processing leads to the alteration of cellular topography and the formation of cellular cavities. As a result, mass transfer and diffusion of moisture improve, drying rate surges and activating energy decreases, and more phyto-constituents nutrients are preserved [5].

4.3.9 Ohmic Heating (OH)

In recent years, ohmic heating (OH), as an advanced thermal method, has been one of the most advanced and promising pre-dehydration treatments. The basis of this technique is the generation of heat within the sample's internal structure via the passage of an electric current. As an aseptic processing technique, it delivers products of higher quality. Due to the material's electrical conductivity, heat is uniformly generated within herbal matrices. Compared to conventional thermal processes, heat is distributed through a whole volume of herbal mass. Therefore, this approach is time-saving and produces high-quality and energy-efficient products. Previous studies indicated that if ohmic heating pretreatment is applied, further heat will not be required for the subsequent inactivation of microbes. Hence, OH brings about the rupture of a cell membrane rupture, and mass transfer occurs [18].

4.4 Diverse Drying Approaches

In post-harvest processing, the most important stage is drying. Drying is paramount since dried herbal material is used in diverse industries: toiletry, medical, and perfume manufacturing. Moreover, dried herbs and herbal material can be preserved longer. Drying could improve the quality of dried herbs and their products by preserving their biologically active compounds. The drying process prevents the decomposition of phytochemicals by some enzymes and microbes. It even avoids decomposition and biochemical reactions, such as oxidation and hydrolysis. Some enzymes are deactivated by drying, such as polyphenol oxidase (PPO) [19].

Moreover, their desirable phytochemicals are conserved, and their shelf life is extended [20]. Due to reducing the volume by drying, storing dried herbal products is much easier and the cost of transportation is much lower [21]. If heat is involved, volatile phytochemicals and essential oils could be volatilized. Thus, raw herbal materials such as latex should be dried below 40°C. For the betterment of the process, air circulation should be established. Each raw material requires a special method of drying. Various herb-drying approaches have been developed in recent decades (Fig. 4.2). New drying methods have been introduced to improve the efficiency and quality of herbal materials and products. Since phyto-constituents are sensitive to high

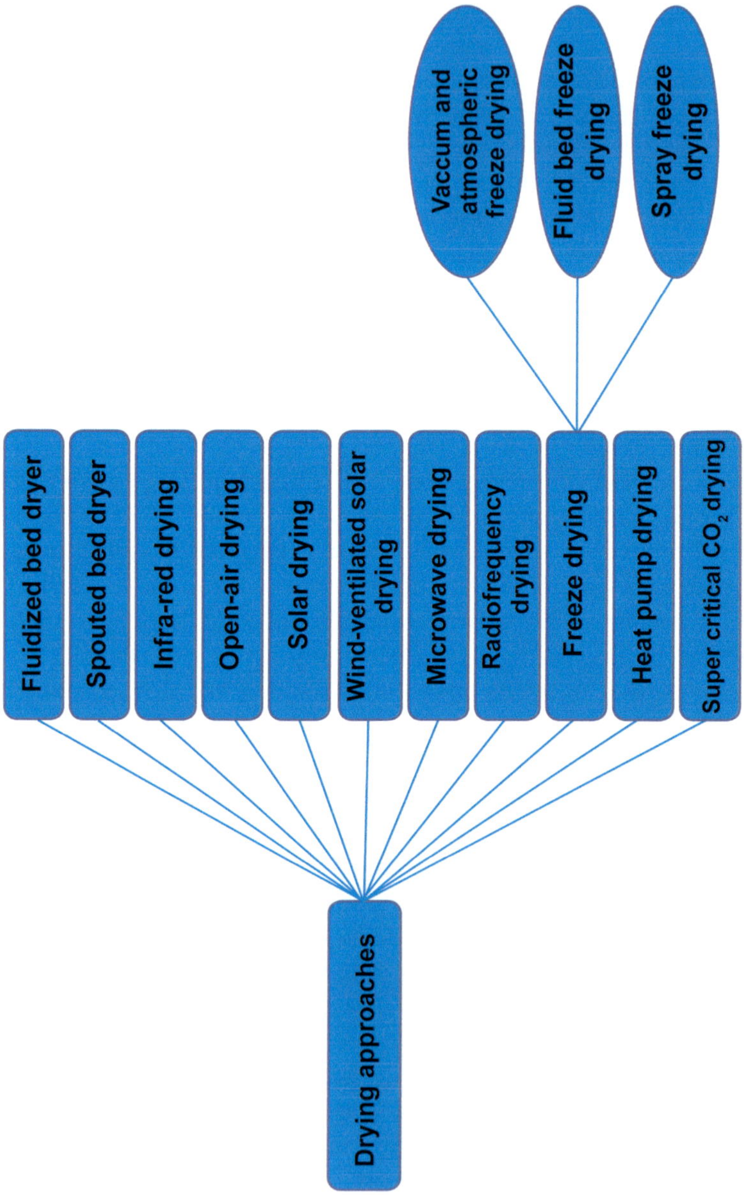

Figure 4.2 Different drying methods.

dehydration temperatures, the lower the drying temperature, the higher the biologically active phytochemicals could be obtained. It goes without saying that at low dehydration temperatures, herbal materials and products are exposed to oxygen long-term, giving rise to a significant drop in active ingredients [22]. Thus, the reduction of drying duration and temperature are equally substantial.

Recently, a plethora of herb-drying research was conducted to introduce new and efficient techniques. Alongside old methods of drying, in the last decades, new methods of drying have been developed. Moreover, from another perspective, drying methods could be either convective or conductive. Supercritical carbon dioxide and heat-pump-assisted drying have received the utmost attention recently [23].

Diverse drying methods could be low-cost or high-cost. Most developing and low-income countries are the major source of herbs and herbal products and export them. One of the fundamental post-harvest processing methods for herbal material is drying to reduce moisture content. As state-of-the-art drying systems require costly equipment, companies with high revenues can afford them; therefore, small-scale and low-income corporations cannot afford costly drying systems. Thus, low-income countries and companies dealing with farmers opt for cost-effective and easy-to-use approaches. Such dryers should meet some requirements and be easily fabricated via natural material and operated effortlessly. They ought to provide superior quality of product and drying kinetics than sun-drying. Moreover, maintenance and replacement should be done with less effort [23]. The following techniques are suitable and meet the mentioned criteria.

4.4.1 Fluidized Bed Dryer (FBD)

This method is primarily used in the pharmaceutical industry to decrease the samples' moisture, mainly powders and granules. This processing unit has a chamber where solid and moist particles are fed and stay on a perforated sheet. Hot air with high pressure and velocity flows into the chamber through the bed, fluidizing and suspending the product causing constant motion.

Thus, the interplay between hot gas and particles as well as efficient dehydration occurs. Finally, the dried material is discharged from the product outlet. This unit could be easily designed and constructed via simple and low-cost ingredients, such as steel or wood. This approach has several advantages; this is a fast-paced method that can be easily controlled, resulting in a reasonable mass transfer rate. Consequently, the duration of the process is shortened and drying kinetic is improved. This easily implantable method imposes lower capital and maintenance charges. Due to applying lower temperatures, the degradation of thermally sensitive phytochemicals is minimized. Uniformity of heat transfer brings about evenly dehydration. The velocity of hot air could be elevated to increase the efficacy of this process. However, a high airflow velocity could trigger gas passage on the material's surface area, causing less diffusivity and moisture transfer. To reduce this issue, an irregular non-continuous manner of fluidization could be suggested to reduce this issue. Implementing infrequent dehydration could preserve the biologically active ingredients, enhance the quality of dried products, and minimize operating costs [24].

4.4.2 Spouted Bed Dryer (SBD)

This approach and fluidized bed dryer are efficient methods to dehydrate seeds with medium and high moisture content. Also, this is a proper approach for herbal material which cannot be easily fluidized. Like FBD, it consists of a cylindrical chamber and a nozzle introducing hot air into the processing unit. An entrance port on one side of the chamber lets the wet material in. A draft tube in the center of the cylindrical vessel elevates the height of the spouting for the material. The spouting air enters the conical chamber. The material particles interact with spouting air arise irregularly and rapidly through the vessel and draft tube and fall slowly. During this cyclic motion, the removal of moisture from the surface of particles occurs. This process provides efficient contact between air and wet particles. This method offers some advantages; pertinent to superior contact between particles and air, the drying rate is improved and the duration of dehydration is shortened. In addition, the dehydration temperature is lower; thus, phytochemicals of thermally sensitive material are preserved. If a low temperature is required for some herbal products, intermittent spouted drying is applied to yield high-quality products via consuming lower energy [24].

4.4.3 Infrared Drying

Infrared drying (IRD) is a thermal method deemed equivalent to sun-drying artificially. Infrared radiation is between visible and microwave on the electromagnetic spectrum, with the wavelength from 0.8 to 1000 μm. This radiation is categorized into three groups; short wave (0.8–2 μm), medium wave (2–4 μm), and long wave (4–1000 μm). A diverse type of infrared is chosen according to the texture of the wet substance to reach optimized dehydration. In contrast to the conventional sun-drying method, it could be undertaken around the clock. Hence, it is cost-effective, has lower energy consumption, is versatile, requires simple equipment installed in any dehydration vessel, and has a high drying rate. In this process, the radiation energy generated from a hot IR lamp is transferred to the substance. Energy is directly delivered to the material without requiring any transfer medium such as water or air. Radiation brings about the heating of the internal part and vibration of molecules giving rise to moisture transfer. The temperature difference between the substance's internal part and the medium yields a pressure gradient causing moisture transfer [25]. IR dryer has a simple design that can be easily constructed and applied. It comprises a manual conveyor aligned to an infrared lamp, where the wet herbal product is placed on it through an inlet hopper. There is a knob of a voltage regulator at the upper part of the processor to regulate the level of irradiation. An intermittent drying mode is also possible using a knob named timing relay. The latter mode is the best option for thermally sensitive material, yielding a high-quality product with acceptable color and phytochemicals and superior drying kinetics at a short duration [26].

4.4.4 Open-Air Sun Drying and Solar Drying

Open-air sun drying and solar drying are traditional ways of dehydration, especially practiced in rural areas or low-income countries with warm climates. Pieces of material are laid on gauze and exposed to sunlight and air for 9–15 h daily. Generally, dehydration takes 5–7 days. Then the dried material is collected and packaged. Despite this method's disadvantages, it is still applied in some tropical and subtropical regions. Compared to other energy sources, solar energy does not cause pollution; it is an abundant source of energy, cost-effective, eco-friendly, and limitless [27].

Notably, as an environmentally friendly method, it is independent of fuel supplies. It goes without saying that this process is time-consuming, and a considerable amount of phytochemicals are lost. Also, the quality of the product is reduced due to insect invasion, the growth of microorganisms, fungi, and mycotoxin, and the reaction of enzymes. In undesirable weather conditions, such as rain, high humidity, wind and dust, products' spoilage, deterioration and decomposition could occur. This method requires large space, no automated or in other words, hand operated, with a low rate. Despite the drawbacks, it is still applied in many rural areas. The other terminology of this method is a passive or natural-circulation system. Industrial and mechanical drying devices or solar dryers came into practice instead of drying material in the open air and under the sky to improve this natural process. These devices are not labor intensive and could efficiently conserve energy and time. They can also overcome the main drawbacks of the classical approach [28].

Drying devices fall into two categories according to heating style and the manner of solar energy utilization; passive or natural circulation and active or hybrid solar energy drying systems. The operation of passive or natural circulation dryers completely depends on the radiation of sunlight. In these devices, the air is heated by solar energy and circulated through the wet material by wind pressure. This approach is named passive due to not applying fans to circulate the air. This approach is deemed an appealing option to rural farmers of remote areas. This option is superior to open-air sun drying in terms of function and economy; firstly, smaller space is required than the mentioned traditional choice to dehydrate equal quantities of product. Secondly, this method yields a product of higher quality and excessive quantity since the material is unlikely to be infested by rodents, insects, and fungi. Thirdly, compared to open-air drying duration of dehydration is shorter. Thus, dehydrated material throughput is elevated. Fourthly, the product is protected from sudden rainfall. Finally, the device's design and material are rudimentary and could be constructed effortlessly with low capital and maintenance costs by rural farmers. There are three passive dryer types: integral, mixed, and distributed [27].

Regarding the design of the distributed mode of the solar-energy or indirect passive dryer, it comprises a drying chamber with insulated walls where heated air circulates and a chimney. This chamber has shelves or trays to place wet material on them. Air heated via solar energy passes through the wet product and leaves the drying chamber through a chimney. Since sun rays do not directly irradiate the material, the herbal product does not caramelize (the process of disintegration of sugar and its reformation into complex polymers) or is thermally damaged. This method is recommended for heat-sensitive products that lose active ingredients due to direct sunlight exposure. Furthermore, this is the best option when the product's color retention is desirable and sunlight could transform the color of highly pigmented products. This dryer is deemed a superior choice when deep-layer drying and better quality are preferred owing to the higher temperature of the chamber. Considering drawbacks, the operating condition is not constant in the drying chamber of the air temperature fluctuation. Hence, this dryer has a rather sophisticated structure that requires higher capital and maintenance costs. Also, wooden drying chambers are vulnerable to climate conditions. Therefore, it is recommended to construct them from metal or waterproof material [29].

Direct-passive or integral-type dryers consist of a drying chamber with opaque walls allowing sunlight transmission.

Drying occurs by direct radiation of sunlight and via circulation of heated air. The heat of sunlight and air removes the relative humidity of the material. Notably, direct sunlight radiation could assist in the dehydration and ripening of some herbs and herbal products. In that case, the required color and flavor of the dehydrated product are developed. In terms of cost-effectiveness and easier application, these dryers are superior to distributed-type dryers. It goes without saying that integral-type dryers cause local overheating and product failure. Furthermore, the drying rate is relatively slower than other dryers, pertinent to incomplete vapor removal. To deal with these boundaries, a solar chimney could be utilized to elevate the air stream's velocity and remove moisture. In mixed-type solar dryers, features of integral and distributed modes dryers have been combined. The structural feature is the same as distributed mode consisting of a dehydration chamber with a chimney (to surge buoyancy) and solar air heater; however, the chamber walls are transparent, allowing sunlight to irradiate directly on material similar to integral-type dryers [30].

4.4.5 Wind-Ventilated Solar Drying

The wind-ventilated solar dryer is a mixed-type dryer in which solar energy and wind are employed to dehydrate herbal material. Some ventilators are applied to circulate air within the system to improve air ventilation. This dryer consists of a chimney with rotary vans powered by wind that draws air into the system. Wind rotary ventilator also exhausts the air constantly. Stationary ventilators could be employed as well that depend on wind speed. Considering the boundary of these ventilators, they do not operate between wind peaks; therefore, they remain inactive during lulls. This dryer contains a damper to control the airflow rate and temperature. Furthermore, in the dryer chamber, warm air, which is heated by solar energy, is circulated via a wind ventilation system. At the bottom of the drying chamber, an outlet of warm air is connected; thus, extra ducting and insulation are not required, which is considered a cost-cutting drive. Also, this design brings about less loss of heat. Besides warm air flow, wet material is heated via direct solar energy radiation passing through transparent sheets covering all sides of the dehydration chamber. Therefore, airflow is of utmost importance in this design [31].

4.4.6 Active Solar Drying Systems

These systems are solely dependent on solar energy as a heating source. Besides solar energy, fossil fuels and electricity could be employed for dehydration. Motor-driven pumps or fans circulate drying air; thus, it is also called a forced convection dryer. In large-scale dehydration systems, dehydration occurs via drying air heated by solar energy and fossil fuel. This approach leads to controlling drying conditions and reducing energy utilization. If the drying air is warm enough, solar energy will solely be used for dehydration; otherwise, dehydrators fired by fossil fuel as the supplemental heating source would be used alongside solar energy to elevate the drying air temperature. These systems are named "hybrid solar dryers" bringing about an optimum drying condition. Active solar energy dryers are grouped into integral, distributed, and mixed-type dryers [31].

Regarding the integral type, particular ducting of air is not required. In distributed type, the solar energy collector and dehydration chamber are separated. Mixed-type designs incorporate several features of integral and distributed types; glazed dehydration chamber and air ducting [31].

Active solar dryers could be supplied with some devices to store solar energy, such as rock or gravel beds; therefore, when the insulation level is low, the drying process of wet material is continued. In some types, desiccants decrease the humidity levels of drying air. Moreover, desiccants could be incorporated in forced-convection designs due to causing resistance toward the flow of air. All in all, for the betterment of the efficiency of solar dryers, fans and fossil fuels as heating sources could be utilized; however, they trigger a surge of operational, capital, and maintenance costs. Thus, their usage is limited by farmers in low-income countries [31].

In conclusion, comparing active and passive solar dryers indicated that forced convection or active modes are more efficient and controllable than passive types. On the other hand, in active dryers, fans derived from electricity or fossil-fuel surge the costs. As a result, the mentioned criteria should be considered carefully in small-scale and large-scale operations [31].

4.4.7 Microwave Drying Techniques

As mentioned in pre-drying methods, microwaves are generated by a rotating electromagnetic field, generating heat all over the substance, creating friction among water molecules, and eventually evaporation and transfer of moisture to the surface. This method is more adventurous than conventional techniques; however, to overcome the shortcomings, such as damage to substance texture, non-uniformity, and low penetration of microwave radiation, this technique is merged with another technique named microwave-assisted hybrid drying. For instance, according to previous studies the combination of convective, microwave, and ultrasound has worked efficiently. The drying rate reached an optimum level, and energy consumption dropped significantly. Also, the number of phyto-constituents, aroma, and color of herbal substance was preserved [32].

If infrared radiation is employed with microwave and convection methods simultaneously, drying kinetics and quality retention will be improved. Another new method is a microwave fluidized bed comprised of a tray for herbal substances and a vibrational structure. The vibrational motor starts working in the drying chamber, and the tray and the raw substance start rolling. Arranged magnetrons on top of the chamber radiate and further assist in dehydration. Nuclear magnetic resonance could detect the drying stages of material and the state of water. In fresh substances, free water dominates; after a while, free water converts to immobilized water, and at the final stage of dehydration, both types of material water drop. In one of the studies, microwave dehydration technology was merged with other methods and called pulsed microwave-assisted hot air dehydration. This dehydration method resulted in dried products with superior color retention compared to other methods. In the microwave-vacuum dehydration technique, the quality of the material and energy efficiency are enhanced. If the power of radiation and the vacuum intensity are elevated, the drying rate will be improved. However, that may trigger the loss of volatile phytochemicals. For dehydration of sensitive herbal products, vacuum levels should be intensified, and microwave power should be decreased. According to previous literature, another employed hybrid system was a combination of microwave power and hot air spouted bed. Microwave-pulse-spouted-bed freeze drying is another hybrid system applied for the dehydration of herbal material. It goes without saying that each herbal material requires a specific device and method for drying. Therefore, more investigation should be performed to determine the optimum option and condition for each substance [32].

4.4.8 Radiofrequency Drying (RF)

RF is a novel method of drying, also named dielectric heating, electronic heating via high-frequency electromagnetic radiation. In this process, an external field is exerted on the non-conductive material; the molecules get polarized (positively and negatively charged ions), giving rise to molecular friction and raising the temperature of the product (frictional heating). Regarding the process, herbal product is placed between two electrodes. High-frequency voltage is applied to the product when the current is drawn. This method is advantageous for the following reasons; offering uniform heating, easily controllable, deeper penetration, and a high dehydration rate. However, the drawbacks of this method should be considered, such as the non-homogenous increase of temperature in material, the dependence of heat production on the type of substance, product size, and power density. Moreover, in general, the efficiency of this method is roughly 50%. Therefore, a combination of conventional dehydration techniques (parallel drying) could be employed to surge drying efficiency. To illustrate, hot air-assisted RF, RF-vacuum, and RF-assisted fluidized bed drying were employed to dehydrate substances [33].

4.4.9 Freeze Drying

In this method, also called lyophilization or cryodesiccation, dehydration occurs at low temperatures to produce powders or solids of high quality. The four operational stages of the process are freezing, vacuum, sublimation, and condensation. First, the product is thoroughly frozen. The freezing rate is important since it determines ice crystals' shape, size, and formation. Second, the pressure is lowered by placing the substance under a vacuum to evaporate water or other solvents. This process is called primary drying due to ice sublimation at below barometric pressure. Then, at the secondary drying stage, bound or unfrozen water molecules are desorbed from the matrix of the material [34].

Considering types of herbal material employed for freeze drying, they fall into three categories: solids, homogenous solutions, and suspensions. In contrast to colloidal systems and solutions, solid herbs have a mechanically stronger cellular structure that is pertinent to the cellular arrangement, material, and geometry. The herbal cell wall comprises polymers of polysaccharides, namely, cellulose, hemicellulose, and pectin, which play an important role in the structural and mechanical strength of plant tissue. These intrinsic characteristics bring about material integrity and avoid the disintegration of herbal cells during freeze drying. In solid herbal material, the turgidity of cells could give rise to collapse, which is not restorable. Structure and composition of herbal tissue are associated with the prevention of collapse. Considering the liquids, such as suspensions, solutions, and emulsions, the concentration of phytoconstituents in the matrix is of utmost significance. Their level could also influence the temperature of the glass transition. The matter appears solid in a glass state; its viscosity is high, with limited molecular movements. While glass transition occurs, glassy material becomes more mobile and less dense. Suppose the temperature increases over the glass transition temperature (collapse temperature), the risk of deterioration in physical properties. Thus, the suitable temperature for freeze drying varies for various herbal substances [34].

This method is carried out at low temperatures; it is preferable for heat-sensitive material, vulnerable to degradation and oxidation. During dehydration, high temperatures, enzymes, and oxygen give rise to the degradation of biologically active compounds. Vitamin C, carotenoids, and phenolic compounds are further prone to degradation during this process. Ascorbic acid is known as the reference index of nutrient retention due to being an easily degraded molecule. Therefore, for preserving nutrients and phytochemicals, the lyophilization technique is suitable and pertinent to the absence of liquid water and oxygen in the dehydration environment and low temperature. Although shortcomings of this method are long duration and costly, it is a preferred approach to yield high-value products in terms of physical, chemical, and nutritional aspects. moreover, the yielded dry powders have pores that make them prone to rehydration [34].

Several innovative methods, intensification of FD process, and pretreatments could be carried out to overcome challenges and elevate process and mass transfer efficiency. Pretreatments could be thermal, chemical, and mechanical. In chemical pretreatment methods, the material is immersed in oleate ester solutions (acidic or alkaline) prior to dehydration. Immersion of substance in an alkaline solution forms cracks on the surface of the cells. However, if the temperature of the solution or the duration of the immersion is high, material texture degradation and taste deterioration would happen. Chemical pretreatments could be complemented or replaced by mechanical ones. For instance, friction and peeling of the substance's surface or skin, or cutting the substance into pieces could be done prior to dehydration. Thermal pre-drying techniques could be chosen before FD, such as blanching of material by soaking it in steam or hot water. It goes without saying that thermal approaches trigger the loss of thermally sensitive ingredients, namely, vitamins, minerals, proteins, or other nutrients [35].

Slow freezing brings about ice crystal formation in the extracellular environment, assisting in the fracture of tissues. On the other hand, rapid freeze drying causes ice crystal formation in the intracellular environment, leading to tissue destruction and mass transfer. The permeability of herbal structure depends on several criteria; composition and microstructure of herbal material, the physical state of the matrix, and glass and lipid transition temperatures. Consideration of the mentioned factors could be used to induce rapid drying and increase the quality of the product [35].

Some pre-drying approaches were introduced in previous studies; repetition of some cycles, such as periodic immersion of substance in liquid N_2, could markedly accelerate the process of dehydration. However, if the mentioned cyclic process is repeated 3–5 times, the fragility of the epidermis will occur. The drawback of this method is the distribution of particles with large sizes in the final dehydrated powder. If the herbal material is liquid with high concentrations of sugar or lipid, dilution or addition of water could be helpful before FD. However, it could increase the cost of the process since FD is not an affordable method [35].

There is an approach called freeze granulation technology, spray freeze drying, or sprays freezing in which droplets of suspension or a liquid are sprayed into an N_2 in the liquid state. Liquid N_2 brings about the instant freezing of droplets and their transformation into granules. Subsequently, the substance is dehydrated pertinent to sublimation of the ice of granules via freeze drying process. This method could be advantageous since the yielded granules are homogeneous in structure and active ingredients. Moreover, the stability and the solubility of the material are improved. On top of that, the density of granules could be determined and controlled based on the solid content of the slurry. The fewer cavities granules have, the less their density will be, offering them good aerodynamic characteristics. Furthermore, as this procedure is mild, the final product will be more, yield less waste, and the solvent could be recycled [35].

Foaming is another pretreatment method applied prior to freeze drying, for liquids cannot be frozen easily. This so-called approach is named foam-mat freeze drying. The drying rate of foamed substances is higher than non-foamed ones. As the total interfacial area of the foamed product increases, the FD duration is decreased. Hence, foamed material is porous with lower density. Thus, the mass load of the dryer device is elevated offering, higher output. Despite several pros of this technique, a few studies have been conducted to combine foaming as pretreatment and FD as the main drying process [36].

Heat sources, such as Infrared and microwave, could be used proceeding FD. Infrared, as a source of heat, could surge the vibration of molecules and offer uniform heating to the material. Therefore, it could accelerate the process of FD. This method is called infrared-assisted freeze drying. Microwave radiation applied before FD could increase the dehydration rate. Since microwave radiation assists sublimation, it could accelerate the drying up to approximately 70%. This technique is called microwave freeze drying. In contrast to infrared radiation, microwave energy is not uniform, giving rise to non-uniform heating of the substance, melting of ice, and overheating of the material. This is the major obstacle to commercializing the microwave freeze-drying method. In previous studies, some non-thermal pretreatment techniques were suggested to tackle shortcomings of conventional freeze drying; to illustrate, pulsed electric field, ultrasound, and high-pressure application were recommended, typically for thermally sensitive products [34].

4.4.9.1 Vacuum (VFD) and atmospheric freeze drying (AFD)

Vacuum freeze drying is a benchmark and established approach in which the products are of higher quality with preserved nutritional and phytochemical ingredients. This method makes the product highly porous and of lower density with maintained color and flavor. The most important challenge of FD is its high energy, manufacturing, and apparatus maintenance costs. To overcome this challenge, AFD was developed, combining the pros of FD and convective dehydration, where FD yields higher quality products, and convective drying is a more affordable approach. In this technique, FD is carried out at air pressure without a vacuum [37].

Compared to vacuum freeze drying, the operational temperature is higher, ranging from −3 to −10°C. This temperature range fulfills both cost and quality requirements since establishing a lower temperature than the mentioned one requires more energy consumption. Furthermore, the AFD approach improves mass transfer between the dehydration medium and the frozen product. Thermally sensitive and volatile ingredients are better preserved by this method. Above all, the powder yielded as the final product will have a larger surface area, freely floating, and superior solubility characteristics. As a result, AFD is an advantageous process that could be commercialized and employed in the pharmaceutical industry. Moreover, AFD is preferable over VFD pertinent to the elimination of vacuum apparatus, which is costly and imposes a considerable financial burden. A heat pump system could be applied to elevate the temperature, to reduce energy consumption, and duration of the dehydration process [37].

Hence, in the AFD process, inert gas, such as nitrogen and helium, could be applied in a dehydration environment to decrease oxidation which triggers substance degradation. Even though vacuum freeze drying offers superior quality products than atmosphere freeze drying, both processes are similar regarding hygroscopicity and kinetic rehydration. Previous studies indicated that AFD gave rise to a deterioration of antioxidant activity compared to VFD by 15%. Investigations demonstrated that employing AFD in fluid beds and spray freeze dryers leads to superior products [37].

4.4.9.2 Fluid bed freeze drying (FBFD)

AFD could be performed in fluidized beds resulting in better mass transfer and dehydration. This process could be used for liquid substances, transforming them into dried powder. First, high pressure and low temperature (−30 to −40°C) with the aid of liquid CO_2 are imposed on the substance. Afterward, that substance will be exposed to normal temperature and pressure that will bring about the foaming of the substance. Then the foamed product will be exposed to freezing temperature to be frozen and dehydrated. About solid and small pieces of products freezing process is performed directly in fluidized beds [38].

4.4.9.3 Spray freeze drying

Spray drying is widespread, yielding freely flowing dehydrated powders. Although this is a fast-rated method, it triggers the loss of thermally sensitive and volatile substances owing to the employment of high temperatures during the process. On the other hand, freeze drying was developed to retain heat-sensitive material. Since sublimation occurs within the freeze-drying process, it leaves porous powders of high quality as an end product. High operational costs of this approach led to the combination of several techniques to reduce the costs. Spray freeze drying contains three steps combining spray and freeze drying techniques. First, the feed liquid, solution, or suspension is transformed into countless numbers of droplets by atomizer (atomization), then the droplets are solidified due to exposure to cold fluid (freezing); afterward, sublimation occurs at low temperature and pressure (dehydration) [38].

Diverse cryogenic materials, gas or liquid could eliminate the sprayed droplets. Freeze drying could be operated under vacuum at either atmospheric or sub-atmospheric pressure. In this method, the size of droplets yielded in the atomization process greatly impacts the freezing rate. Atomization conditions should be optimized; for instance, the geometry of the atomizer nozzle should be adjusted to yield mono-disperse liquid droplets. The energy of atomization, viscosity, flow rate, and surface tension of the feed influence the process rate and end product quality. The more the liquid flow rate is, the more the droplet size will be. The higher viscosity of liquids brings about the consumption of more energy to break and yield smaller size droplets; therefore, the droplets will be larger. Also, surface tension greatly impacts the atomization process as the feed with higher surface tension requires more energy to be atomized. In these cases, atomization could be unchallenging by homogenization and the application of an emulsifier [38].

After the atomization process, with the aid of a cryogen, droplets are transformed into frozen particles. This process is a series of steps; first, preliminary cooling means cooling the substance below the temperature of normal freezing. The second stage is termed recalescence, in which during the cooling of the substance, there will be a subsequent rise in temperature to evolve the structure of crystals and make them larger. At the next stage, the growth of crystals will be slowed by transferring heat from the cryogenic fluid. Eventually, the completion of freezing occurs by decreasing the temperature below the temperature of the cryogen. This process is advantageous due to yielding spherical particles with fine pours and powder of lower density than FD, which produces highly porous particles. In terms of oxidative stability of products, SFD was a superior method to FD and SD due to causing less percentage of oxidation in substances. Therefore, SFD yields products of higher quality and stability. Hence, the dehydration rate in SFD is higher than that of SD and FD as different temperatures are employed on particles during the process. In addition, in SFD, thermal damage is effectively limited, which is one of the major cons of the SD method [38].

Recently, the SFD method has progressed in controlling the particles' size and microstructure. In addition, numerical and computational modeling and simulations have been developed to predict the optimum required temperature and dehydration rate. The simulation tools will reduce the cost of large-scale trials [38].

Concerning the limitations and challenges of SFD, biological products such as proteins are less stable during atomization. Moreover, the rapid cooling of liquid droplets containing proteins triggers denaturation and unfolding. Hence, in this method, large quantities of cold, dry gas, and vacuum conditions are required which makes this method debatable regarding economic viability and employment on an industrial scale. All in all, further studies are needed to realize influential factors on the microstructure of particles to further optimize and control the quality of end products by this method [38].

4.4.10 Heat Pump Drying (HPD)

Heat pump drying and heat pump-assisted drying technologies have attracted attention as promising methods because they are energy-efficient and cost-effective. Heat pump dryers are principally convective, in which dehydration occurs by heat transfer through air convection. In drying, heat source, moisture evaporation, and airflow are paramount. Conventional drying methods employing fossil fuels and electricity impose high operational costs and trigger gas emissions. In contrast to other drying technologies, this method incorporates the capacity to remove moisture from the drying air. Since moisture decreases the dehydration rate, dehumidification could accelerate the dehydration rate. In addition, the dehumidified air is heated repeatedly to provide heat to the dryer. Thus, heat recovery is the superiority of this process over conventional drying systems where heat and energy are wasted. Hence, controlling the temperature and moisture of drying air makes it an environmentally friendly and cost-effective method [39].

Additionally, it could be combined with other types of dryers; especially dryers initially based on convection could be combined with a heat pump. On the other hand, heat pumps are unsuitable for systems requiring a large amount of dehydration air, for instance, spray drying. This method is advantageous for thermally sensitive substances containing ascorbic acid and volatile compounds that retain aroma, texture, flavor, color, and heat-sensitive phytochemicals are retained. However, considering the limitations of this process, it is more applicable for solid substances rather than liquid or semi-solid ones. Hence, as a continuous dehydration system, special modeling, design, and engineering are required, which increases costs [39].

There is heat pump-assisted drying technologies in which the heat pump is in conjunction with solar, microwave, infrared, fluidized bed, radiofrequency, chemical heat pump, and atmospheric freeze drying [39].

4.4.11 Super Critical CO_2 ($scCO_2$) Drying

In the modern age, having green media and an environmentally friendly dehydration approach are necessary. This requirement brought about the introduction of supercritical CO_2-based methods into the realm of science. In this approach, $scCO_2$ is used as a medium for dehydration. The main merits of this process are the employment of ambient temperature and low amount of oxygen that provide a mild operation environment for the product. Furthermore, this method makes the rehydration of dried substances far better than other methods. In terms of end product quality, color, odor, and biologically active ingredients are much better preserved by this process [40].

In comparison with conventional methods of dehydration, for instance, freeze drying and air drying, which take four days and 26 h, respectively, the duration of dehydration via this process is considerably less compared to other methods (2–3 h). Above all, $scCO_2$ dehydration inactivates the microorganisms (yeast, mold, and some bacteria) efficiently [20]. Another distinct advantage of this method is the elimination of vapor-liquid interfaces from the process. The structure of the substance is preserved due to the avoidance of tensile stresses imposed by air drying. Since the critical temperature of the CO_2 is low (31.1°C), dehydration occurs at a much lower temperature than conventional approaches. It goes without saying that $scCO_2$ as a nonpolar solvent, can modestly solve water. A commonly applied approach to tackle this issue is the addition of small amounts of co-solvent to $scCO_2$ to increase the solubility of polar substances in it. Commonly used solvents in conjunction with $scCO_2$ are organic solvents such as acetone, ethyl acetate, and ethanol. They modify the solubility of products in $scCO_2$ by involving several interactions, including dipole-dipole bonding and hydrogen attraction. According to previous studies, the more ethanol is added to the system, the more moisture removal occurs, and the rate of dehydration soars [40].

4.5 Drying Effects on Physical and Chemical Properties

Inevitably, dehydration culminates in the physicochemical transformation of herbal products, whether sensorial deterioration or phytochemically active ingredients. Likewise, various thermally sensitive nutritional ingredients degrade in the heating and dehydration process. As mentioned in the preceding parts, conventional dehydration approaches further trigger the products' physicochemical decomposition. In the following sections, principal factors of quality assessment will be discussed.

4.5.1 Physical Features

4.5.1.1 Structure

Dehydration could alter the physical properties of the herbal material. The structure of the herbal product alters within the drying process which could be characterized by several determinants in which shrinkage, bulk density, and porosity are the most common factors mentioned in scientific manuscripts. The drying process gives rise to a decrease in the material's hydrophilic properties owing to the irreversible rupture of the cell. Due to breaking down and shrinking, the integrity of the cells or the internal structure is lost, and their density is altered, which causes resistance in water absorption or rehydration. Rehydration is the ability of the dried material to absorb water after drying and gain its initial properties. Therefore, dried material should be capable of water reabsorption or rehydration. However, reconstitution capacity, or the ability to regain the initial features, is dependent on the degree of the changes in internal structure. Also, the degree of deterioration of water-holding ingredients, for instance, starch and proteins, is of utmost significance. Structural rupture, the degree of porosity, and texture could be analyzed by taking 3D images and X-ray graphs from the dried product [41].

4.5.1.2 Porosity

Porosity is another physical determinant that determines the texture of the dried product. The texture is appraised by the number of empty spaces or porosity, which depends on the drying method. A well-qualified drying approach could yield highly porous products. Prior studies indicated that FD, microwave, and vacuum drying produce highly porous material among the mentioned drying methods [41].

4.5.1.3 Bulk density

Bulk density is employed in porous and powdered products. Overall, dehydration triggers the growth of bulk density. However, the lower the bulk density of the material, the better its sensory aspects will be. Several elements affect the bulk density of the dehydrated product; however, among the important parameters, porosity is of utmost importance. Alongside porosity, the type of drying method, temperature, and duration of dehydration are of great significance. If the duration of dehydration is longer, the degree of cell destruction, and shrinkage of material will be higher, which gives rise to higher bulk density. Dehydration conditions influence the bulk volume, particles' shape, and the final product's packing density. Prior studies indicated that freeze drying, osmotic dehydration, and microwave drying reduce the bulk density of the material by 50%, thus they are known as the most effective and reliable methods in terms of bulk density reduction when compared to conventional drying approaches [42].

4.5.1.4 Shrinkage

As a physical phenomenon, shrinkage is defined as a reduction in the shape and dimensions of the dried material, pertinent to the evaporation and decrease of the capillary water of the product. Pertinent to the collapse of the structure, it affects the texture and taste of the final product. The optimum approach of dehydration leaves the end product with lower shrinkage. Experimental data illustrated that hot air drying causes more shrinkage and denser products with hard textures. At the same time, freeze drying produces lower shrinkage and higher-quality products. Furthermore, previous reports showed that materials dehydrated by the vacuum microwave method have lower shrinkage degree. This method leads to a shorter drying duration compared to convective approaches and leaves puffed products. Puffed products are yielded when the dried product's structure is porous, and the dehydration duration is shorter. The shrinkage degree of dehydrated products yielded by a vacuum-microwave method is higher when compared to the end products produced by freeze drying. All in all, new drying approaches lead to lower collapse and shrinkage compared to traditional approaches [43].

4.5.1.5 Color analysis

Preservation of the dried product's natural color is of great momentousness. Therefore, the measurement of color changes is used to predict herbs' phytochemical content and quality. Similarly, color assessment is one of the important quality control parameters that immediately affects the percipience and judgment of customers from the product. Parallel to the color, shape, gloss, and size of the end product fabricate and represent the physical features of the dried substance. The color of herbal products is mainly changed and destroyed by conventional dehydration methods. Thus, it is of utmost importance to opt for a suitable dehydration approach and conditions for every product to decrease color degradation and leave an end product with a more attractive color. Browning is a gradual transformation of the color of a dried product to brown or dark brown, which can be considered both desirable and undesirable relative to the kind of herb [44].

Browning reactions are categorized into two groups enzymatic browning and non-enzymatic reactions. In non-enzymatic browning, a combination of reactions occurs since the substance contains several ingredients, such as caramelization, oxidation of ascorbic acid, and Maillard reaction. In enzymatic browning, some enzymes are involved, for instance, PPO, which exists in the product. This reaction could occur in plant harvesting, post-harvesting, and processing stages. Any mechanical action on herbs, for instance, cutting and dicing or their storage at severe temperatures gives rise to exposure of PPO to oxygen. It could be a developing incident to oxidize phenolic compounds into quinones. Afterward, several reactions take place to polymerize the yielded quinones and their derivatives to yield melanin, a brown-colored and insoluble pigment. The more active the PPOs are, the more the rate of enzymatic reaction would be [44].

4.5.1.6 Changes in the number of phytochemicals

The amount of several phytochemicals is highly influenced by the method of dehydration. The content of phyto-constituents is affected by several determinants, for instance, temperature, the duration of exposure, oxygen level, and the presence of light. The mentioned phytochemicals include carotenoids, vitamins, and polyphenols. Prior experimental data showed that the content of phytoconstituents is highly influenced by temperature in the drying process; the higher the temperature, the more destruction of phytochemicals. Moreover, these ingredients are far better retained if oxygen and light levels are reduced. Substantially, prior reports showed that applying combined drying methods assists in retaining phytochemicals [45].

4.5.1.7 Determination of total phenolics and flavonoids

Secondary metabolites are biologically active natural products manufactured by herbs to defend against environmental perils such as the invasion of parasites and other infections. The ingredients are not necessary for herbal growth and reproduction of herbs like primary metabolites (carbohydrates, lipids, amino acids, and nucleic acids); however, they fulfill a crucial function in preventing and treating human ailments. Secondary metabolites are classified into three parts on the basis of their chemical structure: phenolics, terpenoids, and alkaloids [46]. Among the mentioned groups, phenolic compounds occur in a plethora of plants. These ubiquitous phyto-constituents are health-protective and have indicated many biological and pharmacological effects. Therefore, preserving and maintaining phenolic compounds during the post-harvesting process is of great significance, for instance, drying [47]. Furthermore, phenolic compounds are important phytochemicals from various perspectives, notably potent antioxidants. These phytochemicals are found mainly in fresh herbs, and while dehydration, they are decomposed to an extent. Therefore, selecting the superior drying method for each herb is one of the challenging aspects of the post-harvesting process. According to previous studies, air drying was introduced as a better method than oven drying [45].

4.5.1.8 Changes in antioxidant activity

The antioxidant activity of the dried product is related to the stability and retention of some phytochemicals during the dehydration process. For obtaining a dried product of higher antioxidant capacity, it is important to opt for an optimum dehydration method for each herbal material. Among the mentioned active bio-ingredients, polyphenols fulfill a crucial function in the antioxidant activity of products. According to prior scientific evidence, a combination of novel drying methods works better in retaining dried products' antioxidant capacity, while convective methods trigger the degradation of mentioned bio-ingredients. Hence, the drying approach in which the environment is oxygen-free and the temperature is low, loss of polyphenolic content and antioxidant activity is dropped. Since photo-oxidation and thermal oxidation account for lowering antioxidant activity [45].

4.5.1.9 Volatile compounds

The volatile oil composition in aromatic herbal products has been the research subject for many years. Generally, the level of volatile ingredients is decreased in the dehydration process. However, in some of the prior reports, it was indicated that microwave and hot air drying cause greater loss of essential oils. Therefore, to diminish the loss of volatile oil while using the mentioned methods, it is recommended to employ pretreatment methods, or to combine two or three dehydration methods. Therefore, assisted drying technologies, such as employing convective methods of pre-drying (as a traditional method) and vacuum-microwave (novel approach) as finishing approach were introduced as promising approaches for aromatic herbal products [45].

4.6 Storage and Preservation Conditions

Storage of medicinal plants is a part of the post-harvest chain after dehydration that affects the quantity and quality of the phyto-constituents. Generally, herbs and herbal products are stored for long durations before utilizing or manufactured. Therefore, they should be stored in appropriate conditions to minimize physical changes, degradation of biologically active ingredients, fungal, and bacterial infections. Environmental factors, such as humidity and temperature, fulfill a crucial function in the degradation of herbs. There is no constant preservation period for herbal products. Thus, various stability tests should be carried out to determine the proposed period. During storage, medicinal plants are susceptible to metabolic activities that give rise to deterioration. Thus, these alterations should be minimized. The proposed options are reducing metabolic changes, decreasing water content, and cooling the product [48].

Herbal products could be stored at different temperatures; ambient temperature (25°C), refrigerator (2–8°C), and freezer (–20°C). According to prior reports, the storage temperature influences the volatile oil composition. At the same time, the polyphenol content of herbs is less affected if they are stored at a temperature of 4°C [49].

Packaging is another substantial factor that should be addressed in the storage of herbs. Paper and polyethylene bags are suggested to store leaves. Paper bags and cardboard boxes crafted with an inner polyethylene box are recommended. Previous studies showed that storing dry herbs in polypropylene bags influenced the quality and quantity of the volatile oil. There was a considerable decrease in the amount and composition of volatile oil within storing for one year. Apart from the type, the method of packaging, typically the type of packaging gas, greatly impacts the quality and quantity of volatile oils. There are three methods of packaging based on packaging gas: vacuum, nitrogen, and air packaging. Previous studies showed that the herb packaged with nitrogen was higher regarding volatile oil composition and the percentage of active ingredients [50].

4.7 Factors Affecting the Stability of Herbal Material

Stability refers to the duration in which any pharmaceutical substance endures various environmental factors with the lowest effect on its physical, chemical, microbial, and therapeutic properties. Stability testing and determination of expiry date are substantial criteria in the process of herbal drug development. Several requirements should be met to ensure the stability of the herbs and herbal products for proposing shelf life and expiration date. There are some analytical methods according to regulatory guidelines of drug regulatory agencies, whether country-specific or international ones, such as the World Health Organization (WHO) [51].

Evaluating the stability of the herbal preparations imposes some challenges compared to synthetic drugs. Biologically active phytochemicals are composed of several constituents which play a part in therapeutic activities. Hence, herbal preparations comprise several herbs that impose further analytical challenges and costs. Also, herbal substances are unstable due to some criteria, for instance, physical and chemical instability. Physical instability occurs owing to a reaction with the container and some impurities. In addition, environmental conditions trigger variability in complex mixtures of herbs. First and foremost, light is a dominant criterion that triggers the formation of free radicals. Hence, the process of manufacturing, for instance, dehydration, purification, extraction, and genetic variability, affects the stability and the content of phytochemicals. In other words, the decomposition and deterioration of herbal substances occur during storage. Enzymatic degradation and chemical reactions with additives and excipients cause the degradation of phytochemicals and phyto-formulations. Chemical reactions accelerate with the rise of moisture and temperature. Hydrolysis, oxidation, reduction, crystallization, and formulation breakdown occur due to storage in high amounts of moisture and temperature. pH is another determinant for durability testing. Acidic and alkaline pH affect the degradation rate in herbal material [52].

Despite the mentioned challenges, fingerprint chromatograms could assist. Also, the proportion of the analytical and active markers should be assessed. Eventually, the quantity and proportion of active ingredients will be compared to the initial chromatographic fingerprint. If the content of the specific phyto-constituent and therapeutic effects are changed in an herbal product during stability studies, it shows the instability of the herbal product [53].

Since there are phytochemicals in an herbal product that belong to various chemical classes, several intra-molecular reactions may occur. These reactions alter the profile of herbal drugs or products. They could be pertinent to humidity, light, air, and heat during manufacturing, storage, and transportation. To illustrate, tannins make complexes with proteins, heavy metals, and polysaccharides. Also, tannins are incompatible with alkaloids and precipitate. Tannins also interact with glycosides. Heavy metals form complexes with polysaccharides. Moreover, polyphenols make complexes with polysaccharides, and proteins via hydrogen bonding or hydrophobic reactions decrease the number of free polyphenols. These interactions and complexations among active ingredients trigger toxic, less active or even more active products. Thus, it is important to undertake analytical methods and acquire chromatographic fingerprints according to pharmacopeial protocols to ensure the phytochemical profile of the herbal drug is not changed with storage [54].

Another factor that causes biological instability is microbial contamination which gives rise to both health threats and the decomposition of the active ingredients due to the microbial enzymes. Therefore, checking microbial load is an important part of the stability testing of herbs and herbal products. In addition, mold growth triggers the interaction of active phytochemicals with each other, packaging material, and other formulation components. Besides, residues of pesticides and contamination with heavy metals are the other undesirable elements that should be monitored [54].

Several enzymatic activities within herbal material cause changes in the chemical profile, taste, consistency, texture, color, and taste of herbal material. Different types of enzymes include catalase, peroxidase, ligase, chlorophyllase, polygalactouronase, isomerase, lyase, hydrolase, esterase, and transferase catalyze several reactions such as conjugation, interconversion, degradation, and polymerization. These enzymes' activities depend on the condition of processing and storage. Although hydrolytic and microbial decomposition could be reduced by decreasing the content of moisture or dehydration, during the shelf life, enzymatic decomposition could not be prevented. Even though most enzymes are inactivated by heat and water exposure, heat treatment cannot inactivate some enzymes. For instance, peroxidase and catalase are the ones that can activate at the temperature of 60°C in which most enzymes are degraded [54].

Browning is a possible alteration observed commonly in herbal material. Peroxidase, PPO, and Phenylalanine ammonia-lyase are the main reasons for the reaction that produces quinone from phenolic compounds. Quinone could yield complex products, as it is highly reactive [54].

Glycosidases are the other group of enzymes that generate phenolic compounds from terpenoid and flavonoid glycosides. In some cases, a combination of enzymatic activities gives rise to browning. To illustrate, a flavonoid glycoside is hydrolyzed by glycosidase and then catalyzed to quinone by polyphenyl oxidase. The other enzymes accountable for browning are superoxide dismutases, oxalate oxidases, and amine oxidases. Therefore, enzyme assays are required to identify active enzymes in the end product. Also, these assays should detect enzyme levels and elucidate their roles during the shelf life [54].

All in all, stability testing requires assessing physicochemical stability, biological activity, and toxicity that form chemical and biological shelf life. Therefore, an assessment of the criteria above guarantees the safety and therapeutic potency of the herbal material.

4.8 Regulatory Bodies' Recommendation for Testing Stability

There are several regulatory bodies throughout the world that provide guidelines for controlling the stability of herbal products. The Food and Drug Administration (FDA) and European Medicine Agency (EMEA) provide guidelines and analytical methods to prove the stability of products. Stability testing recommended by regulatory guidelines accelerates product chemical decomposition and physical alteration. These tests are aimed at imposing exaggerated stress conditions on herbal products. Stability tastings fall into three categories: accelerated stability testing, performed for six months at 40°C and 75% relative humidity. Second, intermediate stability testing for 6–12 months was accomplished at 30°C and 75% relative humidity. Third, long-term stability testing with the duration of proposed shelf life and employment of 25–30°C and 60–75% relative humidity. Prior studies revealed that a wide range of conditions is employed for stability studies of herbal material. For illustration, temperatures of –80 to 100°C, relative humidity of 0 to 100%, and varied pH conditions were performed within several hours to three years. In polyherbal or traditional herbal medicinal products, stability studies are far more tedious than monoherbal products pertinent to containing several phytochemical markers. Thus, physical stability studies are performed for the majority of polyherbal products. Although regulatory bodies recommend general guidelines for stability studies, herbal materials should be tested case-by-case to ensure their applicability to a specific material [52].

4.9 Recommended Approaches to Handle Instability of Herbal Material

There are some approaches to increase shelf life and produce high-quality end products. Using several typical substrates to produce coatings of nanoparticles could extend shelf life. These coatings could act as a good barrier to protect natural products from moisture and microbial deterioration. The coatings should have typical characteristics: They should be resistant to water to remain intact. They should not produce carbon dioxide and deplete oxygen. Since a limited amount of oxygen should surround the product to prohibit anaerobic respiration, oxygen should not be completely depleted. It should limit permeability to water. It should maintain volatile components active ingredients, such as antioxidants and vitamins, and enhance the appearance. It ought not to react with the product produced. It should not decompose below 40°C, be cost-effective, and of low viscosity [55].

Biopolymers, such as proteins, lipids, and polysaccharides, are edible coatings employed to preserve foods and natural products. These eco-friendly methods are proper alternatives for synthesized polymers. Herbal polysaccharides are excipients that comply with the required conditions. They are non-toxic, stable, available, and easily modified to suit typical conditions. They are cost-effective pertinent to be easily extracted from herbs. As a barrier, they prevent moisture loss and maintain the moisture content of herbal material. Being less permeable to oxygen, they are efficient gas barriers. The main polysaccharides used for coating are chitosan, cellulose, starch, and gums [55].

Chitin is a biopolymer abundant in nature. Chitosan is a polysaccharide derived from chitin which yields semipermeable coatings with antibacterial effects [56]. Chitosan-coating is tough, flexible, clear, and a promising method that forms a thin layer on the product's surface. It extends shelf life by anti-browning activity, inhibiting the loss of volatile components and moisture.

It prevents oxygen penetration into the product, microbial and fungal growth, and increases safety. This coating could be enriched with antimicrobial and antioxidant agents loaded into the chitosan films. These agents could be essential oils, including lemon, lemongrass, cinnamon, mentha, and bergamot oils [55].

Furthermore, some nanoparticles, such as silicon dioxide, could be loaded on chitosan films. The incorporation of acids in chitosan-coating could surge anti-microbial effects. These acids include acetic, ascorbic, citric, and lactic acids. Via these post-harvesting techniques, color change, the growth of yeast, mold, and bacteria could be efficiently prohibited, and the shelf life could be prolonged [57]. The other polysaccharides used as coating are modified starches, alginates, gums, cellulose derivatives, and pectin.

Cellulose and its derivatives are highly abundant polymers with linear structures while being poor barriers for gas and water. These coatings include methyl cellulose, carboxy methyl cellulose (CMC), methyl cellulose (MC), hydroxy propyl methyl cellulose (HPMC), cellulose ether (CE), and hydroxy propyl cellulose (HPC). They have outstanding film-forming characteristics, are water-soluble, resistant to oils, transparent, flexible, tasteless, and odorless, but not cost-effective. The barrier characteristics of cellulose coatings depend on their molecular weight. The ones with higher molecular weight indicate superior coating properties. MC with the lowest hydrophilic characteristics is the least permeable to water. In comparison, CE is highly permeable to water. Prior studies showed that incorporating essential oils in the CMC could prolong the shelf life and further prohibit microbial growth. More studies are required to produce cellulase derivatives with optimum properties [58].

Starch-based coatings are low-cost, biodegradable, and could prevent gas exchange up to an extent. Therefore, they are preferable over plastic polymers. However, a major challenge of starch-based coating is its brittle nature, typically in dry conditions. The addition of biodegradable plasticizers addresses this challenge. Commonly used plasticizers are urea, glycerol, and polyether. Furthermore, previous studies indicated that a combination of starch coating with citric acid could prevent color change and reduce the deterioration of the product's respiration rate. Moreover, incorporating carrageenan, fatty acid esters, chitosan, and starch coating could prolong the shelf life and overall quality [55].

Alginate is a polysaccharide isolated from brown seaweed belonging to the family Phaeophyceae, which acts as a proper barrier for moisture. Combined with volatile oils, it indicated a significant decrease in microbial load, anti-browning effects, and extended shelf life. It could be enriched with citric acid and ascorbic acid [59].

Protein-based coatings could be isolated from various sources, including animals and plants. Considering their characteristics, generally, they are hydrophilic and vulnerable to absorption of moisture, brittle, vulnerable to cracking, and poor water barriers. Proteins fall into two categories: fibrous proteins and globular proteins. The commonly used animal-based ones are insoluble fibrous proteins, including whey protein (a byproduct of cheese production), collagen, milk protein casein, keratin, gelatin, and egg albumin. The common plant-based coatings are water soluble, such as wheat gluten, corn-zein, peanut protein, cotton seed protein, and soy protein. Acid/base solvents and heat should denature a generation of cohesive protein films to yield an extended form. In this extended form, the chains of proteins are involved in several interactions, for instance, covalent, hydrophobic, hydrogen, and ionic bindings. The quality of protein films is highly affected by the sequence and the nature of amino acids. The more chain-to-chain interactions bring about stronger protein films, while they are less flexible and less permeable to gases and liquids [55].

Lipid-based edible coatings categorized as traditional biopolymers are efficient moisture barriers. They include animal wax, such as spermaceti wax, beeswax, shellac wax, lanolin wax, and Chinese insect wax. Vegetable wax is also used, such as Carnauba wax, candelilla wax, bayberry wax, rosin wood wax, sugarcane wax, palm wax, esparto wax, cotton seed wax, rice bran wax, Japan wax, paraffin wax, mineral and synthetic waxes such as paraffin wax, and microcrystalline wax. Fatty acids and vegetable oils extend shelf life owing to reduced respiration and enhanced surface appearance. In addition, due to being hydrophobic agents, they act as proper barriers to water loss. However, their hydrophobic nature causes the formation of more brittle and thicker films. They also do not give palatable organoleptic properties causing a greasy surface and undesirable taste. Therefore, combining lipid substances with proteins and cellulose derivatives could lead to a film of higher quality [55].

Other herbal-originated coatings such as *Aloe vera* gel with anti-fungal and anti-bacterial properties can extend the shelf life, reduce weight loss, and maintain the product quality during storage. Also, they drop respiration rate and oxidative browning and maintain organoleptic and physical properties. Some volatile oils can be used as coatings, such as lemon, orange, ginger, clove, mint, lemongrass, and turmeric essential oils. They can not only surge physicochemical characteristics but also act as anti-microbial agents. In addition, they can reduce weight loss and delay the rate of gas exchange [60].

Overall, the coatings mentioned above are promising environmentally friendly carriers that are preferred over synthetic films. Hence, they are renewable, biodegradable, and edible resources. They decrease gas exchange and weight loss, giving the product a fresh and shiny look. They retain volatile compounds, pigments, ions, vitamins, and antioxidants. They are an efficient post-harvesting approach to extend shelf life, delay the deterioration of herbal material and preserve them from heat, moisture, or other degradation conditions to improve their stability [61].

4.10 Packaging and Durability

Packaging is one of the approaches for improving the quality, safety, efficacy, and stability of herbal material. Since light, moisture, heat, and air could impact herbal materials' stability and shelf life, packaging could enhance their stability. Each herbal material requires typical and proper packaging material. A thorough investigation is required to opt for a suitable and compatible packaging material to preserve the product's properties from chemical, physical, and biological influences. Even though packaging retains the quality of the product from the mentioned factors through time, it is of great importance from the aesthetics point of view. It conveys a positive sense and quality perception to the consumers. Besides compatibility with the plant's physical characteristics and product protection from the environment, it should provide specific information through labeling and brochures. Suitable labeling in combination with legal requirements is crucial to attracting consumers, purchasing decisions, and product marketing. Packaging is communication between the consumer and the final product. Labels should provide information regarding the identity of the product, its ingredients, weight, and the manufacturer [62].

Various packaging materials are used to provide a barrier for herbs and herbal products. Glass and metal packaging are approximately superior barriers against environmental and chemical factors. However, closure devices made of other materials, such as plastic caps, could cause permeability to gas and vapors up to an extent. Plastic packaging is a proper barrier agent and is permeable compared to glass and metal. Packaging should provide biological protection to prohibit spoilage and ailments; for instance, it should be a barrier against microorganisms, insects, and rodents. In addition, it should provide physical protection against mechanical damage, such as shock, vibration, and crushing [63].

Glass, being chemically inert and odorless, is not permeable to vapor and gas. Therefore, it retains the product long without affecting its aroma or taste. Due to its resistance to high temperatures, it is a good choice for the sterilization of the product. Glass is transparent, rigid, offers proper insulation, is environmentally friendly, and recyclable. However, its heaviness results in a high cost of transportation. The other drawbacks of glass packing are its brittleness and vulnerability to breakage [63].

Metal-based packaging materials are the most versatile and widely applied as sound barriers providing physical protection. They are employed as both packaging form and closure as they are formable, recyclable, and greatly accepted by consumers. Metals in both forms, rigid and semi-rigid, mostly made of aluminum, tinplate, tin-free steel, and stainless steel, can be applied [63].

Aluminum is a silvery, and lightweight material mostly combined with oxygen to yield alumina. If magnesium and manganese are combined with aluminum, their physical strength is enhanced. In contrast to most metals, aluminum provides good barrier properties against chemicals, moisture, air, microorganisms, odor, and temperature. Hence, it is flexible and formable and is applied in various forms, such as foils, cans, laminated material, tubes, caps, and closures. Its main drawbacks are its high cost and using it merely as a seamless container. Moreover, it cannot be welded and has limited structural strength. However, aluminum could be rolled to yield thin sheets defined as aluminum foils with diverse thicknesses. One strategy to increase the barrier properties of aluminum foils is the lamination or formation of laminated aluminum, which involves binding aluminum foil to plastic films or papers. Since laminated aluminum is not cost-effective, they are applied to high-value package material. A good substitute for laminated aluminum is metalized film, a polymer film coated with a thin layer of aluminum [63].

Tinplate is a thin layer of steel coated with a thin tin covering on both sides formed by immersing steel sheets in the molten tin. Tinplate is corrosion resistant; however, it can react with the product. Therefore, coating is required. It is lacquered to result in an inert surface between the product and the metal package. Widely used lacquers are made of oleoresins, vinyl resins, and phenolic epoxy groups. It is a formable material and can undergo heat treatment, thus can form containers of various shapes. It is recyclable, low weight, and low cost, making it a suitable alternative to aluminum. Tin-free steel is yielded by applying electrolyte chromium/chrome oxide over steel sheets. It is less expensive than tinplate, formable, paintable, and mechanically strong, and it can withstand heat processing. However, it is difficult to weld and is less corrosion-resistant [63].

Plastics are polymers produced by poly-condensation and poly-addition of monomers. Poly-condensation is defined by the condensation of monomers with at least two functional groups, such as carboxylic, amine, and alcohol groups. Also, in these reactions, byproducts of low molecular weight, for instance, water and methanol, are formed. In poly-addition, a combination of molecules occurs with no liberation of byproducts. Monomers involved in these reactions are unsaturated, with double and triple bonds broken to combine monomers. Plastics are moldable and formable, which offers flexibility in designing containers with diverse sizes and shapes. They are low-weight, inexpensive, corrosion-resistant, thermo-sealable, and printable. The main drawback of plastic is its permeability to molecules of low molecular weight, gas, vapor, and light. More than 30 types of plastics are used for packaging, and the most commonly used ones are polyolefins and polyesters [63].

Paper is another packaging material that cannot protect food for long. It is mostly used for primary packaging as it has poor barrier properties and is poorly heat sealable. Therefore, its functionality and barrier properties should be improved by treating and coating it with lacquers, waxes, and resins. Paperboard is also applied for packaging because it has multiple layers, is thicker, and is heavier than paper. Paper laminates are made of multiple layers to enhance several properties of paper. They consist of papers coated with plastic, aluminum, and polyethylene. Although lamination of paper triggers barrier properties or heat sealability, it considerably increases the cost of paper [63].

Successful packaging involves consideration of several criteria; selection of the proper packaging material compatible with the product's physical, chemical, biochemical, and microbiological characteristics, cost, environmental conditions, consumer requirements, and management of waste issues. Ideally, the nature of each product, properties of packaging material, and interaction between packaging material and product should be assessed to predict the shelf life and stability. The interaction between the package and food takes center stage among the prerequisites for properly selecting packaging material. In addition, the product's acidity, alkalinity, and light sensitivity should be considered. Hence, the rigidity and permeability of packaging material are of profound importance. Low-molecular-weight compounds, whether volatile (flavors and aromas) or non-volatile compounds (fat and pigments), water, and gases can be transported through the package to the product or environment and vice versa. This transportation could trigger chemical changes in product and package, product contamination and loss of quality, package integrity, and its leakage. Furthermore, regarding plastic packaging material, the migration of substances, such as plasticizers, stabilizers, and monomers into the product could occur [62].

Plastic, synthetic polymers, and unrecyclable packaging materials pose enormous environmental and health-related issues. Moreover, the non-degradability of synthetic material arouses the concern about finding an environmentally friendly packaging material. Biodegradable and recyclable polymers made of renewable sources and could be degraded by naturally occurring microorganisms to yield natural products are alternative materials to address the abovementioned concerns. Biopolymers fall into three categories; the first group could be extracted from naturally occurring material such as polysaccharides (starch, chitosan, cellulose), proteins (wheat gluten, whey, soy, fish, pea proteins), or a combination of polysaccharides and proteins. The second group is yielded from the synthetic condensation of naturally occurring monomers, and microorganisms produce the third group. Packaging made of biopolymers is the proper alternative for synthetic polymers or petroleum-based plastics. They have similar properties as synthetic polymers and lead to fewer gas emissions. Hence, several additives could be included in the structure of biopolymers, such as antimicrobial, antioxidant, and anti-browning agents, nutrients, and colorants, which enhance the stability and shelf life of the product [63].

Active or intelligent packaging refers to packaging and product protection by including antimicrobial and antioxidant compounds to decrease the utilization of chemical additives and preservatives that are detrimental to human health. The mentioned compounds are extracted or isolated from herbs and spices. Essential oils are commonly used preservatives, and alkaloids are widely used anti-microbial phytochemicals. Moreover, phenolic compounds, due to having antioxidant and anti-microbial properties, are classified as important additives to biopolymers. Active packaging technology could be beneficial in reducing the number of anti-microbial agents added directly into the product since they slowly release from biopolymer packaging onto the product. Further investigations are required to introduce new natural sources as additives to flourish active packaging, an alternative to traditional packaging systems [64].

4.11 Conclusion

The shelf life of the herbal product depends on the quality of post-harvest practices. Therefore, the producers should adhere to post-harvest guidelines to prevent loss and increase the maintenance of herbal products. In some cases, simple practices could be beneficial, while sometimes high-tech postharvest technologies are required to address the problem of herb product preservation. Furthermore, due to the changes in global trends in the production of plant-based material, new inventions and innovations are needed in processing techniques to maintain the quality and quantity of herbal materials.

References

1. Zhu S, Shirakawa A, Shi Y, Yu X, Tamura T, Shibahara N, et al. Impact of different post-harvest processing methods on the chemical compositions of peony root. *Journal of Natural Medicines*. 2018; 72(3): 757–67.
2. Alamgir A, Alamgir AJTUoMP, Pharmacognosy TEV. Herbal drugs: their collection, preservation, and preparation; evaluation, quality control, and standardization of herbal drugs. In Therapeutic Use of Medicinal Plants and Their Extracts: Volume 1. *Progress in Drug Research*, 2017, Springer, pp. 453–95.
3. Chantuma P, Lacote R, Leconte A, Gohet EJFCR. An innovative tapping system, the double cut alternative, to improve the yield of Hevea brasiliensis in Thai rubber plantations. *Field Crops Research*. 2011; 121(3): 416–22.
4. El-Ramady HR, Domokos-Szabolcsy É, Abdalla NA, Taha HS, Fári MJSARV. Postharvest management of fruits and vegetables storage. In Lichtfouse, E., eds. *Sustainable Agriculture Reviews. Sustainable Agriculture Reviews*, Vol. 15. Springer, Cham. https://doi.org/10.1007/978-3-319-09132-7_2 2015: pp. 65–152.
5. Shishir MRI, Karim N, Bao T, Gowd V, Ding T, Sun C, et al. Cold plasma pretreatment–A novel approach to improve the hot air drying characteristics, kinetic parameters, and nutritional attributes of shiitake mushroom. *Drying Technology*. 2020; 38(16): 2134–50.
6. Thamkaew G, Sjöholm I, Galindo FG. A review of drying methods for improving the quality of dried herbs. *Critical Reviews in Food Science Nutrition*. 2021; 61(11): 1763–86.

7. Mohapatra D, Mishra S. Current trends in drying and dehydration of foods. In *Food Engineering*. 2011; NOVA Publishers, USA, pp. 311–351.
8. De Corcuera JIR, Cavalieri RP, Powers JR. Blanching of foods. In *Encyclopedia of Agri, Food and Biological Engineering*, Marcel Dekker, New York City, NY, USA. 2004, pp. 1–5.
9. Neri L, Hernando IH, Pérez-Munuera I, Sacchetti G, Pittia P. Effect of blanching in water and sugar solutions on texture and microstructure of sliced carrots. *Journal of Food Science*. 2011; 76(1): E23–E30.
10. Dibanda RF, Akdowa EP, Tongwa QM. Effect of microwave blanching on antioxidant activity, phenolic compounds and browning behaviour of some fruit peelings. *Food Chemistry*. 2020; 302: 125308.
11. Feng Y, Tan CP, Zhou C, Yagoub AEA, Xu B, Sun Y, et al. Effect of freeze-thaw cycles pretreatment on the vacuum freeze-drying process and physicochemical properties of the dried garlic slices. *Food Chemistry*. 2020; 324: 126883.
12. Xu X, Zhang L, Feng Y, Zhou C, Yagoub AEA, Wahia H, et al. Ultrasound freeze-thawing style pretreatment to improve the efficiency of the vacuum freeze-drying of okra (Abelmoschus esculentus (L.) Moench) and the quality characteristics of the dried product. *Ultrasonics Sonochemistry*. 2021; 70: 105300.
13. Lebovka NI, Shynkaryk NV, Vorobiev E. Pulsed electric field enhanced drying of potato tissue. *Journal of Food Engineering*. 2007; 78(2): 606–13.
14. Ade-Omowaye B, Rastogi N, Angersbach A, Knorr D. Combined effects of pulsed electric field pre-treatment and partial osmotic dehydration on air drying behaviour of red bell pepper. *Journal of Food Engineering*. 2003; 60(1): 89–98.
15. Kostaropoulos A, Saravacos G. Microwave pre-treatment for sun-dried raisins. *Journal of Food Science*. 1995; 60(2): 344–7.
16. Ashtiani S-HM, Rafiee M, Morad MM, Khojastehpour M, Khani MR, Rohani A, et al. Impact of gliding arc plasma pretreatment on drying efficiency and physicochemical properties of grape. *Innovative Food Science & Emerging Technologies*. 2020; 63: 102381.
17. Niemira BA. Cold plasma decontamination of foods. *Annual Review of Food Science and Technology*. 2012; 3(1): 125–42.
18. Hosainpour A, Darvishi H, Nargesi F, Fadavi AJFS, International T. Ohmic pre-drying of tomato paste. *Food Science and Technology International*. 2014; 20(3): 193–204.

19. Asekun O, Grierson D, Afolayan AJ. Effects of drying methods on the quality and quantity of the essential oil of Mentha longifolia L. subsp. Capensis. *Food Chemistry*. 2007; 101(3): 995–8.
20. Thamkaew G, Sjöholm I, Galindo FG. A review of drying methods for improving the quality of dried herbs. *Critical Reviews in Food Science and Nutrition*. 2021; 61(11): 1763–86.
21. Bhatia S, Al-Harrasi A, Kumar A, Behl T, Sehghal A, Singh S, et al. Anti-migraine activity of freeze dried-latex obtained From Calotropis Gigantea Linn. *Environmental Science and Pollution Research*. 2021; 29(18): 27460–78.
22. Sledz M, Wiktor A, Rybak K, Nowacka M, Witrowa-Rajchert D. The impact of ultrasound and steam blanching pre-treatments on the drying kinetics, energy consumption and selected properties of parsley leaves. *Applied Acoustics*. 2016; 103: 148–56.
23. Hasan Masud M, Karim A, Ananno AA, Ahmed A, Hasan Masud M, Karim A, et al. Conditions for selecting drying techniques in developing countries. In *Sustainable Food Drying Techniques in Developing Countries: Prospects and Challenges*. Springer; 2020, pp. 21–40.
24. Chua K, Chou S. Low-cost drying methods for developing countries. *Trends in Food Science & Technology*. 2003; 14(12): 519–28.
25. Ratti C, Mujumdar AS. Infrared drying. In *Handbook of Industrial Drying*: CRC Press; 2020. pp. 567–88.
26. Salehi F. Recent applications and potential of infrared dryer systems for drying various agricultural products: a review. *International Journal of Fruit Science*. 2020; 20(3): 586–602.
27. Doymaz I. Sun drying of figs: an experimental study. *Journal of Food Engineering*. 2005; 71(4): 403–7.
28. VijayaVenkataRaman S, Iniyan S, Goic R. A review of solar drying technologies. *Renewable and Sustainable Energy Reviews*. 2012; 16(5): 2652–70.
29. Gilago MC, Mugi VR, Chandramohan V. Energy-exergy and environ-economic (4E) analysis while drying ivy gourd in a passive indirect solar dryer without and with energy storage system and results comparison. *Solar Energy*. 2022; 240: 69–83.
30. Sontakke MS, Salve SP. Solar drying technologies: a review. *International Journal of Engineering Science*. 2015; 4(4): 29–35.
31. Ekechukwu OV, Norton B. Review of solar-energy drying systems II: an overview of solar drying technology. *Energy Conversion and Management*. 1999; 40(6): 615–55.

32. Hnin KK, Zhang M, Mujumdar AS, Zhu Y. Emerging food drying technologies with energy-saving characteristics: a review. *Drying Technology*. 2019; 37(12): 1465–80.

33. Zhou X, Wang S. Recent developments in radio frequency drying of food and agricultural products: a review. *Drying Technology*. 2019; 37(3): 271–86.

34. Bhatta S, Stevanovic Janezic T, Ratti C. Freeze-drying of plant-based foods. *Foods*. 2020; 9(1): 87.

35. Shanmugam S. Granulation techniques and technologies: recent progresses. *BioImpacts*. 2015; 5(1): 55–63.

36. Darniadi S, Ifie I, Luna P, Ho P, Murray BS. Foam-mat freeze-drying of blueberry juice by using trehalose-β-lactoglobulin and trehalose-bovine serum albumin as matrices. *Food and Bioprocess Technology*. 2020; 13(6): 988–97.

37. Claussen I, Ustad T, Strømmen I, Walde P. Atmospheric freeze drying—A review. *Drying Technology*. 2007; 25(6): 947–57.

38. Ishwarya SP, Anandharamakrishnan C, Stapley AG. Spray-freeze-drying: a novel process for the drying of foods and bioproducts. *Trends in Food Science & Technology*. 2015; 41(2): 161–81.

39. Goh LJ, Othman MY, Mat S, Ruslan H, Sopian K. Review of heat pump systems for drying application. *Renewable and Sustainable Energy Reviews*. 2011; 15(9): 4788–96.

40. Brown Z, Fryer P, Norton I, Bakalis S, Bridson R. Drying of foods using supercritical carbon dioxide—investigations with carrot. *Innovative Food Science & Emerging Technologies*. 2008; 9(3): 280–89.

41. Witrowa-Rajchert D, Rząca M. Effect of drying method on the microstructure and physical properties of dried apples. *Drying Technology*. 2009; 27(7–8): 903–9.

42. Chegini G, Ghobadian B. Effect of spray-drying conditions on physical properties of orange juice powder. *Drying Technology*. 2005; 23(3): 657–68.

43. Lv H, Lian C, Xu B, Shu X, Yang J, Fei B. Effects of microwave-assisted drying on the drying shrinkage and chemical properties of bamboo stems. *Industrial Crops and Products*. 2022; 187: 115547.

44. Moreira R, Chenlo F, Sineiro J, Arufe S, Sexto S. Drying temperature effect on powder physical properties and aqueous extract characteristics of Fucus vesiculosus. *Journal of Applied Phycology*. 2016; 28: 2485–94.

45. Calín-Sánchez Á, Lipan L, Cano-Lamadrid M, Kharaghani A, Masztalerz K, Carbonell-Barrachina ÁA, et al. Comparison of traditional and

novel drying techniques and its effect on quality of fruits, vegetables and aromatic herbs. *Foods*. 2020; 9(9): 1261.

46. Kabera JN, Semana E, Mussa AR, He X. Plant secondary metabolites: biosynthesis, classification, function and pharmacological properties. *Journal of Pharmacy and Pharmacology*. 2014; 2(7): 377–92.
47. Wallis CM, Galarneau ER-A. Phenolic compound induction in plant-microbe and plant-insect interactions: a meta-analysis. *Frontiers in Plant Science*. 2020; 11: 580753.
48. Raya KB, Ahmad SH, Farhana SF, Mohammad M, Tajidin NE, Parvez A. Changes in phytochemical contents in different parts of Clinacanthus nutans (Burm. f.) lindau due to storage duration. *Bragantia*. 2015; 74: 445–52.
49. Rasul A, Akhtar N, Khan B, Mahmood T, Zaman SU, Khan H. Formulation development of a cream containing fennel extract: in vivo evaluation for anti-aging effects. *Die Pharmazie*. 2012; 67(1): 54–8.
50. King K. Packaging and storage of herbs and spices. In *Handbook of Herbs and Spices*. Vol. 3, Elsevier; 2006. pp. 86–102.
51. Kumadoh D, Ofori-Kwakye K. Dosage forms of herbal medicinal products and their stability considerations—an overview. *Journal of Critical Reviews*. 2017; 4(4): 1–8.
52. Thakur L, Ghodasra U, Patel N, Dabhi M. Novel approaches for stability improvement in natural medicines. *Pharmacognosy Reviews*. 2011; 5(9): 48.
53. Xie P, Chen S, Liang Y-Z, Wang X, Tian R, Upton R. Chromatographic fingerprint analysis—a rational approach for quality assessment of traditional Chinese herbal medicine. *Journal of Chromatography A*. 2006; 1112(1–2): 171–80.
54. Bansal G, Suthar N, Kaur J, Jain A. Stability testing of herbal drugs: challenges, regulatory compliance and perspectives. *Phytotherapy Research*. 2016; 30(7): 1046–58.
55. Dhall R. Advances in edible coatings for fresh fruits and vegetables: a review. *Critical Reviews in Food Science and Nutrition*. 2013; 53(5): 435–50.
56. Ruan C-Q, Kang X, Zeng K. Preparation of water-soluble dialdehyde cellulose enhanced chitosan coating and its application on the preservation of mandarin fruit. *International Journal of Biological Macromolecules*. 2022; 203: 184–94.

57. Xing Y, Xu Q, Li X, Chen C, Ma L, Li S, et al. Chitosan-based coating with antimicrobial agents: preparation, property, mechanism, and application effectiveness on fruits and vegetables. *International Journal of Polymer Science*. 2016; 2016: Article ID 4851730 | https://doi.org/10.1155/2016/4851730.

58. Liu Y, Ahmed S, Sameen DE, Wang Y, Lu R, Dai J, et al. A review of cellulose and its derivatives in biopolymer-based for food packaging application. *Trends in Food Science & Technology*. 2021; 112: 532–46.

59. Song Y, Liu L, Shen H, You J, Luo Y. Effect of sodium alginate-based edible coating containing different anti-oxidants on quality and shelf life of refrigerated bream (Megalobrama amblycephala). *Food Control*. 2011; 22(3–4): 608–15.

60. Kumar S, Bhatnagar TJIJoA, Technology FS. Studies to enhance the shelf life of fruits using Aloe vera based herbal coatings: a review. *International Journal of Agriculture and Food Science Technology*. 2014; 5(3): 211–18.

61. Jafarzadeh S, Nafchi AM, Salehabadi A, Oladzad-Abbasabadi N, Jafari SMJAiC, Science I. Application of bio-nanocomposite films and edible coatings for extending the shelf life of fresh fruits and vegetables. *Advances in Colloid and Interface Science*. 2021; 291: 102405.

62. Marsh K, Bugusu B. Food packaging—roles, materials, and environmental issues. *Journal of Food Science*. 2007; 72(3): R39–R55.

63. Deshwal GK, Panjagari NR. Review on metal packaging: materials, forms, food applications, safety and recyclability. *Journal of Food Science and Technology*. 2020; 57(7): 2377–92.

64. Ghoshal G. Recent trends in active, smart, and intelligent packaging for food products. In *Food Packaging and Preservation*, Elsevier; 2018. pp. 343–74.

Chapter 5

Pharmacology of Natural Latex

Laleh Khodaie,[a,b] Saurabh Bhatia,[c,d] Ahmed Al-Harrasi,[c] and Ajay Sharma[e]

[a]*Department of Traditional Pharmacy, Faculty of Traditional Medicine, Tabriz University of Medical Sciences, Tabriz, Iran*
[b]*Department of Pharmacognosy, Faculty of Pharmacy, Tabriz University of Medical Sciences, Tabriz, Iran*
[c]*Natural & Medical Sciences Research Center, University of Nizwa, Birkat Al Mauz, Nizwa, Oman*
[d]*School of Health Sciences, University of Petroleum and Energy Studies, Dehradun, Uttarakhand, India*
[e]*Department of Pharmacognosy & Phytochemistry, School of Pharmaceutical Sciences, Delhi Pharmaceutical Sciences and Research University, Mehrauli-Badarpur Road, New Delhi, India*

khodaiel@gmail.com

Herbal material fulfills a crucial function in drug design and discovery. Medicinal plants possess versatile pharmacological effects due to their secondary metabolites. The pharmacological survey of herbal latex is a multi-directional approach evaluated in vitro and in vivo. In vitro assays are high-throughput studies with practical laboratory procedures to prognosticate pharmacological potentials. In vitro assays are important to predict the efficacy of new candidate medications. Since the interaction between the substance and the target is complicated and involves multiple

Introduction to Plant-Based Latex: Phytochemistry, Ethnopharmacology, Biological, and Pharmacological Activities
Edited by Laleh Khodaie, Ajay Sharma, Saurabh Bhatia, and Ahmed Al-Harrasi

ISBN 978-981-5129-12-0 (Hardcover), 978-1-003-57071-4 (eBook)
www.jennystanford.com

levels, molecular testing models cannot predict. Thus, several cellular assays are designed for drug-discovery screening applications. Cell lines are suitable and valuable models applied to reach the mentioned targets. Animal in vivo experiments provides stronger evidence regarding the physiological level. Despite developing animal models for drug discovery, scientists witness the failure of drug candidates in clinical trials. Physiological differences influence the results. The results of the mentioned assays alongside each other could predict the biological activity of the herbal products. Different pharmacological activities were demonstrated from latex-producing plants, which were discussed in this chapter, including wound healing activity, antiviral, estrogenic anti-estrogenic, antileishmania, antimicrobial, antioxidant, antihyperglycemic, larvicidal, anthelminthic, anti-inflammatory, proinflammatory, cytotoxic, antitumor, anticancer, antimalarial, homeostatic, cognition-enhancing, and sedative effects. Pertinent to demonstrating various pharmacologic effects, latexes have the potential to be commercialized. However, many active ingredients are yet to be isolated and identified from herbal latexes. Furthermore, various pharmacological activities of latex are yet to be explored. Since latex is an environmentally friendly natural product, it is promising to witness more latex-derived medicines in the near future.

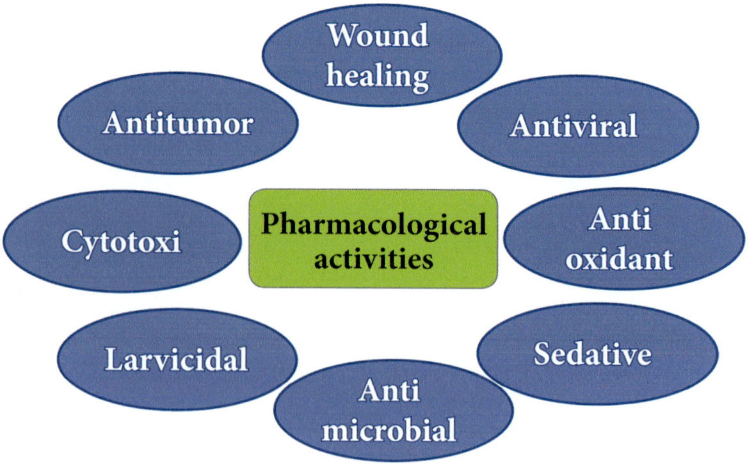

Figure 5.1 Graphical abstract.

5.1 General Introduction

Throughout history, herbs have fulfilled an important function in the life of humankind to alleviate various ailments. Approximately 40% of prescribed medicines are obtained from herbs and herbal products [1]. Even though allopathy has tremendously developed, medicinal plants and their products take center stage to be used as primary health care in many world regions. Due to the mentioned reasons, many researchers have launched a survey on the pharma-cological activities of herbal products. Generally, pharmacological investigations of herbal materials are designed to evaluate their traditional uses. If modern sciences affirm ethnopharmacological uses of herbal material, they will be credentialed to enter higher phases of studies, such as preclinical and clinical trials.

Similarly, the pharmacological evaluation of herbal products assists in transforming herbs from folk to modern medicine. Thus, herbal material is vital in drug discovery and development procedures. Medicinal plants possess versatile pharmacological effects due to their secondary metabolites. Latex exuding from laticiferous plants has shown various pharmacological effects, as a natural and secondary metabolite-rich material. The pharmacological survey of herbal latex is a multi-directional approach evaluated in vitro and in vivo. The biological and pharmacological potential of latex includes antioxidant, antimicrobial, enzyme inhibitory, anti-HIV, cytotoxic, analgesic, anti-inflammatory, anti-diarrheal, etc. This chapter will comprehensively review the pharmacological potential of latex obtained from various medicinal plants.

5.2 In vitro and in vivo Assessments

Drug design and discovery is a lengthy and expensive process that involves preclinical and clinical trials. The preclinical stage includes the sequence of in silico, in vitro, and in vivo assays. In silico studies are computer-based ones to assess the dynamical interaction between the pharmacophore of ligands and 3D structured targets. The simulations by computational tools provide data on the interaction mechanism between ligands and receptors [2]. In silico screening of test chemicals requires further validation to ensure the candidate compound has the potential for pharmacological effect. For the completion of in silico experiments, in vitro studies are carried out.

In vitro assays are high-throughput studies with practical laboratory procedures to prognosticate pharmacological potentials. These models are also useful for comparing the biological activities of candidate agents. These experiments are conducted outside the living organism, namely cell, tissue, or organ-based. In crude terms, these tests are conducted on components of the living organism in a laboratory environment using labware, such as test tubes, plates, etc. These experiments are carried out to decide on the clinical use of drug candidates or proceed with the preclinical drug development process. These studies are advantageous since they are more convenient and cost-effective than experiments on living organisms. In addition, in vitro studies could generate precise, reliable, timely, and reproducible data on candidate agents' safety, potency, and efficacy. In vitro methods could promote our precision from the mechanism of action of drugs, herbal products, and phytochemical compounds. By these methods, cells could be directly exposed to different concentrations of compounds, and manipulation of their physicochemical environment is possible. Also, environmental conditions affecting the cell response could be controlled. The response of cells and sequence of events could be measured in the absence of desired compounds. However, these methods have some drawbacks; properties of test chemicals, such as solubility and volatility, could affect the bioavailability and concentration of the components in the microenvironment.

Moreover, culture conditions could be altered due to factors, for instance, pH, binding to the elements of the culture medium, and osmolality, which could trigger the formation and gradual precipitation of insoluble particulates and nutrients [3]. However, in vitro assays are important to predict the efficacy of new candidate medications. They can facilitate drug discovery by the decrease of exploitation of animals and guidance to drug dosing for clinical studies [4].

In vitro studies fall into cell-free (molecular testing models or biochemical assays) and cell-based. The advantages of cell-free assays are being fast, high-throughput, providing stable and accurate results, and microscale [5]. Cell-free assays are limited to typical targets and provide restricted data about the interactions of test molecules and targets. In contrast, the interaction between the substance and the target is complicated and involves multiple levels, which molecular testing models cannot prognosticate. Thus, several cellular assays are designed for drug-discovery screening applications. By these tests, the responses of cells to special chemicals, for instance, cytotoxicity, biological activity, and biochemical pathways and mechanisms, could be detected and measured. Tumor cell lines are suitable and valuable models applied to reach the mentioned targets. They are immortalized cells derived from different sources, such as human tumors resected during biopsies and surgeries with mutations or chromosomal abnormalities, and divide continuously. Even though they are cost-effective and provide information regarding the functionality of test substances, they offer restricted data compared to testing it on various organ systems [6].

On the other hand, animal in vivo experiments provide stronger evidence regarding physiological levels. Due to several ethical and physical issues with undertaking experiments on humans, model organisms are employed to assess the biological effects of novel drug candidates. These experiments provide information regarding the overall impact of the test substance on a living system. In vitro experiments of drug candidates could give erroneous results and be irrelevant to the results of in vivo tests. Various animal models are employed for in vivo experiments in drug screening; the zebrafish screening model assists in identifying new drug candidates with biological activities and discovering novel pharmacological impacts of existing medicines. Xenopus frogs are other frequently-employed animal models in the process of drug development. Although zebrafish and Xenopus frogs are cost-effective, successful, and extensively used animal models, they differ from mammalian models regarding genetics and metabolic pathways. Dog, rabbit, rat, mouse, mice, guinea pigs, canines, and monkeys are employed as non-human mammalian models to bridge the gap between preclinical and clinical studies and provide data regarding biological processes. Among the mentioned vertebrate models, the mouse is a popular choice owing to genome identity to humans and similar physiological systems. Despite developing animal models for drug discovery, scientists witness the failure of drug candidates in clinical trials. Even though there are genetic similarities between vertebrate models and humans, physiological differences influence the results. Therefore, for discovering effective drug candidates, novel models should be developed for comprehensive understanding and the progress of the drug development process [4].

5.3 Wound Healing Activity

A wound occurs when the integrity and functionality of skin, mucosal tissue, or organ tissue are disrupted. Simple wounds involve skin, but complicated wounds involve body tissues, and internal organs, such as vessels, nerves, and muscles. Skin wounds fall into two groups of acute and chronic. Acute wounds are classified as accidental wounds and surgical ones. Traumas, incisions, excisions, abrasions, and burns trigger acute wounds. Generally, they heal in a sequential and routine process. A timely and orderly manner of wound healing is inflammation, formation of tissue, and remodeling. If any of these phases is prolonged, or proper treatment of an acute wound does not occur, a chronic wound could be formed. In chronic or non-healing wounds, the healing process is impaired, and they do not follow the normal phases of the wound healing process; they remain in the phase of inflammation. They are mostly associated with pathological conditions like diabetes, obesity, and vascular diseases. These wounds give rise to morbidity, mortality, disability, and socioeconomic problems [7]. Hence, wounds exert increasing pressure on healthcare expenses and the economy, especially when prolonged hospitalization period is required.

Even though the biology of wound healing has been investigated for a long time, the precise molecular mechanism has yet to be understood and unraveled. To conduct wound research, unraveling the mechanisms and regulators involved in wound healing is necessary. Various types of cells interplay, such as inflammatory and endothelial cells, platelets, fibroblasts, and keratinocytes which undergo alterations in the phenotype and the expression of their genes. Moreover, cytokines, chemokines, and growth factors contribute to this complicated physiological process. Thus, wound management is challenging due to the numerous factors, a single treatment method and guidelines can be insufficient [8].

Since the skin is a protective barrier of the body against microbes and pathogen entry, the wounded site may undergo microbial infections. As the superficial part of the skin is never sterile, depending on the function of the patient's immune system and the wound's depth and size, microbial colonization and wound infections may occur. Therefore, one of the principal aims of wound-healing investigations is restricting microbial growth and improving local defense against microbes. Besides antimicrobial activity, wound healing agents should have anti-inflammatory, analgesic, and antioxidant properties to assist in wound healing. Prolongation of inflammatory response and extended generation of cytokines and chemokines inhibit injuries to reach the phase of remodeling and synthesis of the matrix. Consequently, wound closure is delayed, and the pain sensation increases. Thus, the anti-inflammatory activities of therapeutic agents are of utmost importance. Hence, inflammation triggers an influx of neutrophils to the injury site, producing free radicals and giving rise to oxidative stress. Thus, antioxidants are considered an integral part of therapeutic agents in wound healing aids. Furthermore, pain sensation is an indispensable part of injuries, which causes discomfort, the release of stress factors, and decreasing quality of life. Therefore, wound healing aids should have analgesic activities [9].

The potential of herbal products and their preparations have attracted attention since modern trends in the healthcare system could not solely address health issues. Typically, wound management, a multifaceted healthcare system issue, further requires the contribution of alternative medicine. Herbs and herbal products were used in traditional and modern medicine since they stimulate the skin's natural repair and regeneration mechanism of skin. Therefore, herbal products captivated the attention of phytopharmaceutical laboratories to isolate and identify active ingredients to assist in wound management [9].

Several reports regarding wound healing activities of herbal latex indicated antibacterial, antifungal, antioxidant, and anti-inflammatory effects. For the assessment of wound-healing activities, various in vitro and in vivo wound-healing models are employed to both validate traditional claims and develop efficient herbal drugs for wounds.

Wound healing is an orchestrated and dynamic process that involves different cell types. In this process, fibroblasts and keratinocytes play a crucial role in wound healing. Keratinocytes are the predominant cellular component of the epidermis, representing the re-epithelialization process by migration, proliferation, and differentiation. Fibroblasts break down the fibrin clot to create an extracellular matrix, a structural and biochemical network comprised of collagen, glycoproteins, and enzymes. Fibroblasts are responsible for tissue contraction in wound healing. The fibroblast in vitro model is an assay for determining the wound healing activity of plant products and phytochemicals. These in vitro assays are useful, fast, and inexpensive to screen various herbal products and phyto-constituents in various conditions. However, these tests can only reprogram some of the determinants involved in complicated wound healing [10].

5.3.1 In vitro Assays of Wound Healing

These assays are useful tools to provide evidence regarding products and differentiate them by their potency. The earliest in vitro and animal model is chick chorio allantoic membrane (CAM) Assay. The assay can evaluate the potential angiogenic properties of test compounds or biomolecules. The CAM assay is developed due to the fusion of allantois and chorion, which is extensively vascularized to make an interface to exchange nutrients, waste, and gases. Since having a dense network of capillaries and functionality, it is exploited to conduct wound healing research. It is a technically simple, accessible, inexpensive, versatile, and fast model that does not require approval from the ethics committee. Regarding this assay's drawbacks, embryo survivability is low [11].

All steps of this experiment should be carried out in a laminar flow hood to maintain sterility. Before experiments, the equipment should be sanitized with 70% alcohol. Nest, a 3-day-old pathogen-free chicken embryonated eggs are selected and a hole cut of 1.0 cm^2 is produced on the eggshell. A sterile, methylcellulose circular disk loaded with a test component is placed on the region where two main vessels join. Then the tape is used to reseal the window, and the eggs are incubated at 37°C in a humidity-controlled chamber for 72 h. After that period, the eggs are opened, and new blood vessels are formed in a membrane treated by test components are observed. The mentioned CAM is compared with the one not impregnated with the test drug. Observing new blood vessels is carried out by taking images via a special camera to evaluate angiogenesis. Morphometric and quantitative analysis of images is carried out by special software [12].

Another laboratory technique to evaluate wound healing is **scratch assay**, in which cell migration, proliferation, and cell-cell interactions are studied. To imitate an injury is done by making a scratch or wound gap on confluent monolayer cells of keratinocytes or fibroblasts. Cells will migrate toward the created wound to close till the establishment of new cell-cell interactions. Cells' motility and growth, which bring about injury healing, are monitored and quantified by capturing images at regular intervals until the scratch is closed. Then, the images are compared to evaluate the rate of migration of cells.

This method has several advantages; it is technically simple, inexpensive, and adjustable for various purposes. It is executable in a laboratory equipped with cell culturing supplies. It imitates cell migration and behavior up to some extent. Also, this method is well-suited to assess the regulation of the migration of cells. Another advantage of this method is its compatibility with time-lapse microscopy to take images of live cells to visualize, track, and analyze intracellular signals. The three main steps of this assay are preparation of cell culture, scratch wound assay, acquisition and analysis of data. As a starting point, a thin layer of epithelial and endothelial cells is grown to obtain a confluent monolayer of cells. Another important component for this test is a coating of extracellular matrix, for instance, collagen, fibronectin, and gelatin, as an adhesive surface for cell types. The stiffness of this component has a great impact on the modulation of the dynamics of the migration. Multi-well plastic cell culture dishes containing 6, 12, or 24 wells. The second step is creating a cell-free gap on the cell monolayer by using a pipette tip, needle, or sharp tool, defined as a scratch assay. Cell-free regions could be created by physically excluding the cells, such as commercial or silicon inserts. The inserts are advantageous over the scratch tools; when seeding, fewer cells are required. Also, the sizes of gaps are more reproducible. The third stage is the image and data acquisition from the biological specimen by capturing images to keep track of the wounds. Once the digital images are acquired, the size of the cell-free area should be measured utilizing image analysis software. For tracking the width of the cell-free area, lines are drawn to measure whether the distance of the lines is decreased to pinpoint the closure of the wound. This phase could be carried out both manually and automatically [13]. Generally, the results of in vitro assays could only partially indicate and recreate the biological process of wound healing in the human body. Moreover, they cannot indicate the interaction between wounded tissue and the wound dressing.

5.3.2 Evaluation of Wound Healing

Various approaches could evaluate the process of wound healing. Wound closure or contraction is a wound healing assessment criterion defined as wound size reduction. Measurement of wound contraction initiates from the day of applying treatment until the day of observing complete epithelialization. The hydroxyproline content, a major component of the protein collagen, could analyze wound tissues. When collagen, a fundamental component of the extracellular network, disintegrates, free hydroxyproline and other peptides are liberated. Thus, the measurement of hydroxyproline is considered a biochemical marker to indicate the tissue collagen content and its turnover [14]. Another key component to monitoring the process of wound healing is **hexosamine**. It is an amino sugar and a biochemical marker since being a crucial substance for the biosynthesis of the extracellular matrix in granulation tissue. After inflicting the wounds, the amount of hexosamine generally surges on day seven and reaches the maximum level on day 12, and gradually decreases. On day 11, the granulation tissue is obtained [15].

Tensile strength is another determinant to measure dermal wound healing. It is related to biomechanical strength and the skin resistance ascribed to the collagen and elastic fiber network. To assess the tensile strength, the force level for splitting the wound is measured. This criterion is reprehensive of the wound healing degree and the strength of wound tissue [16].

The **Percentage of wound contraction** is another indicator of progress in wound healing. It is defined as the decrease of wound area measured by initially tracking the boundaries of the raw wound with a transparent paper in mm^2. This measurement is repeated at 2 to 4 days to record wound healing and contraction progress. Complete epithelialization is deemed as no remaining raw wound. Next, the endpoint of wound healing and incision contraction percentage are measured [16].

Histopathological studies include evaluating the level of re-epithelialization, development of epidermal squamous cells or the flat cells in the outer part of the epidermis, and the level of tissue formation in the wounded area. The mentioned factors are compared with those of normal tissue. Additionally, for qualitative analysis of the remodeling of epidermal and dermal tissue, further factors are evaluated: the level of epithelialization, inflammation, necrosis, angiogenesis, the proliferation of fibroblasts, and the pattern of collagen depositions. Specifically, in burn wounds, histological studies are performed at the end of the experiment or after 25 days. For tissue samples, the wound area is excised; the tissue is processed and stained, then observed under a light microscope to take photomicrographs [17].

5.3.3 Animal Models for Wound Healing

Diverse animal models are used to conduct wound-healing research, whereas small mammals are the proper choices for these investigations, pertinent to multiple reasons. They require less space, water, and food. Moreover, they are accessible, inexpensive, and could be modified genetically. Compared to humans, their wound healing process is accelerated, which lasts several days instead of weeks or months. The commonly used ones are pig and rodent models, pertinent to availability, ease of handling, and being cost-effective. Moreover, mice and rats are favorable models due to the exertion of genetic manipulations to realize the role of a specific gene in a typical biochemical process [18].

Rabbits have been used to conduct wound healing research defined as rabbit ear excisional wound model. The ears of rabbits are wounded by a biopsy punch in which the epidermis, dermis, and cartilage are damaged. In contrast to murine models that wounds heal through a contraction, in this model, healing occurs by re-epithelialization, which is considered a merit of this model. Another advantage of this model is using a small number of rabbits since up to 6 ulcers could be inflicted per ear, which provides enough data [18].

Although these studies do not precisely reflect the wound healing process, they are considered prerequisites for basic studies and development. The developed models for acute wound healing are burn, full thickness, incisional, and excisional. Despite the mentioned developments, acquiring the goals and benefits of this research requires further sensitization of the model. In other words, various species have different anatomical factors, physiological responses, biochemical composition, and processes. The type of receptors in humans is different from the animal species. Typically, in the case of skin, the rodents' skin is covered with fur and attached loosely to the subcutaneous tissue (the deepest layer of skin). In addition, the epidermis and dermis of rodents are more delicate than that of humans. Significantly, rodents' wounds cure by contraction rather than re-epithelialization, while in humans, re-epithelialization, granulation, and tissue formation are the basic healing steps. Thus, the wound healing pattern in animals considerably differs from that of humans. Among the animals, pigs are the closest to humans regarding the biology and skin structure.

To illustrate, the epidermis-to-dermis tissue ratio and the skin function are similar in pigs and humans. For instance, pigskin is not covered with dense hair. There is a similarity in the duration of epidermal turnover, the type and composition of keratinous proteins, and lipids in the outer layer of the epidermis of the two species. In the case of blood vessels, their size, and orientation are identical. Moreover, in pigskin, myofibroblasts, responsible for wound contraction, partially cause wound healing, while re-epithelialization is the major process in wound healing which resembles humans. Yet, pigs are pricier than rodents. Although their healing pattern is similar, there are anatomical differences between the skins of the two species [18].

Several models and species should be selected to evaluate wound healing effects and obtain a precise understanding. That is due to the differences in the anatomy and physiology of animals and humans. Thus, humans are the optimum species for understanding wound-healing cascade to avoid inaccurate understandings and interpretations. Conducting both laboratory and clinical studies is necessary [18].

The drug administration route greatly affects the efficiency and speed of wound healing since the drug delivery route to the target tissue influences therapeutic effects. Two routes of drug administration include topical and systematical, that systematical administration encompasses oral and parenteral routes.

Surgical interventions are required to create the wounds under localized or general anesthesia and sterile conditions. First, hair should be removed from the back of the animal. Next, anesthetized animals are placed on the operation table. Several wound models are employed for wound healing studies [18].

In Excisional Wound Model (EWM), the wound area is prepared with 70% ethanol. Wounds are inflicted on depilated back of the animal with an area of about 200–500 mm^2 diameter and a depth of 2 mm. Both oral and topical administration of drugs could be carried out for therapeutic intervention. The wounds will be checked at regular intervals till observation of wound closure, re-epithelialization, and no residue of the raw wound. In addition, further parameters, such as estimation of collagen, percentage of wound contraction, and duration of epithelialization could be studied [18].

Incisional wound Model (IWM): the wound area is prepared by ethanol 70%. Two longitudinal incisions are inflicted on the skin and cutaneous muscles of two sides of the vertebral column. The length of each incision is 4–6 cm stitched with sutures till tightening of the edges of the wound and its closure. Oral or topical treatment could be administered from the first day of operation till 7 to 9 days. Afterward, the sutures are removed to assess the strength of the healed wounds [18].

The Dead Space Wound Model (DSWM) studies granuloma tissue in terms of changes in physical and mechanical properties. A pouch is made through the infliction of a small nick on the skin of either side of the axilla or lumbar region. The sterilized glass cylinder is introduced into the pouch, and the wounds are sutured. For ten consecutive days, either oral or topical therapeutic medication is administered. Afterward, granuloma tissue is studied to determine wound tensile strength [19].

In Burn Wound Model (BWM), under anesthesia and aseptic conditions, thermal wounds are created by a metal plate heated up to 60°C on the back skin of rats. The plates were kept in that area for 30 s to induce a partial thickness burn injury. To create second-degree wounds, 90°C metal plate will be placed on the dorsal area for 10 s. A 100°C metal plate is applied to the mentioned area for 30 s, for full-thickness burn injuries. After the simulation of the burn injuries, oral or topical treatment is administered. The factors, such as the percentage of wound contraction, hydroxyproline content, and duration of epithelialization will be assessed.

5.3.4 Impaired Wound Models

These models aim to replicate the condition in which nonhealing wounds are developed. These models provide useful data regarding delayed wound healing. The frequently observed conditions that cause impaired wound healing are diabetes, malnutrition, and obesity.

Impairment of diabetic wound healing is due to multifactorial reasons such as irregular inflammatory processes and impairment in the natural formation of new blood vessels. Experimental type 1 diabetes wounds is induced by chemicals or the administration of some drugs, for instance, alloxan or streptozotocin. They are both toxic to pancreatic beta cells giving rise to necrosis and insulin deficiency. Even though this procedure is not thoroughly representative of diabetes in humans, it provides a simple and cost-effective model to acquire data. Insulin resistance and beta cell failure increase the possibility of cutaneous ulcerations.

A growing demand exists for the nutrients assisting in wound healing. Protein-Energy Malnutrition or PEM triggers impairment in restoring ulcers' functional and structural integrity. It causes a delay in the healing process due to several factors, such as alteration in immune function, inflammatory responses, and surging pro-inflammatory cytokines that give rise to the higher possibility of infection. Histological studies on animals assigned to the experimental model of PEM indicated that malnutrition caused a lower percentage of collagen and thinner dermis in malnourished animals than in well-nourished ones [18]. For instance, two groups of rats were assigned to a model to investigate the impact of protein malnutrition on wound healing. In contrast, 8% and 17% of the casein diet were allocated to the malnourished and nourished ones, respectively [20].

Metabolic diseases, such as obesity, are systemic factors to surge the level of ROS or reactive oxygen species that affect the process of wound healing. Due to the prolonged disturbances in the inflammatory phase and angiogenesis process and a decrease in the granulation tissue, epithelialization is delayed. Furthermore, altering the cellular environment gives rise to oxidative stress and cellular changes. In experimental models, a high lipidic diet is utilized for animals to develop and replicate animal metabolic changes. Thus, the healing process of chronic wounds could be investigated in patients suffering from obesity or other metabolic disorders. In this model, two groups are assigned, which receive normal and high-fat diets successively for 45 days. In hyperlipidemic rats, elevated triglycerides, total cholesterol, and blood glucose are the determinants to indicate metabolic disorders. In addition, reduced hydroxyproline and glutathione (GSH) content and increased levels of malondialdehyde (MDA) are indicators of oxidative stress, hyperlipidemia, and metabolic diseases. Afterward, excisional wounds are created on lateral parts of the vertebral column to investigate the process of wound healing in animals with the dysmetabolic condition [21].

5.3.5 Clinical Studies

In the case of acute wound pathology, clinical trials provide an adequate understanding of the wound pathology in contrast to chronic ulcers. Clinical trials provide in-depth identification, and understanding of the pathology and physiology of injuries, the wound healing process and its mechanisms. However, the clinical study of human wounds has encountered some challenges. Various underlying wound etiologies exist, and some ulcers, such as pressure injuries, have more complex etiologies. Various causative factors which lead to varying clinical selections are moisture, immobilization, chronic vascular insufficiency, diabetes, malnutrition, malignancies, autoimmune, and infectious diseases. Typically, the patient's age immensely affects the wound-healing process since the etiology of wounds is different in youngsters and the elderly. Thus conducting research with a coherent study group suffering from similar ulcers is a cumbersome task. To address these challenges, more participants are required, which gives rise to several practical difficulties [22].

A common procedure for wound healing study is partial thickness injury. It is a superficial one that involves the epidermis and the superficial part of the dermis. In this model, the dermis and dermal blood vessels are not damaged. There are several approaches to conduct this research [23].

Tape stripping is a technically simple, standard method, and a common practice. In this method, an adhesive tape impairs the skin by removing the stratum corneum. This model is performed by applying multiple strips on a selected part of the skin to remove its upper layers. It causes loss of trans-epidermal water, activation of the repair process of the epidermis, and proliferation. Various methodological criteria affect this procedure; therefore, standardization of this procedure is mandatory before employing it. These factors include the number of adhesive tapes, pressure applied on the skin while sticking the tape, the stickiness of the tape strips, direction, and velocity while removing them. Besides methodological determinants, other factors influence the result; the body site and the quantity of removed skin, skin hydration, cohesion among the cells, and most importantly, inter-individual response variability [23].

The suction blister model is another well-standardized and controlled procedure to assess wound healing. A blister cavity is filled with interstitial fluid when the epidermis and dermis are separated. Mechanical devices and heat could separate the epidermis from the dermis to form blisters for examination purposes. Various chemical and biological vesicants could also be used to induce blisters. Afterward, the roof of the blisters is excised to obtain open wounds in which partial closure occurs on day four and complete healing on day 26 without leaving scars. In clinical studies, this approach is used to measure various factors effective in acute wound healing using quantitative image analysis. These images determine the loss of transepidermal water and wound area before and after the intervention. By this research tool, migration, proliferation, and differentiation of keratinocytes, or in other words, the kinetics of healing, could be assessed [23].

The abrasive wound model is more invasive, about the induction of uniform abrasions to assess the healing characteristics of various products. In this procedure, superficial and standardized abrasions are inflicted on disinfected skin by scrubbing it with a surgical brush until glistening and bleeding are observed. The entire epidermis layer is removed since their cells are loosely attached, and the basal membrane will be kept intact. In this approach, the depth of wounds resembles the suction blister model in which healing occurs without leaving any scar. Since the infliction of wounds triggers less pain, anesthetics are not required. The major merit of this approach is that it is similar to superficial and minor ulcers. Typically, the efficiency of topical preparations is evaluated by this approach. Moreover, this method could assess topical antimicrobial preparations' efficacy [23].

Wounds could be induced by laser and microderm abrasion, which are wound research tools. Both techniques are advantageous as they are highly reproducible, and ulcers with various depths could be formed. Furthermore, the wounds inflicted by these methods are superficial compared to blister and abrasive wound models. However, induction of wounds by laser and microderm abrasion is not cost-effective compared to the other procedures [23].

Another approach is **split-thickness injuries**, in which a dermatome, a surgical instrument used to cut the skin surface, is utilized to cut the skin at a specific depth. This model removes a layer of 100–1500 microns involving the epidermis and upper dermis. This model is employed to determine re-epithelialization and age's impact on wound-healing. Moreover, data provided by this model reveals the genes contributed to re-epithelialization [24].

In **full-thickness wound models**, the epidermis is removed completely, and the dermis is removed to the deepness of subcutaneous fat to disrupt blood vessels of the dermis. Various surgical instruments, such as dermatome, scalpel, biopsy punches, and laser, could inflict a standardized lesion. As an outstanding research tool, it provides comprehensive data involving all the components contributing wound-healing. The rates of epithelialization, involvement of dermal components, and scar formation are investigated. Furthermore, the macroscopic healing rate and granulation tissue formation could be assessed. Taking biopsies of the wound site assists in assessing collagen and proteoglycans content, angiogenesis, cell population, and histological structure of connective tissue. Typically, wounds inflicted by punch biopsy are useful for assessing wound healing in human and animal models since their depth and size can be managed [24].

Angiogenesis or vasculogenesis is a physiological process leading to the formation of new blood vessels, an important event of wound healing. When an injury occurs, hyperpermeability is induced by microvascular endothelial cells, an initiator for the angiogenesis process. Endothelial cells form a thin layer lining the interior surface of blood vessels that regulate vascular leakage and vasculogenesis. Newly constructed blood vessels supply various cytokines and growth factors for injured tissue. Also, angiogenesis provides nutrients and removes unwanted metabolites. The natural product is examined to reveal whether or not they induce vascularization, accelerating re-epithelialization and ulcer repair. A well-established, simple, commonly used in vivo approach, defined as Miles assay, examines vascular leakage. This technique is reliable for studying the activity of substances modulating permeability and identifying mediators improving vascular leakage. Miles, who developed this technique, used guinea pigs to perform this test; however, in recent years, mice have been a good alternative to clarify the molecular mechanism responsible for regulating vascular leakage. Intravenous injection of blue Evans dye is carried out in the organism of choice and permitted to spread for 30 min. Solvent control, as well as natural substance inducing permeability, is injected in several sites of the dermis. Blue Evans dye reversibly binds to the serum albumin, forming a complex that extravagates and assembles in the dermis. For calculating the level of vascular permeability, the dye is extracted from the dermis by formamide, and the absorbance is measured at 620 nm [24].

5.3.6 Microarray Studies of Wound Healing

Orchestrated and synergistic processes, including the kinetics of cells, enzymatic activities, and neurovascular functioning, lead to the physiological restoration of skin. Molecular mechanisms play a crucial role in controlling this sequential process. However, the molecular mechanisms of skin restoration remained unclear and puzzling. Techniques based on DNA and RNA assist researchers in unraveling the process's molecular basis. As PCR and microarray techniques require trivial skin, they could be easily acquired from humans. They make it possible to assess the expression alteration of various skin genes during injuries and trauma. For this purpose, donor wound sites selected as models should be clean, consistently healing, and have standard depth. Biopsies are taken in typical intervals to represent various stages of the wound healing process. The first biopsy is taken from the unwounded skin, and the subsequent biopsies are obtained from the wounded skin in the acute phase on days 3, 7, 14, and 21. Thus, the impact of therapies on the process of skin restoration and alteration of gene expression could be evaluated. Thousands of genes are involved within the first three weeks of restoration. Molecular studies indicate the type and the most dominant genes associated with the healing process. Collecting and managing the samples for microarray studies and sequencing is very precise and challenging. A well-instructed

Table 5.1 The scientific names and responsible products of latex-producing herbs, and their assessment models for wound-healing activity, and the results

Scientific name	Latex product	Responsible phytochemicals	Assessment model	Results	Refs.
Aloe megalacantha	Topical application of 5% and 10% w/w simple ointment of the leaf latex	Flavonoids and tannins	**In vivo:** excision and incision wound models in mice	A significant increment in the rate of wound contraction, reduction in epithelialization time, and higher skin breaking in both concentrations.	[26]
Hevea brasiliensis	–Intradermal injection of latex serum[1] in PBS[2] for Miles assay –Topical application of serum in CMC[3] for rabbit ear dermal ulcer model	Polypeptides of serum	**In vivo:** Miles vessel permeability assay **In vivo:** The rabbit ear dermal ulcer model **In vitro:** Chorio Allantoic Membrane (CAM) assay	–Significant enhancing the microvascular permeability –In dermal ulcers, this material significantly accelerated wound healing –Significant increase in growth of the blood vessels.	[27]
	–Application of 1%, 0.1%, and 0.01% of serum for MTT assay –Topical application of 1%, 0.1%, 0.01% of serum gel/ cream for abrasion model	Polypeptides of serum	**In vitro:** MTT and scratch assays **In vivo:** skin abrasion wound model in rats	–Viability at 1% and 0.1% and migratory/ proliferative activity at 0.01% concentrations –Significantly higher re-epithelialization, the lower thickness of crusts, and greater extent of epidermal layers at 1% concentration.	[28]
Euphorbia neriifolia	Ingestion of 100, 200, and 400 mg/kg leaf ethanolic extract of latex	Saponins and flavonoids	**In vivo:** excision and dead space wound methods	Increasing the tensile strength, DNA content, epithelialization, and angiogenesis, improving hydroxyproline and protein content and wound contraction, increased catalase and decreased superoxide radical scavenging activity in granulation tissue at 200 and 400 mg/kg.	[29, 30]

Scientific name	Latex product	Responsible phytochemicals	Assessment model	Results	Refs.
E. nivulia	–I.P. injection of crude cysteine protease for reduction of wound area –0.1 mL of the crude cysteine protease for in vitro evaluation of coagulation time –1000 µg/mL cysteine protease placed on the cut portion to evaluate bleeding/clotting time	Glycosylated cysteine protease	**In vivo**: Excision Wound Model in mice	–A significant decrease was indicated in the wound area in the group treated by the enzyme. –Cysteine protease significantly reduces the bleeding and clotting time in mice up to 8 and 14 s.	[31]
E. caducifolia	**In vitro:** 10, 20, 30, 40, and 50 mg/ml of latex to evaluate its effect on hemostasis **In vitro:** 2.5, 5.0, and 10 mg/ml of latex to evaluate angiogenesis **In vivo:** topical application of 2.5, 5.0 and 10 mg/g of latex in 100 g of simple ointment	Synergistic effect of PUFA*, methylpalmitate, amino acids, terpenes, and phytoestrols in latex	**In vitro:** Evaluation of hemostasis by clotting of platelet-free plasma **In vitro:** Chorio Allantoic Membrane (CAM) assay to assess angiogenesis **In vivo:** Excision and incision wound models	–A significant decrease in clotting time, and significant increase in angiogenesis in 2.5, 5.0, and 10 mg/ml treated groups. –A significant increase in hydroxyproline and DNA content in 0.5 and1.0% treated mice. –A significant acceleration of the wound closure in mice treated with 2.5, 5.0 and 10 mg/g latex	[32]
E. tirucalii	10 mg/mL of latex aqueous solution	NM	**In vitro:** Chorio Allantoic Membrane (CAM) assay in fertilized chicken eggs to assess angiogenesis	10 mg/mL latex aqueous solution increased CAM vascular network area and activated the inflammatory response and formation of new blood vessels	[33]

(*Continued*)

Table 5.1 (*Continued*)

Scientific name	Latex product	Responsible phytochemicals	Assessment model	Results	Refs.
Calotropis procera	Topical application of 20 µl/wound of 1% sterile aqueous solution of the latex prepared in propylene glycol	NM	**In vivo:** excisional wound model on the back of guinea pigs	By increasing hydroxyproline (48%) and DNA (23%) and protein synthesis, (58%), and epithelialization led to a significant reduction in wound area owing to epithelialization twice daily within 7 days.	[34]
Carica papaya	Topical application of 1.0% and 2.5% carbopol gels from the latex of unripe fruit	Potentially, papain and chymopapain	**In vivo:** mice burn model.	A significant increment in the hydroxyproline content, faster wound contraction, and significant reduction of the epithelialization time.	[35]
C. candamarcensis	Topical application of 0.01%, 0.1%, 1.0% and 1% (w/w) of latex of unripe fruit in the hydrosoluble base	Potentially, cysteine proteinases	**In vivo:** heat-induced, third-degree burn on mice	–0.01% and 0.1% (w/w) latex formulation enhanced epithelialization –0.1% latex formulation significantly increased healing compared to the, respective, control on days 25–27.	[36]

[1]hydrophilic component of latex or protein fraction
[2]Phosphate Buffered Saline
[3]Carboxy Methyl Cellulose
[4]Not Mentioned

surgeon is required to take the samples. A proper storage procedure is required to ensure the quality of RNA. The location of the wound used for obtaining the samples is of great importance since it influences the result. A sample taken from the center of the wound site differs from one taken from the edges of the wound; therefore, samples should be obtained from the same wound area [25].

5.3.7 The Examples of Herbal Latex with Wound Healing Activities

Table 5.1 presents the scientific name of the plants, the latex product and its route of administration, the responsible phytochemicals, the in vitro and in vivo assessment models, and the research results.

5.4 Antiviral Effects

Medicinal plants, their products, and their active ingredients offer possibilities for anti-viral effects. Thus, they are assessed to determine novel antiviral agents. In contrast to antimicrobial studies, antiviral investigations are insufficient. Therefore, in vitro and in vivo antiviral tests should be performed against DNA and RNA, enveloped or non-enveloped viruses to identify lead molecules as antiviral agents. Pertinent to the growing prevalence and the rapid appearance of resistant strains development of new therapeutic agents seems necessary. This requirement is much more urgent, especially in patients susceptible to infections or immunocompromised ones [37].

Plaque reduction assay determines whether a virus is efficient in forming plaques in the presence of various concentrations of test chemicals. In crude terms, normal cells could take up reactive dyes unless viruses infect them. In that case, they will be unable to absorb the dye, and they form colorless plaques. Therefore, this assay determines the level of virus performance to form plaques. The formation of visible plaques varies in different viruses and depends on the viral replication kinetics varying from 24 h to several weeks. This experiment is undertaken on 6 or 24 well plates to accurately score the plaques, as in plates with small holes accurately, the readability and resolution of the plaques will be affected [37].

5.4.1 Cell Culture and Viruses

For virology experiments, Vero cells, a type of mammalian cell line derived from the kidney of a normal African green monkey, are extensively used. Several licensed lines of Vero cells were acquired from the same source, commercially available, and could produce several viruses. Initially, they are grown in a special culture medium. Various virus strains are prepared to infect Vero cell cultures. After undertaking freezing–thawing cycle on the infected cell cultures three times, they are centrifuged, and the supernatant is filtered, collected, titrated, and stored at a very low temperature. The assay is performed on 96-well plates with flat bottom carrying cell monolayers. After making a ten-fold dilution of viral suspension, the culture medium will be discarded, and inoculation of cells with viruses will be carried out. Inoculated monolayers will be incubated in special conditions for several days, and the plates containing medium will be employed as control. Over time, host cell structural changes will be observed due to viral infection, defined as cytopathic effect (CPE). Lysis, death of the host cell, or its inability to reproduce correspond to CPE. In 24 h, the number of wells with CPE should be recorded. The evaluation will be expressed as $TCID_{50}$ or DL_{50}, corresponding to dilutions of the viral suspensions giving rise to 50% cell degeneration [38].

5.4.2 Antiviral Activity Assays

There are several sites of action for test chemicals to apply their antiviral effect. Therefore, three types of tests could be designed; the first assay is a virucidal test, which aims to indicate the capability of the test compound to disrupt the structural integrity of the virus and prevention of exerting infectious process. In this assay, the viral suspension of an assigned density (10 $TCID_{50}$ per milliliter (v/v)) and different concentrations of the test compound are mixed before the cells are inoculated [39].

The second assay is the pretreatment of cells that the virus and test compounds contact before adsorption of the virus to the cells. It evaluates whether the test compound could associate with cellular receptors to prevent the virus from attaching or penetrating the cells. Thus, this assay evaluates the capability of the test compound to act on cellular receptors [40].

Third, assays to evaluate intracellular replication, in which viral suspension and cells are mixed, and the test compound is added. Therefore, the virus could attach and penetrate the cell, and intracellular replication initiates. On the other hand, if the multiplication of the virus is blocked, it illustrates that the virus replication is inhibited by the test compound [41].

5.4.3 Anti-HIV Assays

HIV-1 and HIV-2, or Human Immunodeficiency Viruses 1 and 2, are retroviruses belonging to the genus of lentiviruses, defined as slow viruses. Having zoonotic origin, they originated from SIV or simian immunodeficiency viruses of primates. They are roughly 100 nm in diameter and have a lipid envelope where transmembrane glycoproteins, gp41 and gp120, are attached. These glycoproteins cause the virus to fuse to the membrane of the host cells and its genome gains entry into the host cell's genome and replicates inside the immune system cells, and transcription and production of viral mRNA occurs. In the host cell, the structural proteins of HIV are biosynthesized and assembled. The main receptors that HIV attached to them exist on CD4-positive T (helper) lymphocytes, macrophages, and microglial cells. Consequently, the patient's immune system does not work properly, hindering its ability over infections and ailments. Typically, the depletion of CD4+ cells gives rise to a malignancy defined as AIDS, which causes the onset of opportunistic infections. It is one of the main medical issues resulting in disability and mortality. In HIV infection, there is a latency period, and there are no symptoms of disease owing to a gradual reduction in the number of CD4+ T cells and a slow surge in the amount of HIV particles [42].

As a fatal ailment, enormous efforts were made to introduce new antiretroviral medicines and vaccines to prevent and cure this fatal disease. The main approach for antiretroviral therapy is the inhibition of protease and reverse transcriptase. However, although this strategy has decreased mortality from HIV infection, the therapies have enormous adverse effects. Hence, long-term treatment triggers drug tolerance. Furthermore, antiretroviral therapies are costly, especially for patients from developing countries. Therefore, novel approaches are required to introduce new therapies, typically the ones acquired from natural reservoirs pertinent to their diversified structure [42].

Evaluation of anti-HIV medications could be conducted experimentally using diverse cell lines derived from human leukocytes. In drug discovery, animal models play a crucial role in the preclinical assessment of medicines and vaccines. Moreover, they assist in gaining perspective on transmission, pathobiology, and progression of HIV infection. For this purpose, the virus could infect various animal species to develop animals with immunodeficiency induced by HIV [42].

However, there are obstacles to developing AIDs syndrome that resembles humans. Different host factors are limiting the estimation of the efficacy of drugs in animal models; the capacity of viruses is diminished for infectivity in animal species, HIV replication is not efficient enough, and the protein response is different. Since HIV is a pathogen-specific to humans, the experimental models could be limited to other lentiviruses, such as SIV. Chimpanzees could be exploited for the research of HIV infection due to evolutionarily being close to human beings. However, SIV infects chimps rather than HIV, and exploiting chimps for research is not cost-effective. Furthermore, SIV infects macaques, and their disease course resembles humans infected with HIV. As there are multiple restrictions to the replication of HIV in other species, there are inadequate animal models to investigate the biology of HIV and its interaction with the host cells [42].

Murine HIV models are extensively used to evaluate the safety and efficacy of novel medications. Genetically engineered rodents could express CD4 receptors and chemokine co-receptors that HIV-1 binds to them and gets entry into the host cell. If the coding region of gp120 is replaced with gp80, an altered virus is obtained. This clone of HIV-1 infects the cells of mice but does not infect human cells. This model is highly adopted in HIV infection research, while it does not simulate the progression of the disease seen in humans [42].

Murine HIV models are humanized mice or humice models, which are versatile experimental tools for studying the biology and interactions of HIV with host cells. Humanized mice are of different types and are immune deficient creatures containing various types of immune cells of humans, including b, T, dendritic, NK cells, and monocytes/macrophages. Humice can replicate HIV 1–2 in the human immune cells in the periphery and tissues. Human immune cells in the humice can ignite and develop innate and adaptive immune responses. Humanized mice are prone to be infected by HIV by the same mechanisms to establish infection in humans. HIV could be transmitted to humice by oral, parenteral, vaginal, and rectal exposure. HIV-Infection could be detected in peripheral blood plasma by viral load testing indicating continuous depletion of CD4+ T cells in tissues and peripheral blood. One of the applications of humice is that antiretroviral therapy used for humans could be employed for them since they replicate the human immune responses over HIV. If HIV treatment is successful, systemic recovery of the CD4 T cells occurs. The other applications of humice are analysis of mechanisms related to immune system activation, transmission, and prevention of HIV. All these considered, there are obstacles in employing humice in HIV investigations. A small plasma volume could be obtained from the animal for the viral load test. For ex vivo functional analysis, the number of peripheral blood cells is limited, and their life span is short [42].

5.4.4 The Examples of Herbal Latex with Antiviral Activity

Table 5.2 presents the scientific name of the plants, the latex product and its route of administration, the responsible phytochemicals, the in vitro and in vivo models of assessment, and the results of the research.

Table 5.2 The scientific names and responsible products of latex-producing herbs, as well as their assessment models for antiviral activity, and the results

Scientific name	Latex product	Responsible phytochemicals	Assessment model	Results	Refs.
Ficus carica	Latex of unripe fruits	NM[1]	Western blotting technique	Fig latex considerably downregulates the expression level of HPV oncoproteins.	[43]
—	Hexane and hexane–ethyl acetate extracts obtained from the latex of unripe fruit	Ferulic acid	In vitro antiviral assay against herpes simplex type 1 (HSV-1), echovirus type 11 (ECV-11), and adenovirus (ADV).	Hexane and hexane–ethyl acetate extracts prevented multiplication of DNA viruses (HSV-1 and ADV) and RNA viruses (ECV-11).	[41]
—	Latex of unripe fruits	Latex enzymes	A clinical trial, application of fig latex on warts and comparison of results with standard cryotherapy.	Latex from the fig fruit indicated antiwart activity due to the proteolytic and keratolytic effects, but its efficacy was less than cryotherapy.	[44]
Euphorbia umbellata	Ethanolic extract of latex obtained from the trunk of plant	Deoxi-phorbol esters	in vitro bioassay using infected and resting CD4+T cells, and primary non-human primate CD4+T cells	They increase HIV transcription and reactivate virus from latency models.	[45]
	4-deoxy-4β-phorbol-12-tiglate-13-phenylacetate	4-deoxy-4β-phorbol-12-tiglate-13-phenylacetate	In vitro bioassay using infected T-cell lymphoblasts	HIV-1 latency reactivation	[46]
E. officinarum	Ingol diterpene; 8-methoxyingol 7,12-diacetate 3-phenylacetate	Ingol diterpene; 8-methoxyingol 7,12-diacetate 3-phenylacetate	In vitro bioassay using Jurkat T cell line	The isolated compound induced cell-cycle arrest and HIV-1-LTR promoter activation and could represent therapies against HIV-1 latency	[47]
Calotropis gigantia	Latex from aerial parts	Lignan glycoside	In vitro bioassay against anti-Influenza virus	Lignan glycoside from the latex showed inhibitory activity against H1N1 and H3N2. It showed a stronger inhibitory effect from ribavirin.	[48]

[1]Not Mentioned

5.5 Assessment of Estrogenicity and Anti-Estrogenicity

Several in vitro assays assist in determining the estrogenic activity of test compounds. In vitro assays are developed on the basis of understanding estrogen's molecular mechanism of action. The activity of estrogenic compounds depends on estrogen receptors (ER) grouped as nuclear receptors. Regarding ER's molecular interaction and actions, when a ligand with estrogenic activity binds to it, dimerization occurs, and the homodimer complex indicates high affinity to specific sites of DNA defined as ERE or Estrogen Response Elements. These sequences are located in the promotors or regulatory regions of genes induced by estrogens or estrogen-like chemicals. Thus, binding of homodimer complex to ERE is a prerequisite for recruiting transcription factors to initiate and modulate transcription of the genes responsible for cell growth, function, or further physiological and pathological activities. Transcribed mRNA involves the synthesis of specific proteins which bring about estrogenic effects in special tissues. For example, estrogenic stimuli could give rise to a proliferation of cells and stimulation of tumor growth. There are several in vitro methods to assess the estrogenic activity of the phytoconstituents. These assays are categorized into three groups; competitive binding affinity of test compounds to ER is measured. Second, gene assays determine the transcription and translation activity of chemicals binding to ER. Third, cell proliferation tests measure the surge of target cell numbers during the proliferation phase [49].

5.5.1 ER Binding Assays

The basis of this well-established and widely-used in vitro test is a measurement of the competitive binding of test compounds to ER. In other words, in this competitive hormone-binding assay, the interaction between ER and ligand is determined. It is based on the quantification of the capability of the test compound to bind to ER in competition with radiolabelled 17β-estradiol. Receptors are obtained by extraction of diverse vertebrates or mammalian tissues. To illustrate, cytosol or cell extract is obtained from rat uterine using high-speed centrifugation. They are incubated with radiolabelled 17β-estradiol and different dilutions of unlabeled test chemicals. As long as test chemicals could compete with radiolabelled 17β-estradiol in receptor binding, radiolabelled 17β-estradiol will be replaced by test compounds. There are various methods to separate free estradiol from bound ones for quantification. Test chemicals are not required to pass a cell membrane to achieve the receptors. There are also non-radioactive techniques, for instance, florescent polarization and enzyme-linked receptor assays (ELRAs). Results are demonstrated as IC50, a measure for drug potency or a concentration of test compound, at which there is a 50% decline in the estradiol binding to the receptor. Although these assays determine whether a test compound could be a ligand for estrogen receptors, they could not characterize whether a compound is an agonist or antagonist. Furthermore, in these assays, the capability of test compounds cannot be determined in terms of initiation of the biochemical cascade, such as gene transcription and synthesis of protein. Since these assays are cell-free: thus, physical characteristics of test chemicals in terms of entering the cells cannot be determined [50].

5.5.2 Reporter Gene Assays

These assays could measure a biological response by studying gene regulation and transcription of target cells. These tests determine whether a test chemical can stimulate transcription dependent on estrogen receptors. The results of these tests further demonstrate the estrogenic activity of the test chemicals. For this purpose, genetically-modified mammalian cells and yeast strains are employed. Transfected cultured cells are applied to undertake reporter gene assays produced by introducing DNA plasmid carrying special nucleic acids (reporter gene sequence), which could express proteins that influence transcription. Assays could be carried out employing T47D and MCF-7 cell lines containing endogenous ER or by the ones lacking endogenous ER, for instance, yeast cells or HeLa cells. The reporter genes in human cancerous cell lines and yeasts code for luciferase and β-galactosidase, respectively. Test chemicals enter the cells and bind to the ER receptor, which triggers activation and alteration of the receptor's conformation. The activated ER complex interacts with soluble cell components to yield a complex interacting with ERE on the reporter plasmid. This interaction leads to reporter gene expression and enzyme production. If the substrate has chemical interaction with a newly synthesized enzyme, the reaction products could be easily detected by spectrophotometric methods or by assays based on antibodies. For comparison of the relative potency of the test chemicals as estrogenic agonists, their capability of inducing reporter genes should be compared with that of reference estrogen. For this purpose, after the acquisition of dose-response data, the EC50 of the test chemical is obtained and compared with that of the reference substance. In the test group, the reference estrogen and test compounds will be exposed to the cells to evaluate substances as estrogen antagonists. Cell cultures should be exposed to the reference estrogen in the negative control group. The induction of reporter gene expression should be compared in two groups of cell culture or in crude terms in the presence and absence of the test chemical to evaluate its estrogen antagonism [51].

5.5.3 Yeast-Based Reporter Gene Assay

Saccharomyces cerevisiae (bakers' yeast) strain has been broadly employed to assess receptors' structure and function and test substances' activity. There are no endogenous steroid hormone receptors in yeast cells; however, recombinant yeast strains are used as estrogen-responsive biosensors for rapid screening of the estrogenic activity of the test chemicals. In crude terms, the genome of Saccharomyces cerevisiae is integrated with the human ER gene and a vector containing a reporter gene coding for the enzyme β-galactosidase. If estrogen or other active substances interact with receptors, the complexes cause reporter gene transcription. To obtain a dose-response curve, calibration, and quantification are carried out by dilution series of the reference standard, 17β-Estradiol. The transfected yeast is incubated with the test chemical for about three days. Suppose the ligand is bound to the receptor, and the receptor is activated. In that case, gene expression occurs, and β-galactosidase is produced and secreted into the culture medium, metabolizes the chromogenic substrate, and induces a color change from yellow to red. Spectrophotometric tools measure the intensity of the red color. The estrogenic activity of the test sample is compared with the reference standard. Anti-estrogenic activity of the test compound is assessed by addition of 17B-estradiolto the culture medium at a concentration to yield a sub-maximal response. The substance's capability is determined by preventing the induced color change by 17β-Estradiol [52].

5.5.4 Yeast Estrogen Screen (YES)

Yeast strains as eukaryotic organisms are advantageous compared to other systems. They have commonalities with mammalian cells regarding cellular features, signaling pathways, and molecular mechanisms. Also, they are cost-effective due to rapidly growing in an inexpensive culture medium. They endure various temperatures and can be effortlessly transported, conserved, and frozen. As non-animal models, they do not impose any ethical challenges to screen therapeutic test substances. Above all, these favorable biosensors are devoid of the receptors of endogenous steroid hormones. Therefore, there would not be any interaction between estrogen and other steroid hormone receptors.

In contrast to receptor-positive cell lines, for instance, MCF-7, which may contain variant or mutant receptors, there is no concern about the negative impact of mutant receptors in yeast strains since estrogen receptor is transfected into their genome. However, the shortcoming of yeast biosensors is their cell wall and active transport system, which is an energy-consuming process, and molecules and ions move as opposed to the concentration gradient. This may impact the evaluation of test compounds' activity; thus, YES assay could not detect several anti-estrogenic substances [53].

5.5.5 Cell Proliferation Assays

E-screen assay is an investigative procedure to analyze the estrogenicity of test compounds on growth, reproduction, or the surge of mitotic activity of human breast cancer cell lines (MCF-7), which internally expresses ER-α. For analysis of substances, various concentrations of the test chemical are included in the culture medium where Human MCF-7 is cultivated. A serial dilution of 17-β-estradiol is provided as a positive control to obtain a dose-response curve. A medium with no hormone-like activity is provided as a negative control [54].

There are several methods based on competing for binding to ER receptors. By these methods, the binding capability of the test phytochemicals is determined whether or not they could compete with 17-beta-estradiol to bind to the recombinant human Estrogen Receptor-α (hrERα). The results of these quantitative tests are expressed as IC50, a concentration of test chemical required to supersede half of the radio-labeled ligand from the receptor. The affinity of test compounds for hrERα is also expressed as relative binding affinity (RBA). The compounds that are capable of binding to ER are further assayed employing various concentrations of the mentioned chemical to obtain the binding curve and calculation of IC50 and RBA values [54].

E-screen assay could be modified to identify antagonists by a two-step test. In the first stage, the capability of the test compound is assessed in terms of inhibiting estrogen action. In crude terms, different dilutions of antagonists are included in the medium, which contains a minimum concentration of E2, which can induce proliferation at a maximum level. If the action of estrogen is prohibited, it indicates the receptor-medicated mechanism of action of the test chemical or the interference of the test substance with a region of estrogen receptor where the ligand binds. Furthermore, to further verify its mechanism of action, the concentration of E2 is increased to reverse it. In the second stage, maximum inhibition is imposed by a minimum dose of antagonist in the presence of different concentrations of E2 [54].

One main shortcoming of this test is that this test needs to be estrogenic-specific. The proliferation of MCF-7 occurs in the presence of cytokines, hormones, nutrients, etc. Therefore, it can trigger the determination of false positive estrogenic substances. Moreover, general growth inhibitors and cytotoxic compounds could give rise to the determination of false negative substances [54].

5.5.6 The Examples of Herbal Latex with Estrogenic Anti-Estrogenic Activities

Table 5.3 presents the scientific name of the plants, the latex product and its route of administration, the responsible phytochemicals, the in vitro and in vivo assessment models, and the research results.

Table 5.3 The scientific name, the responsible product of latex-producing herb, the assessment model for estrogenic anti-estrogenic activity, and the results

Scientific name	Latex product	Responsible phytochemicals	Assessment model	Results	Refs
Euphorbia bicolor	Isolation of flavonoids from 80% methanolic extract by UPLC-ESI-MS/MS	Isoflavones: coumestrol, daidzein, genistein	steroid-regulated yeast system	–Coumestrol and genistein significantly induced estrogenic activity in a dose-dependent manner. –Coumestrol, daidzein, and genistein showed significantly higher anti-estrogenic activities by inhibiting the estradiol activity by 70.7%, 70%, and 90.2%, respectively.	[53]

5.6 Anti-Leishmania Activity

Leishmaniasis is a tropical, subtropical infection spread in various geographical regions and caused by multiple species of leishmania parasite. Even though this ailment is prevalent in low-income countries, it is also endemic in Southern Europe nations. Various leishmania parasites trigger pathogenesis in both humans and dogs. As a zoonotic and vector-borne disease, it is caused by a female sandfly's bite transmitting the leishmaniasis parasite to animals and humans. The parasite's lifestyle is divided into two distinct phases: promastigote and amastigote. Promastigote form develops within the intestine of sandflies, where it is mobile due to owning a long motile flagellum. It is inoculated into the host's skin while the sand fly taking a blood meal is phagocytized by the host's macrophages, where they multiply and transform into amastigotes. The amastigotes live and develop in the host's reticuloendothelial system giving rise to both forms of asymptomatic and symptomatic leishmania. In addition, the parasite could trigger cutaneous, mucosal, and visceral forms of leishmania via migration through lymph fluid or the bloodstream [55].

In cutaneous and mucosal leishmaniasis, it is mainly caused by *L. braziliensis*. Moreover, *L. aethiopica, L. amazonensis, L. major, L. mexicana*, and *L. tropica* cause cutaneous leishmaniasis. In this infection, the parasite is localized. It remains in the skin as ulcers and nodules or disseminates to the mucosal sites and other organs via the lymphatic system and the bloodstream. Initially, the area of skin bitten by a sandfly shows erythema, then it develops to form a small pimple and gradually enlarges and ulcerates. Depending on the leishmania species and the host's immune response, the mentioned process may take several weeks to months. Generally, in immunocompetent patients, the ulcers heal slowly and spontaneously; however, leaving scars. On the other hand, in mucocutaneous leishmaniasis, there are destructive and progressive ulcers in the mucosa which are not self-healing and develop from nasal cavity and mouth to the pharynx and larynx [56].

Non-ulcerative form, known as kala-azar or visceral leishmaniasis, is systematic disease and could be fatal. It is developed beyond the initial area of infection. The diffusion of amastigotes characterizes it to other internal organs (usually the spleen, liver, and bone marrow). That disease is not self-cure and is caused by *L. donovani, L. infantum,* and *L. chagasi*. The infected individuals indicate the signs of systemic infection, such as fatigue, fever and chills, anorexia, and weight loss. Moreover, the lymph nodes, spleen, and liver enlarged, and severe bleeding and anemia could occur. Both forms of local and diffuse types of leishmaniasis are not life-threatening; however delayed diagnosis and therapeutic intervention could trigger progression of lesions, scarring, and secondary bacterial infections [56].

Post-kala-azar dermal leishmaniasis is a kind of dermatosis secondary to visceral leishmaniasis; in crude terms, it is observed in individuals previously infected with kala-azar. It is caused by species developing visceral leishmaniasis, including *L. donovani*, *L. infantum*, and *L. chagasi*. *L. amazonensis*, which is the cause of cutaneous and mucocutaneous leishmaniasis, is another casual species. The infection initiates with measles-like increasing-size ulcers on the face, gradually enlarging and disseminating to the neck, back, and extremities. Owing to undisclosed reasons, this infection recurs in certain individuals after treatment. Due to the evasion of a few parasites from destruction, they can infect the skin. Inadequate therapy and reactivation of latent parasites could be another reason for the mentioned disease. Further investigations are required to decipher parasite's persistence and its pathogenesis progression in some individuals [57].

5.6.1 Experimental Models for Leishmaniasis

For the isolation of parasites, culturing could be the best approach. In vitro cell culture is recommended for evaluation of the anti-leishmania activity of the test compounds against promastigotes of different Leishmania species, defined as resazurin reduction assay. Initially, a suitable solvent is used to dissolve a defined concentration of the test compound. The parasites for culturing could be obtained from needle aspirates or the lesions of a punch biopsy. A typical medium, liquid or biphasic, is needed, while biphasic is a more popular medium. Promastigotes of Leishmania spp. are cultivated in cell culture flasks. A counting chamber is used to count cells under an optical microscope to determine the number of promastigotes per ml. Dead and live parasites are differentiated using trypan blue exclusion dye. Promastigotes of leishmania species are seeded in RPMI-1640 medium in a 96-well cell culture plate. Different dilutions of the test compound are added to the wells in triplicate. A reference drug as positive and negative control is added at a proper concentration in triplicate. The wells allocated to negative control contain promastigotes of leishmania species cultured in RPMI-1640 medium. Several wells, as blank, are specified to merely be RPMI-1640 medium, which is free of the parasite. The wells are sealed and incubated in an incubator for 60 h. Resazurin solution at a concentration of 20 μg/ml is added to each well by pipette. The wells are sealed and incubated in an incubator at 26°C till observation of fluorescent red color in the wells of the negative control. Resazurin is blue to purple, a weakly fluorescent dye permeable to cells and non-toxic. Due to being a redox-sensitive dye, it is irreversibly converted to resorufin due to the cellular respiration of the active cells. Resorufin is pink in color and observable by an unaided eye, which could display the viability of cells. An absorbance microplate reader is used to read the plates at 570 nm. Also, in vitro assessment of antileishmanial activity of test compounds against intracellular amastigotes of different species of leishmania could be carried out by resazurin assay. Different concentrations of the test chemical are prepared in a proper solvent. Macrophages are cultivated in RPMI-1640 medium in cell culture flasks at 37°C. After 3 or 4 days, when macrophages are 70% confluent, they are seeded in 96-well cell culture filled with RPMI-1640 medium and incubated for 18 h at 37°C. Promastigotes of leishmania species are seeded in wells and incubated for 48 h at 37°C. Some wells are allocated to promastigote-free macrophages as a negative control. Different dilutions of the test compound and reference drug are added to the rest of the wells in triplicate. After incubation for 48 h and removal of the supernatant, a lysis solution is added to the wells to disrupt the membranes of macrophages. After incubation for 20 min at room temperature, the plates are centrifuged to remove the supernatant or the solution resulting from cell lysis to add Schneider's insect medium. The plate is sealed and incubated for 72 h in an incubator of 26°C to assist in amastigotes transforming into promastigotes. Resazurin solution is added to each well and thoroughly mixed. The plates are sealed and incubated until observation of red-fluorescent color in the group of the negative control. The microplate reader reads the plate's absorbance [58].

Even though several experimental models with typical features have been established, they cannot reproduce the pathogenesis and clinical syndromes in humans and dogs. In laboratory models, parasites could induce infection in a broad range of mammals; however, they are unlikely to reproduce in real-life situations. In commonly-employed laboratory animals, the induced infection is self-limiting, low, and with few symptoms. Typically, rabbits and rats are invulnerable to being infected by leishmania species, and the guinea pigs are reported to be unsuitable hosts for human-specific leishmania parasites. The commonly-used laboratory animals are rodents such as mice, hamsters and cotton rats, primates, and dogs. Several factors play a role in the outcome of infection in each laboratory animal. In crude terms, several criteria influence whether or not the infection by the parasite is limited to the area of inoculation. The nature and virulence of the inoculated leishmania species and the route of inoculation contribute to the infection outcome.

Furthermore, each animal could resist or be vulnerable to the Leishmania species. To illustrate, *L. major* causes asymptomatic infection in gerbils, yielding a cutaneous and self-limiting infection in humans. However, with a very low amount of inoculum, *L. major* produces a lethal visceral infection. Thus, various animal models have variable outcomes of infection [59].

Different strains of the mouse are vulnerable to most of the leishmania parasites infecting mammals. In these models, the parasites yield cutaneous ulcers in the inoculated site, which could give rise to visceralization or yield metastatic ulcers. In contrast, mice are less vulnerable to *L. donovani* since the produced ulcers cure spontaneously, which could be attributed to the immune system and genetics of the host. Therefore, the golden hamster (*Mesocricetus auratus*) is an appropriate animal model for studying visceral leishmaniasis caused by *L. donovani, L. Mexicana, L. chagasi, L. infantum*, and *L. braziliensis*. Hence, the hamster is a useful animal model for isolating parasites and laboratory experiments. It is an appropriate model pertinent to the rapid development of lesions, cutaneous metastasis, and extended pathology. Since Leishmania parasites cause chronic disease in these animals, the impacts of therapeutical and immunological interventions could be investigated. Several criteria should be considered to modulate the infected hamsters' clinical symptoms, including the parasite's dose, strain, and the site of inoculation. These factors could affect the type of lesions, whether nodular or ulcerative, rapidly or slowly developing, resembling the ones affecting humans. Moreover, the persistence or recurrence of disease in humans could be reproduced in hamster models [59].

Table 5.4 The scientific names, responsible products of latex-producing herbs, and the assessment models for anti-leishmania activity and the results

Scientific name	Latex product	Responsible phytochemicals	Assessment model	Results	Refs
Aloe macrocarpa	The leaf latex and isolated compounds; aloin A/B, and Aloe-emodin	Aloin A/B, Aloe-emodin, rhein	In vitro resazurin assay on amastigotes and promastigotes of *L. aethiopica* and *L. donovani,*	Isolated components showed higher antileishmania activity than latex. Aloin A/B, Aloe-emodin,and rhein showed lower antileishmania activity than the reference drug amphotericin B	[60]
Aloe calidophila	The leaf latex and isolated three anthrones; aloinoside, aloin, and microdontin.	Three anthrones: aloinoside, aloin, and microdontin	In vitro antipromastigote activity tests against the promastigote stage of *L. aethiopica* and *L. major.*	The isolated compounds showed antipromastigote activity higher than the standard antileishmanial drug (pentostam) against *L. aethiopica* and *L. major.* The isolated compounds were more active against *L. aethiopica* than *L. major.*	[61]
Himatanthus sucuuba	The ethanolic extract obtained from the latex of barks	NM[1]	In vitro assay; anti-amastigote activity against *L. amazonensis* infected macrophage cultures In vivo assay; oral administration of the extract to the infected mice	The results of in vitro assay indicated a dose-dependent anti-amastigote effect of the latex extract. It exhibited significant leishmanicidal activity against intracellular amastigotes. It decreased the parasite load in established footpad lesions.	[62]

[1]Not Mentioned

Overall, hamsters and mice are extensively used for in vivo studies. Biopsy specimen from a lesion containing leishmania parasites is subcutaneously injected into the septum, nasal tip, the base of the tail, ears, and footpads of the animal models to grow amastigotes of various species of Leishmania. Hence, these animal models could study the pathogenesis of leishmaniasis and the efficacy of anti-leishmanial drug candidates [59].

5.6.2 The Examples of Herbal Latex with Anti-Leishmania Activity

Table 5.4 presents the scientific name of the plants, the latex product and its route of administration, the responsible phytochemicals, the in vitro and in vivo assessment models, and the research results.

5.7 Assessment of the Antimicrobial Effect

There is a wide array of infectious diseases caused by bacterial strains. Laboratory assays are performed to investigate the susceptibility and resistance of pathogens to the candidate test chemicals. The results assist in preventing and controlling infectious ailments, pave the way for drug discovery, and predict clinical outcomes. Currently, microbial resistance has a considerable impact on therapeutic failures. Important antibiotics are losing their effectiveness due to over-usage and the development of antimicrobial resistance. Treatment failures due to multidrug-resistance bacteria are one of the public health issues. Thus, the discovery of new candidates is of utmost importance. There are gram-positive bacterial species, for instance, *E. faecium*, *E. faecalis,* and *S. aureus,* that the usual antibiotics could not affect. Hence, gram-negative bacterial species, namely *P. aeruginosa*, *A. baumannii*, and *H. pylori* are not sensitive to the commonly-used antibacterial agents. Natural products derived from prokaryotic and eukaryotic microorganisms, animals, and herbs are the main source of discovering new drug candidates. A broad class of complex and versatile antimicrobial compounds with diverse structures could be derived from the mentioned sources.

Various tests are performed to investigate the antimicrobial activity of the test compounds [63].

In vitro antimicrobial assays are performed to assess the potential of natural compounds or phytochemicals to kill or prevent the growth of microbes. For this purpose, several approaches were developed, including agar well diffusion and disc diffusion methods, thin-layer chromatography-bioautography, biofilm, dynamic contact, and time-kill assays. For measurement of minimum inhibitory concentration (MIC), Microdilution methods and broth assays are undertaken. Furthermore, minimum bactericidal concentration (MBC), minimum doubling time/ growth curve (MDT), Bioluminescence, and Flow Cytofluorometric assays are performed in laboratory settings. According to the research objectives, the most suitable assay is selected [64].

5.7.1 Diffusion Methods

Diffusion methods are the official approach used in clinical microbiology laboratories for testing the susceptibility of antimicrobials. By using typical culture media and diverse conditions of incubation, the susceptibility of various bacterial pathogens to the candidate compound could be tested.

In the diffusion method, the principal is the seeding of the inoculum of the test microorganism over the surface of an agar plate. In the disk diffusion method, categorized as the agar diffusion method, near 6 mm filter paper disks, the reservoirs of candidate compounds at different concentrations are placed on the surface of the agar. Incubation should be carried out in proper conditions. If diffusion of the antimicrobial agent into the agar prevents the growth of the microorganism, and the zones of inhibition appear, the test compound is an active antimicrobial agent.

The **agar disk diffusion method** is unsuitable for measuring MIC or minimum inhibitory concentration since not being capable of quantifying the dose of the diffused antimicrobial compound. On the other hand, this method is simple and cost-effective, which could determine the susceptibility of many microorganisms to antimicrobial agents. Moreover, the interpretation of the obtained results is simple and time-saving. There is an appropriate association between the agar disk diffusion test and the clinical cases. Antibiogram is an approach to guide clinicians in selecting appropriate antibiotics for patients infected with typical bacterial strains. Due to the mentioned reasons, this procedure is very common to screen extracts, volatile oils, and phytochemicals.

The **Antimicrobial gradient method** combines dilution and diffusion procedures to measure the MIC value. It is used to determine the susceptibility of the bacterial strain to gradient concentration of antimicrobial in an agar medium to measure the value of MIC. Etest is a MIC version employing a strip loaded with surging test chemical concentrations. This strip is deposited on the surface of agar pre- inoculated with the test microorganism.

The **Agar well diffusion method** is an extremely-used approach to assess the antimicrobial potential of candidate drugs. It resembles the disk diffusion method, while in this method, the agar surface is entirely inoculated by microbial inoculum. Afterward, a hole of 6–8 mm in diameter is aseptically punched on the agar surface and an antimicrobial candidate drug, 2–100 µl in volume, is introduced into the hole. Then, depending on the type of test microorganism, the inoculated plates are incubated under typical conditions. If the diffused antimicrobial agent prevents the growth of the tested microorganism, it could be rendered as an antimicrobial agent [64].

TLC-direct bioautography is a coupled method for rapidly screening the mixture of compounds, such as herbal extracts and phytochemicals, to discover novel antimicrobial agents. As a bioassay-guided approach, thin-layer chromatography, an analytical approach, is merged with detecting biologically-active compounds. Initially, a proper solvent system is introduced to the sample on the TLC layer, and the mixture is separated. Then the developed plate is sprayed or dipped into microbial suspension, allowing growing microorganisms on the adsorbent layer to detect antimicrobial effects of isolated compounds [65].

Even though TLC, compared to other chromatographic approaches (GC, HPLC), has lower separation and characterization power, it could have some benefits. It is an optimal method for analyzing plant extracts due to requiring a limited amount of samples, the capability of the parallel and simultaneous running of several samples, being time-saving, cost-effective, and a high-throughput approach. Furthermore, it could be merged with various bioassays to measure the biological activities of multicomponent samples [65].

Agar overlay bioassay or immersion bioautography is inoculated agar medium used to cover TLC plates, allowing the candidate compound to diffuse into the agar medium. Then, the plates are incubated at proper conditions, depending on the type of test microorganism. Tetrazolium dye is employed for staining to provide growth inhibition zones [65].

5.7.2 Dilution Methods

These techniques are employed to measure MIC (minimum inhibitory concentration) values. It is possible to determine the most effective concentration of the candidate antimicrobial compound, as it has various concentrations in agar and broth medium inoculated with various microorganisms. In crude terms, in this in vitro technique, quantitative measurement of antimicrobial activity is possible. MIC value expressed in mg/ml or μg/ml defines the lowest concentration of the test compound, which prevents the growth of the microorganism and could be detected by an unaided eye. In crude terms, observation of the first well containing clear solution with no turbidity indicates the MIC, as microbial growth results in the turbidity of broth. MIC guides clinicians in making treatment decisions regarding the vulnerability of microorganisms to an antimicrobial agent [66].

The **Broth dilution method** is one of the basic and commonly-used procedures to assess the susceptibility of microorganisms to test compounds. In this approach, two-fold dilutions of the candidate compound are prepared in tubes containing a minimum volume of 2 ml liquid growth medium defined as macrodilution or 96 well microtitration plate named microdilution. Then, standardized and adjusted microbial inoculum (0.5McFarland) is used to inoculate each well or tube and is incubated at proper conditions [66].

Microdilution has some merits compared to the macrodilution method; since microdilution is the miniaturized form of the test, space, and economy of reagents are the major advantageous points of it. Hence, the microdilution method facilitates reading and recording the results, and discerning the growth are facilitated.

Furthermore, dye reagents, such as tetrazolium salts and alamar blue dye (resazurin), could be used to determine the endpoint of microdilution assays. MIC values could be affected by the size of the inoculum, the type of growth medium, the incubation time, and the inoculum preparation method [66].

MIC is categorized into MBC, a minimum bacterial concentration, and MFC, a minimum fungicidal concentration, which estimates the bactericidal and fungicidal activity of the test chemicals. MBC is the lowest candidate compound concentration, killing 99.9% of the particular bacterium after a 24 h incubation. MBC could be measured after subculturing the broths from tubes or wells on the surface of fresh and non-selective agar plates. Therefore, the number of surviving microbial cells could be determined and read as CFU/mL (Colony Forming Unit per milliliter) after 24 h incubation. Like MBC, MFC is measured for antifungal agents against *Candida*, *Aspergillus*, or other fungal isolates [66].

The **Agar dilution susceptibility method** is also used to determine MIC, in which serial two-fold dilutions of candidate antimicrobial agents are incorporated into a molten agar medium. Afterward, a defined number of cells are inoculated on the surface of the agar plate. The endpoint of this test is the lowest dilution of the agent, which could prevent microbial strain growth under the proper incubation condition. Therefore, this is an appropriate method for assessing candidate antimicrobial and antifungal agents. For determination of MIC, this method is preferred to the broth dilution method when the test compound is colorful and due to its color, the growth of the microbial strain is not detectable. In this case, antimicrobial activity of the single test compound is tested against several isolates. Also, this method is suggested as a standardized technique for fastidious organisms, such as anaerobes and Helicobacter species [66].

5.7.3 The Examples of Herbal Latex with Antimicrobial Activity

Table 5.5 presents the scientific name of the plants, the latex product and its route of administration, the responsible phytochemicals, the in vitro and in vivo assessment models, and the research results.

Table 5.5 The scientific names and responsible products of latex-producing herbs, as well as the assessment models for antimicrobial activity and the results

Scientific name	Latex product	Responsible phytochemicals	Assessment model	Results	Refs.
Aloe harlana	Latex and isolated compounds, 7-O-methylaloeresin, and aloin	anthrones (aloin) and chromone (7-O-methylaloeresin A)	In vitro antimicrobial activity by disc diffusion method	Latex and isolated compounds compounds could significantly suppress gram-negative bacterial strains, *E. coli, S. typhi*, and *V. cholera*. They showed antifungal activity against *C. albicans* and *A. niger*. Aloin suppressed *E. coli.* 7-O-MA was a potent antibacterial against gram-positive bacterial strains, particularly *S. aureus*.	[67]
Aloe monticola	Two chromones isolated from leaf latex	Two chromones: aloesin and 7-O-methyl-6'-O-coumaroylaloesin	In vitro antimicrobial activity by disc diffusion method	Isolated compounds showed strong activity against *S. typhi* Shigella dysentery and *S. aureus*.	[68]
Aloe weloensis	Leaf latex	NM[1]	In vitro antimicrobial activity by disc diffusion method	Leaf latex significantly suppressed bacterial strains *S. aureus*, *E. coli*, *P. aeruginosa*, and *E. fecalis.*	[69]
Aloe trigonantha	Isolated compounds from leaf latex; **group A**: aloesin, 8-O-Methyl-7-hydroxyaloin A/B, Aloin A/B, Aloin-6'-O-acetate A/B.	Aloesin, 8-O-Methyl-7-hydroxyaloin A/B, Aloin A/B, Aloin-6'-O-acetate A/B (**GroupA**)	In vitro antimicrobial activity by disc diffusion method	**GroupA** isolated compounds significantly suppressed gram-negative bacteria; different strains of *E. coli*, *Salmonella enterica*, *S. typhi*, *S. dysentery*, different strains of *Vibrio cholera*.	[70]

(*Continued*)

Table 5.5 (*Continued*)

Scientific name	Latex product	Responsible phytochemicals	Assessment model	Results	Refs.
	Group B: emodin and emodin anthrone, 8-O-methyl-7-hydroxyaloin A/B, aloin A/B, and aloin -6′-O-acetate A/B.	Emodin and emodin anthrone, 8-O-methyl-7-hydroxyaloin A/B, aloin A/B, and aloin-6'-O-acetate A/B (**Group B**)		**Group B** isolated compounds were significantly active against *E. coli* and *V. cholera.*	
Aloe trichosantha	Leaf latex for isolation of 2 anthrones: aloin A/B (1) and aloin-6'-O-acetate A/B (2)	Aloin A/B (1) and aloin-6'-O-acetate A/B (2)	In vitro antibacterial activity by disc diffusion and broth dilution methods	*E. coli, Salmonella,* and *V. cholera* strains were highly susceptible to both isolated compounds. The zone of inhibition by compound 2 against *V. cholera* was higher than that of the standard drug ciprofloxacin.	[71]
Ficus carica	Hexane extract from ripe dried fruit	Coumarins	In vitro antibacterial activity by disc diffusion and micro dilution methods	The hexane extract inhibited the growth of *S. saprophyticus, S. aureus, S. epidermidis, E. cloacae, E. coli, E. faecalis,* and *P. aeruginosa.* The activity of hexane extract was comparable with tetracycline as standard antibiotic against *S. saprophyticus,* and *S. aureus.*	[72]
Ficus microcarpa	Chitinases obtained from the latex of the herb's trunk and branches	Class I chitinase	In vitro antifungal activity by disc diffusion method	Class I chitinase showed strong anti-fungal activity against *Trichoderma viridae.*	[73]

Scientific name	Latex product	Responsible phytochemicals	Assessment model	Results	Refs.
Hevea brasiliensis	Hevein extracted from the latex	Hevein (cysteine-rich protein)	In vitro antifungal activity by disc diffusion method	Hevein showed antifungal activity against *Candida albicans, C. tropicalis, C. krusei.* (Significance ND)	[74]
	Latex of the stems	B-serum is separated from latex, a rich source of proteins and nucleic acids.	In vitro antifungal activity by disc diffusion assay	Latex B-serum showed anti-fungal activity against *C. albicans.*	[75]
	Crude latex	α-D-mannosidase and N-acetyl-β-D -glucosaminidase	Macrobroth method of dilution	The crude latex showed antifungal activity against *Trichosporon cutaneum.* The latex showed synergistic effect with amphotericin B against *C. albicans*	[76]
Euphorbia antiquorum	Latex of the stems was used to isolate homodimeric lectin	Homodimeric lectin (EantH)	Broth and agar microdilution method	EantH inhibited the growth of the Gram-positive bacteria *S. aureus* and *S. epidermidis*, and gram-negative ones *Propionibacterium* acnes and *Streptococcus agalactiae.* EantH killed *P. acnes* and *S. agalactiae* at higher concentrations. (Significance ND)	[77]
E. characias	The latex of the stem was used to purify proteins	Latex proteins	Macro broth method of dilution	Synergistic antifungal activity of the ketoconazole and latex proteins against *C. albicans.* Minimum dosage of ketoconazole is required.	[78]

(*Continued*)

Table 5.5 (*Continued*)

Scientific name	Latex product	Responsible phytochemicals	Assessment model	Results	Refs.
E. tirucalii	80% hydroalcoholic extract from the latex obtained from branches and trunk of the plant	Perhaps tetracyclic triterpene euphol	Agar well diffusion method	It significantly suppressed gram-negative bacteria (*P. aeruginosa*, *Klebsiella sp*, *Proteus mirabilis*, and *Shigella dysenteriae*).	[79]
Calotropis procera	Stem latex	NM	Agar well diffusion method	The latex inhibited dermatophytes, *Trichophyton* and *Microsporum spp.*	[80]
		Cysteine peptidases and osmotin like proteins	Agar well diffusion method	Antifungal activity against filamentous fungi *Fusarium solani* and *F. oxysporum.*	[81]
C. gigantia	Chloroform extracted fraction of latex obtained from the stem	Perhaps methyl nonanoate, a saturated fatty acid	Agar well diffusion method and minimum inhibitory concentration (MIC)	Chloroform extracted fraction of latex significantly suppressed *S. mutans* and *L. acidophilus*	[82]
Himatanthus articulates	The latex obtained from the bark	NM	Agar well diffusion method	latex inhibited *Candida albicans.*	[83]
Lactuca sativa	The latex obtained from aerial parts	Probably terpenes and cardenolides but also enzymes, in particular, glucanases	Agar well diffusion method	Latex as an antifungal inhibits the growth of *Candida albicans.*	[84]

[1]Not Mentioned

5.8 Evaluation of Antioxidant Activity

Free radicals are unstable and highly-reactive molecules with an unpaired electron and incapable of existing independently. They are a by-product of cellular metabolic processes, and due to having an odd number of electrons, they attack molecules and yield a free radical. This process could result in a cascade and could damage the living organism. ROS and RNS (reactive nitrogen species) are free radicals and by-products of the cellular metabolic process. When the concentration of the mentioned substances is high, they cause oxidative stress, damaging the integrity of organic molecules and the structure of living cells. Therefore, a wide range of chronic and degenerative diseases, such as diabetes mellitus, Alzheimer, cancer, autoimmune, and cardiovascular disease are developed. Living organisms could produce natural antioxidants or endogenous ones to counteract the damage of free radicals. However, supplements and foods as exogenous antioxidants or free radical scavengers could prevent oxidative and nitrosative stress and repair the harm generated by ROS and RNS [85].

Herbal products can scavenge free radicals and are enriched with antioxidant ingredients. Even though there are several phytochemicals known as antioxidants, polyphenols, and flavonoids include the main group of plant-based antioxidants. Herbal ingredients exert their antioxidant activity through several biochemical pathways to protect cells. For instance, they apply scavenging activity against radicals of superoxide. Hence, they could surge superoxide dismutase activity in different body organs [86].

Herbal latex, a complex fluid produced by the secretary system of some medicinal plants, consists of diverse secondary metabolites with antioxidant activity. Coexistence of polyphenols and flavonoids in herbal latex and their synergistic activity take center stage to exert and boost antioxidant effects. Phenolics, namely flavonoids, coumarins, lignans, and tannins, compose secondary antioxidant metabolites and could act against conditions related to oxidative stress [87]. Various experiments were carried out to assess the antioxidant capacity of herbal latex to shed light on the reason for their ethnic uses to prevent and cure ailments.

5.8.1 Methods of Measuring Antioxidant Activity

Various methods and tools exist to evaluate the candidate agents' antioxidant properties. A combination of chemical assays with detection instruments is recruited to assess the scavenging activity of test compounds. In recent years, there has been remarkable progress in assessing antioxidant properties. The basis of the evaluation of scavenging activity is the measurement of the reducing power of the agents against special kinds of free radicals or oxidized substrates. Beyond chemical assays, live cell, and tissue antioxidant assays could be employed for this purpose [88].

Several approaches involve removing free radicals. There are diverse methodologies to assess the antioxidant capacity of natural compounds; they could be categorized based on the lipophilicity and hydrophilicity of the antioxidants, solvents, and the kind of reagents, whether radicalic or non-radicalic. Also, the chemical mechanism is another factor in classifying antioxidant methodologies: HAT (hydrogen atom transfer), ET (Electron Transfer), and chelation-based approaches. DPPH, FRAP, ABTS scavenging, and CUPRAC are ET-based tests. ORAC and TRAP are HAT-based tests. In these methods, the transfer of hydrogen atoms or electrons to free radicals is directly measured in which typical chemicals or biological molecules are involved. These typical chemicals are added to the mixture to remove radical species or ROS.

ROS are highly reactive agents generated from oxygen during cellular metabolisms. They include hydrogen and lipid peroxides, and superoxide. These assays are lipid-free, or in other words, lipid substrate is not required. In these assays, a chemical system is usually employed, which contains one oxidizable ingredient to assess the antioxidant property of the test chemical. These tests are categorized into two groups: reactions based on the transfer of hydrogen atoms and second, ones based on the transfer of an electron. Some reference antioxidant agents, such as ascorbic acid and torolax, are employed to compare the antioxidant activity of test chemicals with them. There are several instruments for detecting and measuring antioxidant properties. The commonly-used ones are based on spectrophotometry [88].

ORAC assay is HAT-based and used to measure the test compounds' antioxidant properties in industry and for academic purposes. ORAC stands for Oxygen Radical Absorbance Capacity, in which the capability of agents is measured based on prohibiting oxidation induced by peroxyl radical. Peroxyl radicals are yielded, while lipids in organ systems or foods are oxidized. The values yielded by this assay are used to infer the antioxidant efficiency of the test compounds. A fluorescent protein probe, such as phycoerythrin, with bright red-orange fluorescence, was initially employed. Later, phycoerythrin was replaced with frequently-used, non-protein fluorescent probes, like fluorescein. ROS gives rise to the quench of the signal of the fluorescent agent. When antioxidant agents are added to the mixture, they absorb the produced ROS, and the signal of the fluorescent agent is preserved. A fluorometer is used to record the fluorescence of the probe. In ORAC assay, ROS producers are azo compounds belonging to lipophilic and hydrophilic groups. Commonly, AAPH (2,2′-azobis (2-methylpropionamidine) dihydrochloride), a hydrophilic agent, is used, which decomposes thermally and generates free radicals of peroxyl. This test measures the capability of the test compounds to donate a hydrogen atom. Trolox, an analog of vitamin E, is used as a reference drug to compare ORAC values of test chemicals with it. Production of peroxyl radicals is highly determined by temperature, which triggers inconsistency in the results. Therefore, during the assay, it is of utmost importance to check the temperature [89].

DPPH assay is categorized as an ET-HAT-based, very simple, commonly-used, and cost-effective test indicating test compounds' potential to scavenge free radicals. UV–Vis spectrophotometer is required to carry out this test. DPPH (2,2-diphenyl-1-picrylhydrazyl) is a chemical compound, a purple-colored free radical chemical with strong absorption around 520 nm. In the medium, the antioxidant agents donate a hydrogen atom or an electron to alter the radical of DPPH to its molecular or reduced form. Conversion of the purple-colored radical DPPH into the yellow-colored and reduced form brings about the spectrophotometric determination of antioxidant activity. The results are communicated as IC50, an indicator of the test chemical concentration causing scavenging of 50% radical of DPPH. There are modifications to reduce the costs of this bioactivity test. Filter papers and TLC plates, as flat surfaces, could be employed to increase the speed and efficiency of the test, defined as bioautography. In this method, the spots of plant extracts and phytochemicals are applied on chromatography plates, or papers are immersed in the mobile phase mixture in a solvent tank. The ingredients of spots are separated and sprayed with radical DPPH solution. Screening of antioxidant ingredients could be carried out by immersing the papers and plates into the solution of radical DPPH to detect the change of DPPH color [90].

Ferric reducing antioxidant power (FRAP) is another assay that the basis of this approach is the transfer of a single electron in an acidic condition (ET-based assay). It was developed to quantify the antioxidant's power to reduce the ferric ion (Fe^{3+}) complex to the ferrous (Fe^{2+}) complex and produce an intense blue color. The acidic condition of the medium assists in iron solubility and induces electron transfer. Initially, tripyridyltriazine (TPTZ) was used as a ligand to bind to iron, while alternative ones were applied to bind to ferric, including ferrozine and potassium ferricyanide. This method is also a spectrophotometric approach to quantify the antioxidant potential of the test compounds. The antioxidants' reduction ability on a ferric ion is deduced as their capability to reduce ROS [88].

ABTS assay is spectrophotometric, and an ET-based assay was developed to determine and rank the antioxidant property of the test compounds by comparing the values with that of the reference drug of Trolax. This simple and commonly-used assay is based on the discolorization of ABTS (2,2′-azinobis(3-ethylbenzothiazoline-6-sulfonic acid). $ABTS^{+\bullet}$ is a chemically stable chromophore and a water-soluble radical cation. The maximum absorption of the ABTS is at 734 nm. When an antioxidant agent donates an electron or hydrogen ABTS, its intense blue-green color decreases. To put it in another way, an efficient antioxidant agent in the test solution neutralizes radical cation of ABTS and, consequently, weak absorption at 734 nm. The amount of discoloration of this chromophore is dependent on the intrinsic antioxidant capacity of the test compound and the duration of the reaction. H_2O_2 or ferryl myoglobin is applied to react with ABTS to yield a radical cation form. One of the advantages of this method is the solubility of ABTS radical cation in both lipophilic and aqueous media. Despite being cost-effective and available in laboratories and industry, several criteria, such as incubation time, the ratio of antioxidants, and ABTS radical cation, could affect the result. Similar to the DPPH assay, the major challenge of this test is that $ABTS^{+\bullet}$ is not biologically relevant due to being an artificial compound and not existing in biological systems [91].

Hydrogen peroxide scavenging activity is a spectrophotometric and ET-based assay to measure the radical scavenging potential of the test compounds. The basis of this test is the determination of H_2O_2 disappearance. The absorbance of this mixture would be taken at 230 nm. First, 40 mM hydrogen peroxide solution in buffer phosphate will be prepared. Then, various concentrations of test compound will be dissolved in distilled water, added to the hydrogen peroxide phosphate buffer and mixed. After 10 min, the absorbance of the mixture will be measured at 230 nm against a negative control solution (phosphate buffer without hydrogen peroxide). Positive control solution is various concentrations of ascorbic acid solved in a hydrogen peroxide solution. The percentage of scavenging activity will be measured according to the following equation: $A_0 - (A_1/A_0) \times 100$, where A0 is the absorbance of the control and A1 is the absorbance of the test compound [92].

Nitric oxide is a free radical, a vasodilator, and a signaling molecule yielded by inducible nitric oxide synthase (iNOS) at the inflammation sites. Release and maintenance of NO at the inflammation sites, such as the intestine, lungs, and joints, trigger immunological responses. NO is involved in immunological diseases and chronic inflammation. Stimulation of inflammation could be mitigated by scavenging compounds preventing NO production. Natural products are alternative agents and radical scavengers which prevent NO production [93].

Nitric oxide (NO) scavenging activity is a spectrophotometric, ET-based method in which Griess reagents are used to determine nitrite ions. Griess reagent is a mixture of 1% sulfanilamide and 0.1% naphthyl ethylenediamine dichloride. Nitrite ions are yielded when an aqueous solution of sodium nitroprusside prepared at physiological pH reacts with oxygen. Different concentrations of test compound and ascorbic acid are mixed with an aqueous solution of sodium nitroprusside. After incubation for 150 min and at the temperature of 25°C, 0.5 ml of the incubated solution would be taken and added to the sulfanilic acid reagent. This process is carried out to finalize diazotization by incubation at room temperature for 5 min. Afterward, 0.1% naphthyl ethylenediamine dihydrochloride at the volume of 1 ml will be added to the mixture and incubated for 30 min. The absorption will be measured at 540 nm. Control or blank contains reagents but the test chemical. Next, the amount of the inhibited could be calculated [94].

The **in vitro assays based on chelating** measure the capability of the test chemicals to yield complexes with different valence metal ions. Free metal ions participate in $OH^{\bullet}$ production reactions. Hydroxyl radicals are extremely reactive species that attack the cell membrane and biological macromolecules leading to mutations and cell death. The reaction of a free metal ion such as Fe^{2+} and Fe^{3+} with H_2O_2 generates hydroxyl radical (Fenton reaction). Compounds with antioxidant capacity could react with unsaturated metals and produce saturated complex structures, preventing oxidation reactions. Therefore, assessing chelating potential of natural products and phytochemicals is a common procedure. Even though iron acts as a coenzyme and is a significant component of macromolecules, free Fe^{2+} ions are toxic prooxidants. The protective effects of chelating antioxidant agents in reducing the concentration of the mentioned ions play a crucial role in surviving living organisms [95].

2-deoxy-D-ribose degradation assay is a spectrophotometric assay designed to evaluate the chelating potency of the biomolecules to protect 2-deoxyribose or DNA against the damage caused by $OH^{\bullet}$. Similarly, it measures the antioxidant potential of the compounds to scavenge hydroxyl radicals. This in vitro assay is simple, cost-effective, and based on the measurement of MDA (malonic dialdehyde), the product of 2-deoxyribose breakdown. It is used to determine radicals of hydroxyl or OH radical species. Hydroxyl species are yielded during the reaction of hydrogen peroxide and iron complex. While heating the mixture, hydroxyl radicals react with 2-deoxyribose, and MDA is formed and measured during the reaction with TBA or thiobarbituric acid. Adding a test chemical with hydroxyl radical scavenging capacity to the reaction media brings about competition between the test chemical and 2-deoxyribose sugar. Antioxidant agents yield stable complexes that do not react with H_2O_2. Therefore, 2-deoxyribose is not degraded. Although the results of this test are prepared quickly, this assay is not carried out in a biological system. Hence, Fe-antioxidant complexes can have absorption at 532 nm, the wavelength of the compound generated by 2-deoxyribose degradation [95].

The results of various antioxidant assays with diverse mechanisms could be significantly different. Furthermore, since oxidizing agents and mechanisms of action are various, there will be a low correlation among the values. Therefore, integrated methods should be developed to evaluate the antioxidant potential of the test chemicals, especially herbal bioingredients.

5.8.2 The Examples of Herbal Latex with Antioxidant Activity

Table 5.6 presents the scientific name of the plants, the latex product and its route of administration, the responsible phytochemicals, the in vitro and in vivo assessment models, and the research results.

5.9 Anti-Hyperglycemic and Anti-Diabetic Activity

Diabetes mellitus, a chronic and endocrine ailment, is featured by hyperglycemia. In this disease, defects in insulin production alter the metabolism of proteins, fats, and typically carbohydrates. Not only does diabetes cause a high mortality rate worldwide, but it is a huge financial strain placed on the healthcare services. The common approach to lowering postprandial hyperglycemia in type 2 diabetes is decreasing the digestion of carbohydrates. If the digestion enzymes of α-amylase and α-glucosidase are inhibited, the postprandial glycemic reaction will be controlled. α-amylase, as a glycoside hydrolyzing enzyme, cleaves the alpha bonds of starch or other polysaccharides. It can predominantly produce oligosaccharides, maltose, dextrin, or shorter-chain

Table 5.6 The scientific names and responsible products of latex-producing herbs, the assessment models for antioxidant activity, and the results

Scientific name	Latex product	Responsible phytochemicals	Assessment model	Results	Refs
Aloe harlana	Leaf latex	7-O-MA (chromone derivative) and aloin (anthrone derivative)	In vitro DPPH assay 2-deoxyribose degradation assay	Significant antioxidant activity due to the phenolic structure of compounds. Chromon derivatives show stronger scavenging activity than the anthrone derivative. 7-O-M A was a stronger antioxidant	[67]
Aloe megalacantha	Leaf latex	Phenolic and flavonoid compounds	In vitro DPPH assay	Statistically significant antioxidant activity compared to the blank solution.	[96]
Aloe monticola	Leaf latex	NM[1]	In vitro DPPH assay	Statistically significant antioxidant property compared to blank control. It showed a half inhibitory concentration compared to ascorbic acid.	[96]
Aloe pulcherrima	Leaf latex	NM	In vitro DPPH assay	Significant and concentration-dependent antioxidant activity.	[97]
Ficus religiosa	Methanol extract from latex	flavonoids, polyphenols, and methionine	DPPH and phosphor-molybdenum radical assays	Methanol extract indicated statistically significant antioxidant property compared to ascorbic acid.	[98]
Euphorbia bicolor	80% methanolic extract of latex obtained from stem, leaf, and inflorescence bracts	Phenolic compounds, flavonoids, and proanthocyanidin	Ferric Reducing Power, ABTS, Hydrogen peroxide, Nitric oxide, and DPPH scavenging activity	The latex extract showed significant and concentration-dependently antioxidant activity compared to the standard drugs of BHT, ascorbic acid, and quercetin	[99]

[1]Not Mentioned

glycosides. α-glucosidase breaks down oligosaccharides into the monosaccharide of glucose. The currently-used enzyme inhibitors for controlling postprandial hyperglycemia in diabetic patients are acarbose, voglibose, and miglitol. However, they are not preferable for long-term usage because they have several side effects, especially in the gastrointestinal system. Thus, natural resources, such as herbs and their bio ingredients are searched to find safer enzyme inhibitors. Several herbal latexes are known traditionally to possess anti-diabetic activity. Their encouraging therapeutic results could assist in the anti-development of natural anti-diabetic medicines with fewer contraindication and unfavorable effects and more therapeutic efficacy [100].

5.9.1 In vitro Assays Concerning Carbohydrate Hydrolyzing Enzymes

α-Amylase Inhibitory assay is a common test employed in laboratories. α-amylases, a predominant digestive enzyme, randomly hydrolase α-1,4-glycosidic bonds of the starch chain. Measurement of amylase inhibitory activity of the test compounds is one of the procedures to evaluate the hypoglycemic effects of test chemicals. In this assay, amylose azure is used as an insoluble substrate that couples to dye and breaks down by the enzyme α-amylase. Color's intensity indicates the inhibitory activity of the chemicals on the enzyme. The absorbance will be measured at 595 nm. Initially, the substrate solution should be warmed up to the 37°C. Then, samples prepared in various concentrations, substrate solution, plus stop reagent will be added to the test tubes. Next, blank control will be prepared by adding substrate, water, and stop reagent. Finally, the test tubes will be centrifuged for a while, and the supernatant will be transferred to the transparent wells to read the optical density [100].

In the second chromogenic method, the pancreatic α-amylase enzyme of porcine, potato starch dissolved in phosphate buffer as a substrate, and a test compound dissolved in DMSO will be mixed to the plastic tube. The incubation is carried out at 25°C for 3 min. Afterward, the mixture is removed and poured into the tube holding solution of chromogenic reagent defined as

DNS, 3,5-dinitro salicylic acid, and incubated in the water bath of 85°C. Determination of enzyme inhibitory activity is undertaken by measuring the absorbance at 540 nm. Control tubes should contain DMSO instead of the test chemical. The absorbance of the test mixtures will assist in calculating the percent of generated maltose. Acarbose was employed as a reference drug [100].

Determination of β-galactosidase Inhibitory Activity is another common test employed in laboratories. β-galactosidase is an enzyme that catalyzes lactose degradation to yield monosaccharides, glucose, and galactose that undergo glycolysis to produce energy. Since inhibition of this digestive enzyme could result in hypoglycemia, researchers are investigating natural products to find novel agents with β-galactosidase Inhibitory Activity. β-galactosidase enzyme solution will be added to the phosphate buffer, and various concentrations of the test compound will be preincubated for 15 min at 37°C. Afterward, the solution of ONPG, 2-Nitrophenyl β-D-galactopyranoside, will be added to the mixture and incubated for 30 min at 37°C. A sodium carbonate solution will be added to terminate this reaction. The spectrophotometer will be used to measure the absorbance of the mixtures at 410 nm. The blank container lacks the enzyme solution. The measured optical density of the solutions will be employed to assess the inhibitory enzyme activity of the test chemicals [101].

Determination of α-Glucosidase Inhibitory Activity is another conventional test. α-Glucosidase enzyme involves the digestion of carbohydrates in the digestive system by hydrolyzing carbohydrates to simple sugars surging the blood glucose level. The intestinal absorption of maltose or other products of amylase enzyme is carried out by α-glucosidase, which further hydrolysis them to glucose. Inhibition of this enzyme in diabetic patients could bring about a reduction of hyperglycemia [100].

An in vitro assay to measure the α-glucosidase inhibitory potential of the test compound involves mixing the α-glucosidase solution with phosphate buffer. Various concentrations of the test compound will be prepared and added to the mixture, and the mixture will be incubated at 37°C for 30 min. The stop reagent of Na_2CO_3 will be added to conclude the reaction, and a spectrophotometer will measure the absorbance at 405 nm.

Acarbose will be used as a reference medicine. The optical density of the test compound, blank and reference drug will be employed to evaluate the enzyme inhibition activity of the test chemicals [100].

5.9.2 In vivo Animal Models to Assess Anti-Diabetic Activity

Diverse animal models have been employed to assess in vivo activity, side effects, the mechanism of action of herbs, and herbal products. Due to the large variation and heterogeneity of symptoms in diabetic patients, several animal models should be used to display a special type of diabetes with its typical features. Two types of nondiabetic and diabetic animals were employed to study herbal material's hypoglycemic and anti-diabetic effects. However, the hypoglycemic agents could be ineffective in man, which could be pertinent to diverse metabolic pathways in animals' and humans' livers. In addition, the species have a substantial sensitivity difference regarding metabolism, absorption, and elimination rates. Small rodents are the commonly used models to study the glucose-lowering effect of the test compounds. They are not costly, available, and in general, they rapidly demonstrate diabetic symptoms. Animal species could be manipulated genetically, surgically, and pharmacologically to induce diabetes. Even though the herbal product could not replace insulin, it could partially mimic the role of insulin or increase the efficiency of low insulin concentrations in the body [102].

Commonly used animal models induced hyperglycemia could be categorized as nutrition-induced, and chemical-induced. In nutrition-induced ones, diabetes could be developed by long-term overnutrition to simulate obesity-dependent diabetes, resistance to insulin, and dysfunction of pancreatic beta cells. In this approach, toxic chemicals are not employed, and other vital organs are not intoxicated. However, considering the drawbacks of this method, it could not develop actual hyperglycemia in normal animals, and it is improper to study the parameters of circulating glucose. In the second group, animals are induced hyperglycemia by the cytotoxic effect of chemicals, such as alloxan and streptozotocin. These chemicals trigger loss of beta cells in the pancreas giving rise to insulin deficiency instead of residence to insulin. This approach is inexpensive and could be easily developed and maintained. However, chemically induced hyperglycemia is unstable and reversible; results could also be variable since pancreatic beta cells are regenerated spontaneously. Thus, long-term experiments may encounter problems. Furthermore, the mentioned toxic drugs could cause toxicity in other vital organs [102].

Diabetes could be induced surgically by reducing the mass of islet beta cells, similar to type 2 diabetes. In this approach, the cytotoxicity of chemical drugs to induce diabetes will be avoided. However, the procedure is cumbersome and increases the risk of mortality. Moreover, the dissection of insulin-secreting cells (beta cells) may occur with alpha cells secreting glucagon affecting regulatory response [102].

Natural products, such as herbal latex, display nonspecific action. They contain several phytoconstituents indicating antioxidant properties, lowering the level of cholesterol and triglycerides alongside decreasing the level of glucose. Natural products could exert anti-diabetic effects by various routes, extrapancreatic effects, or prevention of carbohydrate absorption. Notably, the optimum designed model for studying the anti-diabetic efficacy of herbs includes a combination of beta cell dysfunction and insulin resistance models. In addition, the mentioned models will study the secondary advantages of herbal products and glycemic control. Secondary benefits include anti-hyperlipidemic, anti-obesity, and antioxidant effects. To generate optimum results, diverse animal models should be employed to determine the efficacy of herbal extract and its mechanism of action [103].

5.9.3 The Examples of Herbal Latex with Anti-Hyperglycemic Activity

Table 5.7 presents the scientific name of the plants, the latex product and its route of administration, the responsible phytochemicals, the in vitro and in vivo assessment models, and the research results.

Table 5.7 The scientific names, responsible products of latex-producing herbs, the assessment models for anti-hyperglycemic and anti-diabetic activity, and the results

Scientific name	Latex product	Responsible phytochemicals	Assessment model	Results	Refs
Aloe megalacantha	Leaf latex and preparative TLC isolates	Mainly anthraquinone glycosides could be responsible for this effect.	In vitro α-amylase inhibitory activity by 3,5-dinitro salicylic acid (DNSA) method	The leave latex and two unknown isolates showed dose-dependent and statistically significant α-amylase inhibition activity compared to the blank.	[96]
	Leaf latex	NM[1]	In vivo study on streptozotocin-induced diabetic mice	It exhibited a significant anti-hyperglycemic effect, improved body weight, and lowered cholesterol and triglycerides levels.	[104]
Aloe monticola	Leaf latex and preparative TLC isolates	Anthraquinone glycosides could be responsible for this effect.	In vitro α-amylase inhibitory activity by 3,5-dinitro salicylic acid (DNSA) method.	The leave latex and two unknown isolates showed dose-dependent and statistically significant α-amylase inhibition activity compared to the blank.	[96]
Aloe pulcherrima	Leaf latex	NM	In vitro sucrose and maltase inhibitory activity by glucose oxidase assay. In vitro α-amylase inhibitory activity by dinitro salicylic acid assay. In vivo study on streptozotocin-induced diabetic mice.	Latex significantly inhibited sucrase, maltase, and α-amylase enzymes in a concentration-dependent manner in in vitro assays, however, it indicated a higher inhibitory effect on sucrase. Latex showed significant anti-diabetic activity and improved body weight in mice compared to vehicle control.	[105]
Himatanthus drasticus	EtOH extract of latex obtained from stems and plumieride	Perhaps plumieride	In vitro assay to evaluate α-amylase and α-glucosidase inhibitory activity	EtOH latex extract and plumieride significantly inhibited α-amylase and α-glucosidase.	[106]

[1]Not Mentioned

5.10 Larvicidal Activity

Larvacide is an insecticide affecting the larval phase of the insect's life. There are severe diseases spread by infectious mosquitoes, which are vectors of protozoan diseases, such as malaria and filariasis, and virus illnesses, including dengue fever, encephalitis, and yellow fever. When a mosquito swallows the blood of an infected individual, viruses, and bacteria can be transmitted from one infected person to the next one through its saliva. Vector-borne illnesses account for the suffering and mortality of millions, especially in developing and underdeveloped nations. Also, they are responsible for economic and social burdens in societies [107].

Prevention of mosquito-borne illnesses is based on decreasing and controlling mosquito larvae or the immature stage and reducing their population in their adult stage. Synthetic insecticides, such as organophosphates, carbamates, Organochlorines, and pyrethrins, were developed and used at large-scale to control vector-borne illnesses. However, they jeopardize human health because they can potentially cause endocrine, nervous, and reproductive problems. Being nonbiodegradable and extremely toxic, they are found in soil, water, crops, and food. Furthermore, they can have negative impacts on fetuses during pregnancy. Apart from being hazardous for humans and the environment, their effects on mosquito populations could be temporary since they become resistant to chemical insecticides due to metabolic detoxification and alteration on action sites [107].

Therefore, searching for simple and sustainable alternatives for a healthier lifestyle and avoiding environmental problems is important. Natural active ingredients could act as efficient larvicidal and insecticidal agents. Herbal insecticides have managed pests and insects for thousands of years. Latex is among the herbal exudates that demonstrated insecticidal activities against larvae and adults of insects.

In larvicidal bioassays, mosquito larvae reared in the lab with known instar are kept in water for 24 to 48 h. To assess the toxic effect of substances against larvae, different concentrations of test substances are exposed to larvae at various intervals of 24, 48, and 72 h. This test could be performed in a filter-paper pocket or other containers. If the percentage of larvae mortality is equal to or more than 50%, it will be deemed toxic against larvae.

For evaluation of the larvicidal effect of unknown substances, selected bioassay is performed. This test's starting point would be a maximum concentration of 500 ppm. The mentioned concentration could be diluted serially. Three containers should be allocated to each concentration, and the test should be repeated in triplicate. Each container will have larvae, some powdered food, and a prescribed concentration of the test chemical. After exposure, the number of alive larvae will be counted at 24, 48, and 72 h. If a concentration of the test substance indicated a 50% larvicidal effect, a dose bioassay is performed to calculate LC50 and LC90. The containers will be allocated to five concentrations selected from the first assay and the negative control. In a control container, mortality should not be more than 10%. In this assay, test containers should have larvae, powdered food, and a special concentration of test compound. The negative control ones contained the solvent used to solve the test substance, larvae, and powdered food. For the calculation of LC50, statistical analysis software could be used [108].

5.10.1 The Examples of Herbal Latex with Larvicidal Activity

Table 5.8 The scientific names, responsible products of latex-producing herbs, and the assessment models for larvicidal activity and the results

Scientific name	Latex product	Responsible phytochemicals	Assessment model	Results	Refs
Aloe yavellana	Leaf latex and two isolates aloin A/B and microdontin A/B	Aloin A/B and microdontin A/B and other minor active phytochemicals.	LPT method against *Amblyomma variegatum* tick larvae	Leaf latex, aloin A/B, and microdontin A/B showed significant mortality. Microdontin A/B was 3 times more active than aloin A/B. Leaf latex of the plant was much more active than the purified phytochemicals.	[109]

(*Continued*)

Table 5.8 (*Continued*)

Scientific name	Latex product	Responsible phytochemicals	Assessment model	Results	Refs
Calotropis procera	Latex of aerial parts	Protein content in the water-soluble fraction	In vitro egg hatching and larval development	Crude latex of *C. procera* significantly inhibited egg hatching and larvae development of *Aedes aegypti* (Linn.) in vitro.	[110]

Table 5.8 presents the scientific name of the plants, the latex product and its route of administration, the responsible phytochemicals, the models for studying larvicidal effects, and the research results.

5.11 Anti-Inflammatory and Analgesic Activities

Any mechanical injury of tissue, infection, and autoimmune stimuli involves sophisticated responses defined as inflammation in which leukocytes and fluids accumulate in the irritated site. In the process of inflammation, initially, vasodilation and subsequently increased perfusion occurs. It brings about the infiltration of leukocytes into the inflamed tissue, edema, and swelling. Consequently, local warmth increases due to the elevation of metabolic activity and function of leukocytes at the site. Increased vascular perfusion and swelling trigger stimulation of the nerve endings, causing pain and itching. Chemical mediators released from inflammatory cells and damaged tissue are associated with an inflammation response. Chemical mediators fall into three categories; amines increase the dilation of arteries and permeability of the capillaries, including histamine and serotonin, peptides such as bradykinin, and 20-carbon, polyunsaturated fatty acids defined as eicosanoids, namely prostaglandin, leukotriene, and thromboxane. Histamine is stored in mast cells and released from blood cells, defined as basophils contributing to acute inflammation. Serotonin, as a multifaceted signaling mediator, exists in platelets of humans and exerts its function through serotonin receptors localized in CNS and peripheral organs. Bradykinin, a nano peptide mediator, brings about a surge in prostaglandin synthesis and production of local pain. Eicosanoids, such as leukotriene and prostaglandin, are the products of enzymes 5-lipoxygenase and cyclooxygenase, respectively. Prostaglandins are involved in vascular permeability and reinforce the actions of other mediators, such as kinin, serotonin, and histamine. Leukotriene also belongs to the eicosanoid family and contributes to acute inflammatory responses. Other inflammatory mediators include nitric oxide (NO), lipoxins, growth factors, and platelet-activation factor (PAF) [111].

Based on the process duration, there are two types of inflammation; acute and chronic. Immediately after external stimuli, acute inflammation occurs, which takes a few days. Although acute inflammation is the body's response to external stimuli, if not controlled, it could give rise to chronic inflammation associated with autoimmune and Alzheimer disease, arthritis, atherosclerosis, and allergies. Acute inflammation is associated with the release of soluble mediators, such as cytokines and chemokines, causing macrophages and neutrophils to migrate to the site of inflammation. The mentioned cells play an active role in acute inflammation. If acute inflammation does not recover, it will lead to the migration of T lymphocytes to the area of inflammation, damage of tissue, and eventually fibrosis. Macrophages and monocytes take center stage in both phases of inflammation [111].

Different medicines have been introduced to suppress the crises of inflammation in the body. Anti-inflammatory drugs fall into two groups, steroidal and non-steroidal drugs, with various side effects and toxicity risks. Thus, the main objective of the drug discovery field is unraveling novel drugs with the minimum dose and adverse effects, as well as optimum efficacy. Researchers could get guidance from alternative and traditional medicine as a source of naturally derived medication. The first naturally-derived and clinically-practiced anti-inflammatory medicine was aspirin, isolated from salicylate-containing herbs. Hence, natural herbal products have created a new perspective on isolating anti-inflammatory medications [111].

There are several scientific approaches and guidelines to elucidate the anti-inflammatory effects of natural products in the acute phase of inflammation. In preclinical studies, several agents could be administered to induce inflammation in animal models.

5.11.1 Induction of Acute Inflammation

The first approach concerning chemically induced inflammation is carrageenan induction of paw edema. In this method, the injection of carrageenan subcutaneously gives rise to acute inflammation and edema formation due to the extravasation of plasma proteins, neutrophils, water, and the release of different mediators. The formation of edema comprises two phases; in the initial phase, serotonin and histamine are released, and diminish within 2 h. In the second stage, lysosome, protease, and prostaglandins are released, which takes three to 5 h. The procedure includes dividing animals into three groups: the control receiving vehicle, the second group receiving reference anti-inflammatory drug, and the third, injected test compound. The edema volume should be calculated for each group, and the percentage of edema inhibition should be compared between the treated animals and the reference drug-receiving group [112].

The second method is the **induction of paw edema by histamine,** which affects endothelial cells and evokes the release of eicosanoids and neuropeptides, giving rise to proinflammation and hyperalgesia. The process is similar to the induction of acute inflammation by carrageenan, in which the histamine solution is injected subcutaneously into the left hind paw of animals rather than carrageenan. According to the abovementioned method, the volume of edema and the percentage of inhibition of inflammation are calculated and compared [112].

The third approach is the **induction of vascular permeability by acetic acid**. Since acetic acid increases the dilation and permeability of blood vessels and the release of fluid, plasma proteins, and mediators, such as histamine and eicosanoids. Thus, inhibitory activity of the test chemicals could be evaluated against the induced inflammation. The procedure allocates six rats to each control, test, and treatment group. First, an acetic acid solution is injected intraperitoneally. Right after, Evan's blue is intravenously administrated to the rat's tail vein. The absorbance of dye exuded into the cavity of the peritonea is measured after thirty min [112].

Pleurisy is inflammation of layers defined as pleura covering lungs, which jeopardizes lung tissue and could give rise to diverse respiratory diseases. **Carrageenan-induced pleurisy** is another approach to induce acute inflammation employed to evaluate the anti-inflammatory effects of the test substances. It is about injecting carrageenan 1% solved in saline solution into the pleural sac. After exposure to carrageenan, inflammation, infiltration of leukocytes, and accumulation of ROS occur, triggering lung injury. Thus, in this experimental model, inflammation and oxidative stress fulfill a crucial function in pleurisy. The rodents are anesthetized, and the skin of the intercostal site and the underlying muscle is desiccated to inject carrageenan solution into the pleural cavity. A day after carrageenan administration, the pleural space is opened, rinsed by saline, and aspirated to remove and measure exudate. The leucocytes of exudate are stained and counted by a light microscope [112].

Acetic acid colitis assay is another commonly employed assay. Ulcerative colitis is an inflammatory ailment that afflicts the colon and rectum. It causes the development of small ulcers in colon tissue. This assay is one of the experimental approaches for studying IBD. There are similarities between colitis experimentally induced in animals and the one inflicting humans. In this model, the cascade of the inflammatory process, lipid peroxidation, and enormous production of ROS species cause deterioration of tissue damage. Thus, this test is a tool to investigate test compounds' anti-inflammatory, anti-ulcerative, and antioxidant effects. For the first step, rats of determined age and weight are selected. They are divided into negative and positive control and intervention groups. Then 4% (v/v) acetic acid in 0.9% NaCl is administered into the rectum of the rats by a catheter. The test substance should be administered 5 min after colitis induction and continued for three days. On the fourth day, the rats are euthanized and operated on to exclude the last 8 cm of the colon. The colons should be opened and rinsed to study the colonic lesions visually for histologic evaluation by light microscopy. Antioxidant enzymes, such as superoxide dismutase, malondialdehyde, and catalase could be assessed. Furthermore, the levels of various cytokines could be measured by enzyme immunoassay kits available commercially [112].

Dextran-induced edema in rats; the characterization of this model is induction of acute inflammation in rats. Similar to rat paw edema induced by carrageenan, the anti-inflammatory potential of test compounds could be assessed. However, carrageenan develops biphasic edema in which mast cells release serotonin and histamine in the initial phase. In the second phase, neutrophils are infiltrated, and prostaglandin E2, nitric oxide, and IL-1β are released. In the dextran model, vascular permeability surges, kinins are activated, and mast cells release histamine and serotonin. These incidences evoke osmotic edema, while the level of neutrophils and proteins is low [112].

In this assay, 500 μg of dextran is injected into the rat paws subcutaneously. One group of animals is injected with different concentrations of the test compound before injection of the test compound. By a plethysmometer, the paw volume should be measured before the injection of the test compound and following the injection in predetermined intervals. The difference in the paw volume (ml) before and after the injection is calculated [112].

Cotton-pellet granuloma test in rats is employed to evaluate the efficacy of compounds over subacute inflammation in the proliferative phase. Cotton pellets are implanted subcutaneously in the body of rodents; consequently, the granuloma is formed at the surrounding site of the pellets. Granulation tissue is a reddish-colored mass with multiplied blood vessels accumulated with liquid containing the mixtures of proteins and immune cells. The formation of granulomatous tissue, adsorption of fluid and the mentioned substances affect tissue's dry and wet weight, respectively. Initially, a hot air oven is used to sterilize pellets made of cotton wool. Then, they are implanted into the body of anesthetized rats. Within seven days, different doses of the test substance are administered daily. After that period, the cotton pellets in both control and intervention groups should be excluded for drying and obtaining the mean weight. Moreover, both groups calculate and compare the granuloma formation and prevention percentage [113].

5.11.2 Induction of Chronic Inflammation

Rheumatoid arthritis (RA) is an autoimmune, long-term, and inflammatory ailment in which the immune system attacks healthy body cells. Its manifestation is painful and swollen joints, and finally, mobility loss. The common strategy to relieve RA is using medicines to alleviate its symptoms and prevent the progression. Various approaches to investigating RA are used, and experimental animal models are commonly used. These animal models represent signs and pathogenesis of RA to evaluate potential therapeutic compounds to alleviate and treat RA-related primary and secondary implications. For a generation of RA, the suitable experimental models are rats and mice, susceptible to typical adjuvants for the induction of arthritis [114].

One of the in vivo studies for induction of this chronic inflammation is **formaldehyde-induced arthritis in rats,** in which 0.1 ml of 2.5% formaldehyde solution solved in normal saline is subcutaneously injected into the hind paw of rats. Rats are categorized into three groups: control-receiving vehicle, the second group receiving reference anti-inflammatory drug, and the third group receiving test compound or herbal extract. A digital caliper measures the rat's paw thickness on alternate days. There is another approach to induce arthritis in rats using adjuvant to induce chronic inflammation in the joints of rats through intradermal injection of the *Mycobacterium tuberculosis* bacilli prepared in sterilized paraffin oil and mannide monooleate. The intradermal injection of this suspension at the paw or tail base of the rats triggers inflammation to evaluate the anti-arthritic activity of the test substances [114].

Another in vivo study is **thermal hyperalgesia**. Thermal injury is combined with hyperalgesia and inflammatory pain. In this assay, rodents, such as a contact thermode, are exposed to heat to prompt a first-degree burn on the skin. This method is useful for studying primary and secondary hyperalgesia since heat stimuli induce pain in the inflamed area, defined as primary hyperalgesia, and outside the inflamed site, characterized as secondary. To detect heat sensitivity, Plantar Test or Hargreaves' Method measures rodents' response and pain sensation to heat stimulation. Heat stimulus is generated by the light source and used on the plantar part of the hind paw, and paw withdrawal is detected. If the rodent does not respond to the stimulus, it will be recorded as the termination of the trial. Withdrawal latency is a term used to express the results of this assay and is defined as the amount of time needed to notice the response to the stimuli. Before the induction of stimuli, the intervention group receives an injection of the various concentrations of the test substance at the hind paw, and the control group receives the injection of the vehicle. Paw withdrawal latencies are evaluated at determined intervals [115].

5.11.3 Examples of Herbal Latex with Anti-inflammatory Activity

Table 5.9 presents the scientific name of the plants, the latex product and its route of administration, the responsible phytochemicals, the models for studying anti-inflammatory activity, and the research results.

5.12 Pro-Inflammatory and Immune System Stimulation

The immune system's role is to bring about host defense against pathogens, and its proper and measured response is of utmost importance. The inappropriate and unmeasured response of this system could be pathologic. Similarly, insufficient response and immune system overreaction may trigger ailments and tissue damage. To reach this goal, anti-inflammatory and pro-inflammatory agents should balance dynamically to obtain an acceptable outcome. An immunostimulant agent is a compound that activates and reinforces the immune system's responses. These agents are noteworthy in patients with immune-suppressed ailments, such as the ones suffering from malignancies, AIDS, and SARS. Besides the mentioned conditions, there are newly emerging and naturally occurring infectious diseases and epidemics, which jeopardize the health of humankind worldwide. No typical treatment or vaccine exists for the mentioned ailments. However, non-specific immunotherapy or the employment of immunostimulants could help strengthen natural immune responses as a profitable strategy [128]. In recent years, immune system activation has become a promising field. Significantly, natural products or plant-derived compounds have the potential to be used as immune system stimulants.

Table 5.9 The scientific names, responsible products of latex-producing herbs, the assessment models for anti-inflammatory activity, and the results

Scientific name	Latex product	Responsible phytochemicals	Assessment model	Results	Refs
Aloe megalacantha	Leaf latex	NM[1]	In vivo test by carrageenan-induced inflammation of paw model	Dose-dependent and significant reductions of inflammation as compared to negative control groups.	[26]
Euphorbia bicolor	Methanolic extract of latex from leaves, inflorescence bracts, and stems	NM	In vivo test rat model of hind paw inflammation and induction of thermal hyperalgesia tests.	Intraplantar injection of induced significant thermal hyperalgesia and mechanical allodynia followed by long-lasting, non-opioid peripheral analgesia compared to vehicle injections.	[115]
	80% methanolic extract from the latex of stem, leaf, and inflorescence bracts.	NM	In vivo test by rat model of orofacial pain	It significantly reduced orofacial mechanical sensitivity in rats as compared to the negative control. In addition, it can decrease pain inducted by oxidative stress through a reduction in proinflammatory chemokines and cytokines.	[99]
Euphorbia lactea	Tirucallol, a triterpene isolated from the latex of leaves	Tirocallol	In vivo anti-inflammatory activity by ear edema induced by tetra decanoyl phorbol acetate. Western blot analysis of iNOS expression.	Tirucallol could suppress ear edema by topical application. It prevented neutrophil migration. It reduced Nitric Oxide production by preventing the expression of iNOS.	[116]
E. royleana	The hydrosoluble fraction of latex	NM	In vivo test by edema induced by carrageenan and dextran. Arthritis induced by formaldehyde Cotton-pellet granuloma test in rats	Oral administration of hydrosoluble fraction of latex significantly decreased inflammation induced by carrageenan and dextran. It significantly reduced arthritis. It poorly prevented granuloma. Formation.	[117]

(*Continued*)

Table 5.9 (*Continued*)

Scientific name	Latex product	Responsible phytochemicals	Assessment model	Results	Refs
Calotropis procera	Latex of aerial parts	Proteins of latex	Acetic acid-induced colitis in rats	Significantly ameliorated ulcerative colitis and acute inflammation comparable to the mesalazine.	[118]
Himatanthus drasticus	Lupeol acetate isolated from latex obtained from stems	Pentacyclic triterpene lupeol acetate	In vitro assay; MPO and LDH release from human neutrophils. In vivo assays; oral administration of lupeol acetate to a mouse; the models: paw edema induced by carrageenan and dextran, and the carrageenan-induced neutrophil migration into peritoneal cavities.	Oral lupeol acetate yielded a significant and dose-related inhibition of IL-2, IFN-γ, and TNF-α in pleural exudates of mice. It decreased the number of iNOS cells. Pro-inflammatory cytokines and the NO system participate in lupeol action.	[119]
	Latex from stems	Probably triterpenes, such as α-amyrin, β-amyrin, and lupeol cinnamate	In vivo paw edema model	The latex significantly reduced paw volume and serum levels of malondialdehyde.	[120]
Plumeria pudica	Water-soluble proteins of latex obtained from aerial parts.	Proteinases, chitinases, and proteinase inhibitors	Acetic acid-induced colitis in rats	Intraperitoneal injection of proteins significantly inhibited inflammatory cell infiltration, and the release of cytokine IL-1β, contributing to the reduction of colon inflammation.	[121]
P. rubra	Protein fraction of latex obtained from aerial parts	Latex proteins	Ethanol-induced gastric lesions in mice	Intravenous injection of proteins showed a gastro-protective effect and inhibited gastric lesions by the involvement of prostaglandin and NO/cGMP/K_{ATP} pathway	[122]

Scientific name	Latex product	Responsible phytochemicals	Assessment model	Results	Refs
Calotropis procera	Protein fraction of latex obtained from aerial parts	High molecular weight protein sub-fraction	In vivo assay; Freund's complete adjuvant-Induced monoarthritis in rats	A significant and dose-dependent decrease in joint swelling and improving joint functions.	[123]
	Chitinase isoforms obtained from latex	Chitinase isoforms	In vivo assay; mouse peritonitis model.	Chitinases showed significant anti-inflammatory effect	[124]
	The latex protein fraction	Proteins of latex	In vivo assays; acetic acid-induced abdominal constrictions, formalin, and hot plate tests	The significant and dose-dependent anti-nociceptive effect	[125]
Calotropis gigantea	Lupeol isolated from the latex obtained from arial parts	Lupeol, a pentacyclic triterpene	Arthritis induced by intradermal injection of Freund's Complete Adjuvant (FCA) in the hind paw	Oral administration of lupeol significantly decreased proinflammatory and anti-inflammatory cytokines, primary signs such as edema and pain, and secondary complications such as CVD and diabetes.	[126]
E. tirucalii	Crude extract obtained from branches and trunk	Perhaps triterpenoids euphol and tirucallol	An in vitro study using macro-phages RAW 264.7	The crude latex significantly reduced the production of the proinflammatory cytokines TNF-α, IL-6, and NO; the NO reduction was even more significant.	[127]

[1]Not Mentioned

Prior studies showed that latex yielded from sap-producing plants, especially its protein fraction, could induce an immune response or prompt pro-inflammatory activity. These responses include cellular response, migration of neutrophils, and vascular reaction, the elevation of vascular permeability. Furthermore, they cause the release of diverse pro-inflammatory mediators, namely, cytokines, chemokines, eicosanoids, and leukotriene.

5.12.1 Examples of Herbal Latex with Pro-Inflammatory and Immune System Stimulation Activity

Table 5.10 presents the scientific name of the plants, the latex product and its route of administration, the responsible phytochemicals, the models for studying pro-inflammatory activity, and the research results.

5.13 Anthelmintic Activity

Helminths are parasitic worms that live and nourish in the host's intestine and give rise to malabsorption and loss of nutrients. They are classified into three groups of nematodes (roundworms), cestodes (tapeworms), and trematodes (flatworms or flukes). Contamination by these helminths occurs when their eggs are present in infected human and livestock feces. The mature eggs hatch and yield larvae which grow and produce mature worms. The mature worms could penetrate the skin to enter the bloodstream and are transported to body organs causing various infections. Since helminth parasites affect crops and the production of foods, they affect the economy. Thus, anthelmintic medications are of utmost importance in human and veterinary medicine. These drugs are employed to treat and control diseases caused by parasitic helminths. However, frequent administration of currently-used drugs triggered the development of resistance to anthelminthic drugs [137].

Table 5.10 The scientific names, responsible products of latex-producing herbs, the assessment models for pro-inflammatory activity, and the results

Scientific name	Latex product	Responsible phytochemicals	Assessment model	Results	Refs.
Carica candamarcensis	Latex obtained from unripe fruits	Combination of cysteine proteinases	An In vivo evaluation, mice were injected with heat-killed *Salmonella enterica,* and the proteinase fraction of latex	As an immune system activator, lymphocytes in the bloodstreams of pretreated mice with cysteine proteinases were significantly increased. COX-2 and IL-1 mRNA transcripts were upregulated.	[129]
Cryptostegia grandiflora	Latex obtained from young branches	Proteins of latex	In vivo evaluation of vascular permeability by injection into peritoneal cavity, induction of paw edema, and air pouch models.	As an inflammatory agent, it caused neutrophil migration, vascular permeability, and surged myeloperoxidase.	[130]
E. tirucalii	80% hydroethanolic extract of latex obtained from branches and trunk	Euphol	In vivo test; oral administration of the latex extract to the rats	The extract significantly increased TH1, TH2 cytokines, and chemokines enhancing the immune system.	[131]
	Eutirucallin; a lectin as a carbohydrate-binding protein	Eutirucallin; a lectin as a carbohydrate-binding protein	In vivo test; intra-peritoneal injection of eutirucallin to the male rats	Eutirucallin can produce pro-inflammatory agents, IL12, and TNF-a cytokines and nitric oxide.	[132]

(*Continued*)

Table 5.10 (*Continued*)

Scientific name	Latex product	Responsible phytochemicals	Assessment model	Results	Refs.
Calotropis procera	water-soluble fraction of latex obtained from the aerial parts	NM[1]	In vivo test; intra-peritoneal injection of the water-soluble fraction to rats	Intraperitoneal injection of fraction in mice induced migration of macrophages to the intraperitoneal cavity, confirming the proinflammatory effects.	[133]
	Latex obtained from the aerial parts	Biogenic amines, perhaps histamine in latex	In vivo test; rat paw edema model by injecting dried latex	Dried latex indicated dose-dependent, significant, and intense inflammation of the rat paw.	[134]
	A soluble protein fraction	Proteins of latex	In vivo test; by intraperitoneal injection of protein fraction to mice infected by Listeria monocytogenes	Immunomodulatory and protective properties and prevention of intracellular bacterial infections by increasing leukocyte migration to the infectious area and increasing TNF- α, IL-6, IL-1β and iNOS.	[135]
	Proteins of latex	Proteins of latex	In vivo test; by intraperitoneal injection of protein fraction to mice infected by *Salmonella enterica* [model of typhoid fever)	Prevention of septic shock and salmonella clearance occurred. Murine model of typhoid fever. Neutrophils infiltered into the peritoneal cavity of pretreated mice with latex proteins.	[136]

[1]Not Mentioned

Furthermore, their less efficacy and adverse effects led to the urgent need to develop novel drug classes. Natural products are a valuable source of drug discovery because they contain various compounds with several functional groups at diverse concentrations. Various natural remedies have been traditionally used throughout the world. Typically, diverse herbal latexes are used in folk medicine. They are historically proven to obtain anthelmintic activity, and they are historically proven to obtain anthelmintic activity. Recent studies focus on the scientific validation of ethnobotanical practices and the isolation of active ingredients. There are different approaches to evaluating the anthelmintic activity of the test substances; first, they could kill the parasite or its eggs. Second, they could prevent the growth of parasites. Third, they could indicate their inhibitory activity by paralyzing the helminths. Thus, they cannot attach to the host tissues [137]. The ensuing paragraphs will discuss various in vitro assays proving the anthelminthic activity of test compounds, and the list of herbal latex with anthelminthic effects will be discussed.

5.13.1 Egg Hatch Assay

This test is an in vitro approach to evaluate the anthelmintic activity of the test compounds. The helminth eggs are isolated from the feces of infected humans or animals using the floatation method, and the isolated eggs are washed with distilled water. A 24-well plate, with a volume of 1 ml per well, is used, and 100 to 150 eggs, and the different concentrations of the test compound are added to each well. In wells allocated to the negative control, distilled water is added instead of the test solution. According to the type of helminth parasite, a respective anthelminthic medicine should be used as a positive control. First, the plates are cultured at 20°C for 48 h. Then, hatching should be prevented in each well by adding a solution defined as helminthological iodine. Then, the number of st-stage larvae and the unhatched eggs should be counted using an optical microscope to calculate the hatching percentage for each extract concentration [138].

5.13.2 Larval Motility Assay

In this in vitro test, the criterion for anti-parasitic activity is the prevention of adult worm mobility. Various concentrations of test substances are prepared and included into the wells. Collected and washed mobile adult worms are also placed in plate wells. PBS is used as a negative control, and an anthelminthic medicine should be used as a positive control. The adult worms' mobility should be observed with a magnifying glass every 6 h. The observation ends when the worms are immobile for 5 s, indicating they are all dead. Motility inhibition could be assessed by a percentage indicating the number of immobile ones versus the total number of parasites [139].

5.13.3 Larval Migration Inhibition Assay

Larval migration inhibition assay is an in vitro test based on measuring larvae migration rate across a membrane after exposure to test compounds. After hatching eggs, larvae take ten days to reach L3 or the infective stage. A defined number of L3 larvae are exposed to different concentrations of test compounds. Then, they are incubated at 20°C for three h. Afterward, the larvae suspension should be washed and centrifuged with PBS. A mesh with a diameter of 20 μm should be used to suspend larvae to migrate across it within three h at 23°C. PBS is used as a negative control, and an anthelminthic medicine should be used as a positive control. After incubation, the emigrated larvae should be included in PBS and counted with magnifying glass. Therefore, the percentage of test compounds' inhibitory activity on larvae migration could be calculated [140].

5.13.4 Larval Development Assay

This technique is employed to assess the sensitivity level of parasites to anthelmintic agents and drug discovery. Initially, the parasitic eggs are collected and washed. Then, an equal number of eggs are transferred into containers containing a growth medium and a certain amount of test substance. Next, the eggs are incubated, and the incubation duration is determined according to the time required to attain L3-stage larvae. If the eggs are not hatched, or the larvae do not reach the L3 level, the test substance would have the potential of an anthelmintic effect [141].

5.13.5 In vivo Anthelmintic Test

In this study, the anthelmintic potential of test substances is evaluated against the helminth parasite infecting an animal. The animal is infected orally with L3-stage larvae of a selected helminth parasite. After 3 to 4 weeks, the McMaster technique should verify the animal infection. The infected animal's feces are collected and cultured for seven days at 25°C to allow the eggs to hatch and develop to obtain L3-stage larvae. Rodents could be selected for performing in vivo tests by infecting them by ingesting a solution containing 1000 L3 larvae through gavage. Ten days after infection, the rodents are divided into several groups, and each group orally receives a determined concentration of test substance dissolved in a vehicle through dosing gavage. Rodents in the control group are merely administered by vehicle. On day 13th of post-treatment, the rodents are sacrificed to remove and open their stomachs longitudinally. The removed stomachs are immersed in distilled water and located in a bain-marie at 37°C to transfer larvae into the distilled water. After removing the stomachs and adding formaldehyde solution, they are examined by microscope to detect adult and pre-adult worms [142].

5.13.6 Examples of Herbal Latex with Anthelminthic Activity

Table 5.11 presents the scientific name of the plants, the latex product and its route of administration, the responsible phytochemicals, the models, and the research results.

Table 5.11 The scientific names, responsible products of latex-producing herbs, the assessment models for anthelmintic activity, and the results

Scientific name	Latex product	Responsible phytochemicals	Assessment model	Results	Refs
Ficus carica	latex from unripe fruits and leaves of the fig tree.	Cysteine proteases	In vitro larval motility assay on adult male and female *Trichuris muris* (a rodent nematode).	Exposure to *Ficus carica* latex and cysteine proteases caused decrease in the motility of *Trichuris muris* adult worms in vitro. Damage to the worms' cuticle happened after 45 m and 30 min, respectively (significance ND).	[143]
Ficus benjamina	Latex of apex and branches	Cysteine proteases	In vitro larval motility assay on adult male and female *Trichuris muris* (a rodent nematode).	Exposure to *Ficus benjamina* latex caused decrease in the motility of *Trichuris muris* adult worms in vitro. Damage to the worms' cuticle happened after 45 m. (significance ND)	[143]
	Latex protease obtained from apex and branches.	Latex protease	In vitro larval development assay. In vitro exsheathment inhibition assay	Latex protease prevented development and exsheathment of larvae of the gastrointestinal nematode *Haemonchus contortus.*	[144]
Carica papaya	Fruit latex	Cysteine proteases	In vivo anthelmintic test	Papaya latex significantly decreased both adult worm burden and egg output in the large intestine of mice infected with adult *Trichuris muris.*	[143]
		Perhaps: cysteine proteinases like papain and chymopapain	In vitro tests; egg hatching and larval motility assays	Latex and purified papain were vermicides against gastrointestinal Nematodes, *Strongyloides venezuelensis* eggs and larvae	[145]
	Ethyl acetate extract of latex obtained from	Perhaps 1-(2′,5′-dimethoxyphenyl)-glycerol	In vitro assay, egg hatch test, larval development test, and worms motility test	The extract inhibited egg hatching, larval development and motility of the adult worms of *Haemonchus contortus.*	[146]

5.14 Cytotoxicity, Antitumor, and Anticancer Tests

Cell cytotoxicity or cell stress is generally defined as the capability to kill living cells. Toxic agents exposed to living organisms could affect them negatively. Therefore, it is important to quantitatively determine toxicity of toxic substances to decide how hazardous they are and how they should be exposed to them. Cytotoxicity test is one of the initial steps and a prime indicator to evaluate in vitro biological effects of natural compounds. Physicochemical factors could trigger cytotoxicity, such as the denaturation of proteins. Moreover, cell stress could occur by interfering with typical cellular pathways, such as activating or inhibiting of the enzymes or receptors [147].

Cytotoxic compounds must not target healthy living cells by necrosis or apoptosis. Also, some compounds trigger cytostasis, in which cell viability is decreased, or in other words, growth and division of the cells are actively stopped. Exposure of cells to cytotoxic agents could bring about several fates; in some cases, necrosis occurs in which the cohesion of cell membrane is broken due to the lysis of cells, the contents of the cells are released, the metabolism is shut down, and they are quickly killed. Due to the accelerated embark of in vitro necrosis, apoptotic machinery is not activated, and the related markers are not produced. In the process of apoptosis, cell viability declined. In other words, the cell growth and division are stopped, and it could embark on the process of apoptosis or activation of intracellular cell death. In this process, a series of molecular incidences occur, including a shutdown of metabolism, cytoplasm shrinkage, the nucleus condensed, the DNA cleaved, and membrane cohesion destroyed. Cytotoxicity of test substances could be measured by several tests discussed in the ensuing paragraphs [147].

5.14.1 Brine Shrimp Lethality Assay

As a high-throughput screening approach, this test could be advantageous from various aspects; Since the least or no toxicity is a crucial determinant to developing a novel drug or cosmetics, it assists in identifying chemicals with the cytotoxic effects that could cause unintended health risks. In addition, this test could be beneficial to discover novel pharmaceuticals with potential therapeutic activity against cancerous cells. Brine shrimps or various species of *Artemia* are suitable aquatic invertebrates to screen for toxicity of substances. Their year-round availability and high sensitivity to toxicity make them appropriate species for tests. The procedure of this study involves exposure of *Artemia* nauplii with various concentrations of the test chemical within 24 h to obtain the percentage of mortality. A container containing Artemia and the solvent will be allocated to control. This test will be carried out in triplicate to acquire an accurate LD50 [148].

5.14.2 Cell-Based Tests

Several cell-based tests have been developed to determine the toxicity. For this purpose, various cell lines are used to assess cytotoxicity of the test compounds against both malignant and normal cells regarding cellular division and changes in their morphology. Cell culture is useful, cost-effective, and time-saving to determine the toxicity of substances and cell viability. To this purpose, several tetrazolium-based tests exist: MTT, WST-1, MTS, and XTT. **MTT assay** is a common colorimetric and quantitative test to assess the viability and metabolic activity of the cells. In this test, monolayer cells are plated in multiwall plates or dishes and incubated with MTT reagent or tetrazolium salt (3-(4,5-dimethylthia-zol-2-yl)-2,5-diphenyltetrazolium bromide). Since the MTT reagent possesses a lipophilic structure, it could transfer through the mitochondrial membrane. The oxidoreductase enzyme of metabolically active cells reduces this chromogenic reagent into an insoluble compound named formazan. The optical density of intracellular formazan derived from MTT is measured by a microplate reader, which indicates its concentration and reflects mitochondrial activity [149].

WST-1 Assay is another experimental procedure to determine the antiproliferative activity of test substances using cell proliferation reagent, WST-1(2-(4-iodophenyl)-3-(4-nitrophenyl)-5-(2,4-disulfophenyl)-2H-tetrazolium). This reagent is used to quantify cell viability in the format of a 96-well plate. This test is also a tetrazolium-based and colorimetric approach in which dehydrogenase enzymes reduce tetrazolium salts to formazan. Spectrophotometric equipment is employed to determine the number of living cells. Both MTT and WST-1 assays determine the viability of cells according to the activity of their dehydrogenase enzymes. The difference between this test and MTT is that the MTT reagent is reduced inside the mitochondria, while WST-1 is reduced inside and outside this cellular organelle. The WST-1 reagent yields formazan, which is soluble in water, while in MTT assay, the water-insoluble purple formazan crystals are formed. WST-1 test is more convenient compared to MTT as in the MTT-based test long time of exposure is needed (24, 48, 72 h); in WST-1 assay, the incubation time is 2 h, thus shorter incubation period is required. Furthermore, the results of the MTT assay could be falsely positive due to the reduction of MTT by some compounds without the mediating viable cells [150].

MTS test is another MTT-related and chromogenic assay based on the enzymatic reduction of tetrazolium salt by mitochondrial enzymes of viable cells in culture. When the MTS reagent or 3-(4,5-dimethylthiazol-2-yl)-5-(3-carboxymethoxyphenyl)-2-(4-sulfophenyl)-2H-tetrazolium is reduced, the produced formazan could be solved in culture medium. MTS test is alternatively used instead of MTT, as the reagent toxicity is less and formazan solubilization is unnecessary. Hence, in this assay, the absorbance of produced formazan is at 490 nm [151].

XTT is another quantitative and chromogenic test for the assessment of the proliferation of cells. As per the assays mentioned above, dehydrogenase enzymes of mitochondria in living cells take center stage in this test. XTT is another tetrazolium-based approach in which XTT reagent or yellow tetrazolium salt (sodium 3′-[1-(phenylamino carbonyl)-3,4-tetrazolium]-bis (4-methoxy6-nitro) benzene sulfonic acid hydrate) is reduced to orange-colored formazan. The yielded formazan is soluble in a culture medium and could be spectrophotometrically detected. The more viable cells exist in the medium, the more orange-colored formazan is yielded. The absorbance of yielded formazan could be detected at 450–500 nm [151].

5.14.3 In vivo Antitumor Tests

It is a cumbersome task to deduce the toxicity of chemicals and the health of humans according to the results of an in vitro test. Although in vivo studies are time-consuming, costly, and require infrastructures and facilities, they offer improved anticipation of reactions in the living organism. Different animal models could be selected to assess the chemotherapeutic potential of the test compounds. It is of utmost importance to select an appropriate tumor model. Rodents are common models for preclinical evaluation of their efficacy. For tumor formation, subcutaneous, intravenous, or intraperitoneal injection of carcinogens is carried out. When the tumors reach the proper size, the test chemical is administered to observe the tumor response by pursuing and measuring the tumor size over time. For this reason, the results of in vitro cytotoxic tests demonstrating target cells are the basis of selecting a proper in vivo model to test the efficacy of substances. Furthermore, in vitro tests shed light on selecting sensitive models, the amount of initial dose, and the administration schedule. The endpoint observes the antitumor effect and modulation in cellular pathways in these studies. The tumor size and morbidity associated with the tumor should be decreased. Besides, lifespan should be lengthened, and quality of life should be increased [152].

5.14.4 Examples of Herbal Latex with Cytotoxic, Antitumor, Anticancer Activity

Table 5.12 presents the scientific name of the plants, the latex product and its route of administration, the responsible phytochemicals, the models for studying cytotoxic, antitumor, anticancer activity, and the research results.

5.15 Antimalarial Activity

Malaria is a parasitic disease spread to individuals by infected mosquitos, causing many symptoms. It is a curable ailment, and the symptoms could vary from mild to severe, grouped as uncomplicated and complicated malaria. Malaria parasite development occurs in

Table 5.12 The scientific names and responsible products of latex-producing herbs, as well as the assessment models for cytotoxic, antitumor, anticancer activities and the results

Scientific name	Latex product	Responsible phytochemicals	Assessment model	Results	Refs
Calotropis procera	Hexane, dichlo-romethane, and ethyl acetate fractions of latex obtained from aerial parts	Perhaps anticancer cardenolides: uscharin and voruscarin, and cyclopeptides	MTT assay	The extracts exhibited significant cytotoxic activity against HL-60 (human leukemia), Ovcar-8 (human ovarian adenocarcinoma), HCT-116 (human colon cancer), and SF-295 (human glioblastoma)	[153]
	Latex proteins isolated from latex of aerial parts	Latex proteins	MTT assay	Latex proteins showed significant cytotoxic effect on human peripheral blood mononuclear cells (SF295 and MDA-MB-435 cell lines).	[154]
	Latex proteins isolated from latex of aerial parts	Latex proteins	In vivo anti-tumor assay by oral and I.P. administration of latex proteins to mice transplanted with sarcoma 180	LP significantly reduced tumor growth and sarcoma cell proliferation by intraperitoneal and oral administration routes	[155]
	Aqueous suspension of dried latex obtained from aerial parts	NM[1]	-In vivo assay, oral administration of latex suspension to the mice with hepatocellular carcinoma -In vitro MTT assay	The suspension showed in vivo chemopreventive effect. Vascular endothelial growth factor was decreased in mice. The suspension indicated cytotoxic effect against hepatoma (Huh7) and non-hepatoma (COS-1) cell lines and nontransformed hepatocytes (AML12).	[156]

(Continued)

Table 5.12 (*Continued*)

Scientific name	Latex product	Responsible phytochemicals	Assessment model	Results	Refs
	Isoforms chitinases isolated from latex obtained from aerial parts	Isoforms chitinases	In vitro MTT assay	Highly and significantly cytotoxic against HCT-116 (colon carcinoma), OVCAR-8 (ovarian carcinoma), and SF-295 (glioblastoma) tumor cell lines.	[124]
Calotropis gigantia	Cardenolides isolated from latex obtained from aerial parts	Cardenolides	In vitro MTT assay	The cardenolides exhibited potent and significant cytotoxic effects against human breast cancer cell MCF-7.	[157]
Ficus carica	Latex from unripe fruits and leaves of fig trees.	NM	MTT assay	It significantly suppresses cervical cancer cell (HeLa and CaSki) proliferation without toxic effect on the normal HaCaT cells.	[43]
	Latex obtained from the leaves	Perhaps proteases such as ficin, caseinolytic, and gelatinolytic enzymes	In vivo test by induction of breast tumor in female Wistar rats	Daily and intratumoral injections of fig latex significantly decreased breast tumor volume within four weeks by induction of apoptosis. It did not cause any adverse effect on hematological and histological factors.	[158]
	Water fraction latex obtained from fig leaves.	Synergistic effect of protocatechuic, caffeic, ferulic, p-coumaric acids, and quercetin.	WST-1 assay	Water fraction latex obtained fig leaves significantly decreased the proliferation of T98G, U-138 MG, and U-87 MG, GBM cell lines.	[159]
	Latex from leaves	Perhaps ficin, a cysteine proteinase	MTT assay	The latex significantly prevented the growth of the stomach cell line without any obvious effect on peripheral blood mononuclear and normal cells.	[160]

Scientific name	Latex product	Responsible phytochemicals	Assessment model	Results	Refs
	The latex from fruits and leaves	NM	MTT assay	The latex could significantly reduce the viability of Hella cell line, which was significant compared to the negative control.	[161]
	Latex obtained from the leaves	NM	MTT assay for assessment of cellular phototoxicity	The latex indicated significant antiproliferative activity on the human tumor cell line A375 (melanoma) after irradiation at a specific UVA dose.	[162]
	Latex obtained from leaves and fruits	Combination of 6-O-palmitoyl-β-D-glucosyl-β-sitosterol	MTT assay	The latex indicated cytotoxicity against Raji cells, DG-75, Jurkat, HD-MAR, DU-145, and MCF-7 (significance was not defined).	[163]
	Fruit latex	NM	MTT assay	The proliferation of Human glioblastoma, and hepatocellular carcinoma was significantly decreased. No significant effect was observed on normal liver cells. Latex specifically inhibited the proliferation of cancer cells.	[164]
Hevea brasiliensis	Latex of trunk and branches was used for the preparation of C-serum	C-serum, which was a rich source of proteins and nucleic acids	MTT assay	Significant anti-proliferation properties on hepatocellular adenocarcinoma cell lines (HepG2).	[165]

(*Continued*)

Table 5.12 (*Continued*)

Scientific name	Latex product	Responsible phytochemicals	Assessment model	Results	Refs
Euphorbia bicolor	Latex from leaves, inflorescence bracts and cut stems, and the isolates, resiniferatoxin, and rutin	The combination of latex extract, resiniferatoxin, and rutin	MTT assay	The combination of latex extract, resiniferatoxin, and rutin induced significant antiproliferative activities in ER-positive MCF-7 and T47-D and triple-negative MDA-MB-231 and MDA-MB-469 breast carcinomas in a dose-dependent manner, but not in human normal primary dermal fibroblast cultures.	[166]
Euphorbia dendroides	Jatrophane diterpenoids isolated from the latex of stem	Jatrophane diterpenoids	P-gp inhibiting activity assay by flow-cytometry	Eleven isolated jatrophane diterpenoids were significantly but moderately active against lung, colorectal carcinoma, and glioblastoma cell lines.	[167]
E. umbellata	Euphol: tetracyclic triterpene alcohol	Euphol: tetracyclic triterpene alcohol	MTT assay	Euphol is particularly active against glioma cells (including primary, pediatric, and adult glioma cell lines), much more active than the temozolomide, a clinical drug used to treat some brain cancers	[168]
E. tirucalli	Latex from branches and trunk	Euphol: tetracyclic triterpene alcohol	In vivo study; oral administration of an aqueous solution of latex to tumor-bearing rats	Treatment with 50 μL/mL latex resulted in significant reduction of tumor cell proliferation by arresting the cell cycle of cancer cells.	[169]
	Crude latex from branches and trunk	Perhaps triterpenes such as euphol and tirucallol	MTT assay	The latex was significantly cytotoxic against Gastric Adenocarcinoma cancer cells (AGS) without damaging healthy ones.	[127]

Scientific name	Latex product	Responsible phytochemicals	Assessment model	Results	Refs
	Latex diluted in saline	NM	- In vitro MTT assay on melanoma B16/F10 cells. - In vivo assay by oral administration of diluted latex to the melanoma induced mice	Latex showed a significant decrease in cell viability on B16/F10 murine melanoma cells in vitro Latex showed a significant decrease in the volume fraction of metastatic colonies. The results were not associated with adverse histopathological changes in the liver, kidneys, or spleen.	[170]
E. esula	Latex diluted in ultrapure water	NM	In vitro thiazolyl blue assay	The latex significantly inhibited the proliferation of the human gastric carcinoma cell line.	[171]
E. nivulia	Ingol diterpenes isolated from the latex of the stem and trunk.	Ingol diterpenes	MTT assay	Ingol diterpenes showed significant cytotoxic activity against Colo 205, MT2, and CEM cell lines.	[172]
Himatanthus drasticus	Proteins isolated from latex obtained from stems	NM	In vivo antitumor test by intraperitoneal administration of proteins to mice	A significant anti-tumor agent in Sarcoma 180 and Walker 256 carcinosarcoma	[173]
	Latex obtained from stems	Probably triterpenes, such as α-amyrin, β-amyrin, and lupeol cinnamate	In vivo antitumor assay on S180 solid tumor model	Latex significantly showed anti-tumor activity in Sarcoma 180-bearing mice	[120]

[1]Not Mentioned

infected red blood cells producing toxic compounds, such as the pigment of hemozoin. When the erythrocytes lyse, merozoites, hemozoin, and other toxic compounds are released. They stimulate immune cells and treat cerebral malaria with severe pathology could occur [174].

Recently, resistance to antimalarial agents drove the attention of researchers to explore novel chemotherapeutics. Natural resources, such as traditional herbal remedies and products, are inexpensive and easily administrable. Modern science assists in the identification and authentication of practices based on traditional knowledge to discover plant-derived medications and determine their potential and safety to control malaria. There are several in vivo and in vitro approaches to indicate the antimalarial activity of test compounds [174].

4-day suppressive assay or Peter's test is an **in vivo assay** employed on mice infected by injection of *Plasmodium berghei berghei*. Three days after injection, the parasitemia is tested by taking blood smears to be assured of 30–40% parasitemia. The blood taken from the eye vein of the infected mice is mixed with normal saline and injected intraperitoneally. The mice are categorized into intervention and control groups 3 h after infection. The treatment group orally receives different dilutions of the test compound for four days. The remaining groups are administered chloroquine (reference drug) and the solvent (negative control) to calculate percentage suppression and ID50. Sap-producing plants endemic to different regions indicated antimalarial activity [174].

5.15.1 Examples of Herbal Latex with Antimalaria Activity

Table 5.13 presents the scientific name of the plants, the latex product and its route of administration, the responsible phytochemicals, the models for studying antimalaria activity, and the research results.

Table 5.13 The scientific names and responsible products of latex-producing herbs, the assessment models for antimalaria activity and the results

Scientific name	Latex product	Responsible phytochemicals	Assessment model	Results	Refs
Aloe megalacantha	Leaf latex	NM[1]	Peter's 4-day suppressive test on *Plasmodium berghei* (in vivo)	Significant parasite suppression as compared to the negative control.	[175]
Aloe pulcherrima	Leaf latex, two C-glycosylated anthrones, nataloin, and 7-hydroxyaloin.	Nataloin and 7-hydroxyaloin	Peter's 4-day suppressive test on *Plasmodium berghei* (in vivo)	leaf latex, nataloin, and 7-hydroxyaloin showed antimalarial activity compared to the negative control. 7-hydroxyaloin indicated the maximum suppression of the parasite.	[176]
Aloe citrina	Leaf latex, and two isolated named homonataloin A and B	Homonataloin A and B	Peter's 4-day suppressive test on Plasmodium berghei (in vivo)	The latex and isolated compounds exhibited significant antimalarial activity.	[177]

[1]Not Mentioned

5.16 Homeostatic Activity

Hemostasis includes the formation of fibrin clots and their dissolution, defined as coagulation and fibrinolysis, respectively. Formation and dissolution of clots should be in balance; otherwise, vessels may be blocked, causing severe vascular disorders, such as deep-vein thrombosis [178].

Hemorrhage or uncontrolled bleeding is one of the leading causes of death. Fatal bleeding is mainly triggered by the impairment in coagulation process and vascular injury, leading to failure of organs, coma, and death. Due to the restricted capacity of the intrinsic coagulation pathway, especially in massive bleeding, the generation of novel and effective hemostatic agents plays a crucial role. Furthermore, shortcomings of commercial hemostatic agents, such as gauze, fibrin, cellulose-based compounds, chitosan, alginate, and gelatin, open new avenues of research. Herbal products could facilitate the swift generation and interaction of the blood proteins to form a network and aggregate erythrocytes to induce coagulation [178].

Clotting assays are performed to determine the coagulant activity of test compounds. **Activated Partial Thromboplastin Time (aPTT)** is a low-cost and easily performed approach that measures the duration of blood clot formation. Any injury to blood vessels triggers clotting factors to yield clots and prevent bleeding. aPTT assay monitors how well the clotting factors are functioning. The principal of this approach is the preparation of citrated plasma, including a substitute for platelet, activator of clotting proteins, and $CaCl_2$ results in clot formation. Platelet-Free Plasma (PFP) is initially prepared by mixing the blood sample with trisodium citrate. The sample is centrifuged, and the separated supernatant is PFP. Increasing dilutions of the test substance ought to be incubated with PFP for 5 min (37°C). Next, various dilutions of the test substance solved in Tris-HCl buffer (pH 7.4) should be added to PFP pre-warmed to 37°C. After one min of incubation, $CaCl_2$ is added, and the clotting duration is recorded.

Prothrombin Time (PT) assay is another test to determine the duration of clot formation. Prothrombin is a coagulation factor and a protein synthesized in the liver. In this test, PFP, increasing dilutions of the test substance, tissue thromboplastin, and calcium are incubated at 37°C to initiate coagulation [179].

Current medications used for thrombosis involve thrombolytic representatives with a short half-life and the possibility of bleeding. Thus, a novel research area involves developing more efficient antithrombotic agents. Enzymes, such as proteases in herbal latex, can have antiplatelet/antithrombotic activity by breaking down the subunits of fibrinogen protein. Various approaches were proposed for the evaluation of the fibrinolytic activity. In an assay defined as a **fibrinogenolytic activity**, different concentrations of plant latex are exposed to human fibrinogen, and several hours are given till the termination of the reaction. It is noteworthy that pH and temperature play an important role in the activity of enzymes. Thus the optimum condition of the reaction should be detected. Imaging software monitors enzyme activity or the percentage of degradation of fibrinogen protein. In the **fibrinolytic activity** assay, platelet-free plasma and a solution of $CaCl_2$ are mixed to obtain the fibrin clot. The formed clot is removed, washed, and exposed to various concentrations of test substance solved in sodium phosphate buffer. After three h of incubation, the reaction is ceased by the denaturing buffer, and the sample is applied to gel electrophoresis to examine the degradation of fibrin. Herbal products are used as homeostatic agents [180].

5.16.1 The Examples of Herbal Latex with Homeostatic Activity

Table 5.14 presents the scientific name of the plants, the latex product and its route of administration, the responsible phytochemicals, the models for studying homeostatic activity, and the research results.

Table 5.14 The scientific names and responsible products of latex-producing herbs, the assessment models for homeostatic activity, and the results

Scientific name	Latex product	Responsible phytochemicals	Assessment model	Results	Refs
Carica candamarcensis	Latex from immature fruit	Combination of cysteine peptidases	Clotting assays, aPTT, TT, and PT	Proteolytic fraction of latex, as a thrombolytic agent, with proteolytic cleavage of fibrinogen and fibrin, significantly increased PT, TT, and aPTT, in a dose-dependent manner.	[181]
Ficus carica	Latex from unripe fruits and the leaves of fig tree.	Ficin or the mixture of proteases	Clotting assays, aPTT, and PT	Ficin shortened the activated partial thromboplastin time and the prothrombin time due to the activation of human coagulation factor X.	[182]
Euphorbia hirta	Thrombin-like serine protease (hirtin)	Thrombin-like serine protease (hirtin)	The fibrinogenolytic activity was analyzed by 12% SDS-PAGE.	The enzyme efficiently hydrolyzed fibrinogen and fibrin clot, it showed fibrinolytic activity	[183]
Euphorbia resinifera	A serine protease of latex named EuRP-61	A serine protease in latex	Fibrinogenolytic activity was analyzed by 12% SDS-PAGE.	EuRP-61 could digest human fibrinogen in a dose-dependent fashion. It has the potential to protect against thrombosis.	[184]
Calotropis gigantia	Aqueous extract of latex	Perhaps cysteine proteases	Fibrinogenolytic activity was analyzed by 12% SDS-PAGE.	The latex showed a hemorrhagic effect, hydrolyzing strongly all the subunits of fibrinogen and crude fibrin clot even at lower concentrations.	[185]

5.17 Cognition Enhancing Activity

Cognition impairment involves a gradual decline of memory and cognition, a preliminary sign of Alzheimer's disease. It is a multifactorial disorder associated with reduced transmission mediated by acetylcholine, oxidative stress, and neuroinflammation. Employment of herbal remedies for improving cognition impairment dates back 5000 years ago, and they are gaining more attention for drug discovery. Herbal remedies are multi-targeting agents due to containing multiple ingredients. Likely, the bioactive phytochemicals of herbs function synergistically. Thus, they could prevent and cure cognition decline. Various herbal mixtures, such as herbal latex have been used traditionally to alleviate central nervous system disorders, such as cognition impairment. Modern science confirmed traditional knowledge indicating that active latex phytochemicals showed neuro-protective effects and improved cognitive functions [186]. In vitro and in vivo experiments are performed to verify the cognitive enhancing effect of the herbal latex.

The level of acetylcholine in the brain plays an important role in the pathophysiology of cognitive dysfunction, as loss of cholinergic neurons and decline of acetylcholine could result in Alzheimer's disease. The Ellman method is an in vitro and colorimetric test performed to assess the potential of test compounds to inhibit the activity of acetylcholinesterase (ACHE) and (BCHE) butyrylcholinesterase enzymes. The enzyme hydrolysis acetylcholine into choline and acetic acid. Prevention of this enzyme brings about the accumulation of acetylcholine and hyperstimulation of individual neurons. In this test, acetylthiocholine is the substrate with similar kinetic behavior to acetylcholine and reacts with Ellman's reagent or DTNB (5,5′-dithiobis-2-nitrobenzoic acid), yielding a yellow-colored substance named 5-mer-capto-2-nitrobenzoic acid. The absorbance of the product is measured at 412 nm. Herbal products are used as cognition-enhancing agents [187].

5.17.1 The Examples of Herbal Latex with Cognition Enhancing Activity

Table 5.15 presents the scientific name of the plants, the latex product and its route of administration, the responsible phytochemicals, the models for studying cognition enhancing activity, and the research results.

Table 5.15 The scientific name, responsible product of latex-producing herb, the assessment model for cognition-enhancing activity, and the results

Scientific name	Latex product	Responsible phytochemicals	Assessment model	Results	Refs
Calotropis procera	Latex obtained from aerial parts	NM[1]	In vitro acetyl-cholinesterase Microplate Inhibition Assay In vivo assay, scopolamine- and electroconvulsive shock induced amnesia in rats	Latex inhibited the enzyme in in vitro test Significant and dose- dependent improvement of performance in rats	[188]

[1]Not Mentioned

5.18 Sedative and Anxiolytic Effects

Insomnia is a chronic sleep disorder and a health issue that negatively affects the quality of life. Sedative-hypnotic medicines and anti-depressants are commonly used to facilitate insomnia which has several side effects and causes difficulties for patients. Therefore, people resort to ethnomedicine, herbal products, and complementary therapies. For example, folk uses herbal latex to relieve anxiety and sleep disorders. However, modern pharmacological research is conducted to confirm their anxiolytic and sedative effects. Several tests evaluate the mentioned effects of herbal products; traction, fireplace, hole board, and thiopental-induced sleeping tests are the common assays.

Courvoisier, Rousselet, and Laroche developed a traction test that diazepam is a reference drug, and the test material is intraperitoneally injected into the mice hung from a stretched rope. Normal mice pull their legs to reach the rope, while the sedative mice fail to perform the mentioned action [189].

In the fireplace test, two groups of control and intervention receive diazepam and test material intraperitoneally. One hour after injection, the mice are situated in a glass tube 30 cm long. Their effort to escape from the tube within 30 s is considered normal, while the mice showing no effort within the mentioned time are affected by sedative material [190].

In the hole board approach, curiosity and exploration of the mouse are explored. They are situated in a box-like device (40×40) with a hole in its bottom. The mouse is intraperitoneally administered diazepam and a test substance as a reference and test drug. The more insertion of the head of the mouse into the hole is recorded, the more alert the mouse is. Thus, sedative drugs result in less insertion of the mouse head [191].

In the elevated Plus-Maze Test, a maze apparatus is used to determine whether or not the test substance is sedative. One hour after administrating the test compound, the mouse is situated in the center of the maze apparatus for 5 min comprising two open and two closed arms. The sedative substance triggers a higher number of entries into the open arm. Similar to the tests mentioned earlier, diazepam is the standard drug [192].

The open-field test assesses whether the mouse can spontaneously explore after receiving sedative material. An open-field instrument comprises a white box dividing the area into black and white squares. The mouse will be checked while visiting the squares 1 h after administering the test and reference substances. Then, the number of blocks crossed by the animal is calculated. The lower the number mentioned, the more sedative the mouse is [193].

In the thiopental-induced sleeping test, the animals are initially exposed to thiopental, and after 30 min, the test material is injected intraperitoneally. The results are recorded as the time between injection of test material and sleep, as well as sleep and wake up [194].

5.18.1 The Examples of Herbal Latex with Sedative and Anxiolytic Activity

Table 5.16 presents the scientific name of the plants, the latex product and its route of administration, the responsible phytochemicals, the models for studying sedative and anxiolytic activity, and the research results.

Table 5.16 The scientific names, responsible products of latex-producing herbs, and the assessment models for sedative and anxiolytic activity and the results

Scientific name	Latex product	Responsible phytochemicals	Assessment model	Results	Refs
Lactuca saligna *L. serriola* *L. viminea*	lactucin-type guaianolides isolated from latex obtained from stems	lactucin-type guaianolides: Lactucin, lactucopicrin	In vivo assay; traction, fireplace, holeboard, and thiopental-induced sleeping tests induced in rats	The isolated compounds exhibited sedative activity	[195]

5.19 Conclusion

Herbal products, typically herbal latex, comprise an important part of the medicine market worldwide. They are estimated as the primary source of healthcare. Thus, their long history of use, popularity among people, and biologically active secondary metabolites assigned them to be explored for deriving novel natural medicines. Novel high-throughput technologies pave the pave to reach novel pharmaceuticals. Due to secondary metabolites of latex-producing plants, they inhibit the growth of bacteria, fungi, viruses, tumors, and cancer cell lines. They also showed analgesic and anti-inflammatory effects, larvicidal, anti-sedative, and cognition-enhancing activities. In addition, a wide array of biologically active compounds is responsible for the mentioned activities. Pertinent to demonstrating various pharmacologic effects, they have the potential to be commercialized. However, many active ingredients are yet to be isolated and identified from herbal latexes. Furthermore, various pharmacological activities of latex are yet to be explored. Since latex is an environmentally friendly natural product, it is promising to witness more latex-derived medicines in the near future.

References

1. Patel K, Patel DK. Medicinal importance, pharmacological activities, and analytical aspects of hispidulin: A concise report. *Journal of Traditional and Complementary Medicine*. 2017; 7(3): 360–6.
2. Kapetanovic IJC-bi. Computer-aided drug discovery and development (CADDD): in silico-chemico-biological approach. *Review Chemico Biological Interactions*. 2008; 171(2): 165–76.
3. Harry GJ, Billingsley M, Bruinink A, Campbell IL, Classen W, Dorman DC, et al. In vitro techniques for the assessment of neurotoxicity. *Environmental Health Perspectives*. 1998; 106(suppl 1): 131–58.
4. Saeidnia S, Manayi A, Abdollahi M. From in vitro experiments to in vivo and clinical studies; pros and cons. *Current Drug Discovery Technologies*. 2015; 12(4): 218–24.
5. Contreras-Llano LE, Tan C. High-throughput screening of biomolecules using cell-free gene expression systems. *Synthetic Biology*. 2018; 3(1): ysy012.
6. Katt ME, Placone AL, Wong AD, Xu ZS, Searson PC. In vitro tumor models: advantages, disadvantages, variables, and selecting the right platform. *Frontiers in Bioengineering and Biotechnology*. 2016; 4: 12.
7. Nuutila K, Katayama S, Vuola J, Kankuri E. Human wound-healing research: issues and perspectives for studies using wide-scale analytic platforms. *Advances in Wound Care*. 2014; 3(3): 264–71.
8. Velnar T, Bailey T, Smrkolj VJJoimr. The wound healing process: an overview of the cellular and molecular mechanisms. *Journal of International Medical Research*. 2009; 37(5): 1528–42.
9. Maver T, Maver U, Stana Kleinschek K, Smrke DM, Kreft S. A review of herbal medicines in wound healing. *International Journal of Dermatology*. 2015; 54(7): 740–51.
10. Walter MN, Wright KT, Fuller HR, MacNeil S, Johnson WEB. Mesenchymal stem cell-conditioned medium accelerates skin wound healing: an in vitro study of fibroblast and keratinocyte scratch assays. *Experimental Cell Research*. 2010; 316(7): 1271–81.

11. Ribatti D. The chick embryo chorioallantoic membrane (CAM) assay. *Reproductive toxicology (Elmsford, NY)*. 2017; 70: 97–101.
12. Naik M, Brahma P, Dixit M. A cost-effective and efficient chick ex-ovo CAM assay protocol to assess angiogenesis. *Methods and Protocols*. 2018; 1(2): 19.
13. Jonkman JE, Cathcart JA, Xu F, Bartolini ME, Amon JE, Stevens KM, et al. An introduction to the wound healing assay using live-cell microscopy. *Cell Adhesion & Migration*. 2014; 8(5): 440–51.
14. Mathew-Steiner SS, Roy S, Sen CK. Collagen in wound healing. *Bioengineering*. 2021; 8(5): 63.
15. Chaudhari M, Mengi S. Evaluation of phytoconstituents of Terminalia arjuna for wound healing activity in rats. *Phytotherapy Research*. 2006; 20(9): 799–805.
16. Tekleyes B, Huluka SA, Wondu K, Wondmkun YT. Wound healing activity of 80% methanol leaf extract of Zehneria scabra (Lf) Sond (Cucurbitaceae) in mice. *Journal of Experimental Pharmacology*. 2021; 31(13): 537–44.
17. Thakur R, Jain N, Pathak R, Sandhu SS. Practices in wound healing studies of plants. *Evidence-Based Complementary and Alternative Medicine*. 2011; 2011: 438056.
18. Masson-Meyers DS, Andrade TA, Caetano GF, Guimaraes FR, Leite MN, Leite SN, et al. Experimental models and methods for cutaneous wound healing assessment. *International Journal of Experimental Pathology*. 2020; 101(1–2): 21–37.
19. Kumari M, Eesha B, Amberkar M, Kumar N. Wound healing activity of aqueous extract of Crotalaria verrucosa in Wistar albino rats. *Asian Pacific Journal of Tropical Medicine*. 2010; 3(10): 783–7.
20. Sampaio LAS, Costa JS, Freire TFC, Reis SRDA, Deiro TCBDJ, Medrado ARAP. Influence of protein malnutrition on cutaneous wound healing in rats. *Revista de Nutrição*. 2018; 31: 433–42.
21. Deng L, Du C, Song P, Chen T, Rui S, Armstrong DG, et al. The role of oxidative stress and antioxidants in diabetic wound healing. *Oxidative Medicine Cellular Longevity*. 2021; 2021: 8852759.
22. Wilhelm KP, Wilhelm D, Bielfeldt S. Models of wound healing: an emphasis on clinical studies. *Skin Research and Technology*. 2017; 23(1): 3–12.
23. Kuhlmann M, Wigger-Alberti W, Mackensen Y, Ebbinghaus M, Williams R, Krause-Kyora F, et al. Wound healing characteristics of a novel wound healing ointment in an abrasive wound model: a

randomised, intra-individual clinical investigation. *Wound Medicine.* 2019; 24(1): 24–32.

24. Brash JT, Ruhrberg C, Fantin A. Evaluating vascular hyperpermeability-inducing agents in the skin with the Miles assay. *Journal of Visualized Experiments.* 2018; 19(136): 57524.
25. Swamy SM, Tan P, Zhu YZ, Lu J, Achuth HN, Moochhala S. Role of phenytoin in wound healing: microarray analysis of early transcriptional responses in human dermal fibroblasts. *Biochemical and Biophysical Research Communications.* 2004; 314(3): 661–6.
26. Gebremeskel L, Bhoumik D, Sibhat GG, Tuem KB. In vivo wound healing and anti-inflammatory activities of leaf latex of Aloe megalacantha baker (Xanthorrhoeaceae). *Evidence-Based Complementary and Alternative Medicine.* 2018; 2018: 5037912.
27. Mendonça RJ, Maurício VB, Teixeira Lde B, Lachat JJ, Coutinho-Netto J. Increased vascular permeability, angiogenesis and wound healing induced by the serum of natural latex of the rubber tree Hevea brasiliensis. *Phytotherapy Research.* 2010; 24(5): 764–8.
28. Leite MN, Leite SN, Caetano GF, Andrade TAMd, Fronza M, Frade MAC. Healing effects of natural latex serum 1% from Hevea brasiliensis in an experimental skin abrasion wound model. *Anais Brasileiros de Dermatologia.* 2020; 95: 418–27.
29. Mali PY, Panchal SS. Euphorbia neriifolia L.: review on botany, ethnomedicinal uses, phytochemistry and biological activities. *Asian Pacific Journal of Tropical Medicine.* 2017; 10(5): 430–38.
30. Bigoniya P, Rana A. Wound healing activity of Euphorbia neriifolia leaf ethanolic extract in rats. *Journal of Natural Remedies.* 2007; 7(1): 94–101.
31. Badgujar S, Mahajan R, Chopda M. Wound healing activity of latex of Euphorbia nivulia Buch.-Ham. in mice. *Research Journal of Pharmacology and Pharmacodynamics.* 2009; 1(2): 90–2.
32. Goyal M, Nagori BP, Sasmal D. Wound healing activity of latex of Euphorbia caducifolia. *Journal of Ethnopharmacology.* 2012; 144(3): 786–90.
33. Bessa G, Melo-Reis PR, Araújo LA, Mrué F, Freitas GB, Brandão ML, et al. Angiogenic activity of latex from Euphorbia tirucalli Linnaeus 1753 (Plantae, Euphorbiaceae). *Brazilian Journal of Biology = Revista Brasleira de Biologia.* 2015; 75(3): 752–8.
34. Rasik AM, Raghubir R, Gupta A, Shukla A, Dubey MP, Srivastava S, et al. Healing potential of Calotropis procera on dermal wounds in Guinea pigs. *Journal of Ethnopharmacology.* 1999; 68(1–3): 261–6.

35. Gurung S, Skalko-Basnet N. Wound healing properties of Carica papaya latex: in vivo evaluation in mice burn model. *Journal of Ethnopharmacology*. 2009; 121(2): 338–41.

36. Gomes FS, Spínola Cde V, Ribeiro HA, Lopes MT, Cassali GD, Salas CE. Wound-healing activity of a proteolytic fraction from Carica candamarcensis on experimentally induced burn. *Burns: Journal of the International Society for Burn Injuries*. 2010; 36(2): 277–83.

37. Ellan K, Thayan R, Raman J, Hidari KI, Ismail N, Sabaratnam V. Antiviral activity of culinary and medicinal mushroom extracts against dengue virus serotype 2: an in vitro study. *BMC Complementary and Alternative Medicine*. 2019; 19: 1–12.

38. Kiesslich S, Kamen AA. Vero cell upstream bioprocess development for the production of viral vectors and vaccines. *Biotechnology Advances*. 2020; 44: 107608.

39. Meléndez-Villanueva MA, Morán-Santibañez K, Martínez-Sanmiguel JJ, Rangel-López R, Garza-Navarro MA, Rodríguez-Padilla C, et al. Virucidal activity of gold nanoparticles synthesized by green chemistry using garlic extract. *Viruses*. 2019; 11(12): 1111.

40. Mazzon M, Ortega-Prieto AM, Imrie D, Luft C, Hess L, Czieso S, et al. Identification of broad-spectrum antiviral compounds by targeting viral entry. *Viruses*. 2019; 11(2): 176.

41. Lazreg Aref H, Gaaliche B, Fekih A, Mars M, Aouni M, Pierre Chaumon J, et al. In vitro cytotoxic and antiviral activities of Ficus carica latex extracts. *Natural Product Research*. 2011; 25(3): 310–9.

42. Huot N, Rascle P, Müller-Trutwin M. Contribution of animal models to HIV research. *Virologie*. 2019; 23(4): 229–40.

43. Ghanbari A, Le Gresley A, Naughton D, Kuhnert N, Sirbu D, Ashrafi GH. Biological activities of Ficus carica latex for potential therapeutics in Human Papillomavirus (HPV) related cervical cancers. *Scientific Reports*. 2019; 9(1): 1013.

44. Bohlooli S, Mohebipoor A, Mohammadi S, Kouhnavard M, Pashapoor SJIjod. Comparative study of fig tree efficacy in the treatment of common warts (Verruca vulgaris) vs. *Cryotherapy*. 2007; 46(5): 524–6.

45. Valadão ALC, Pezzuto P, Silva VAO, Goncalves BS, Rossi AD, da Cunha RD, et al. Reactivation of latent HIV-1 in vitro using an ethanolic extract from Euphorbia umbellata (Euphorbiaceae) latex. *PloS One*. 2018; 13(11): e0207664.

46. Tostes JB, Carvalho AL, Ribeiro da Silva AJ, Mourao PJP, Rossi AtD, Tanuri A, et al. Phorbol esters from the latex of Euphorbia umbellata: bioguided isolation of highly potent HIV-1 latency interrupters in virus reservoir cells. *Journal of Natural Products.* 2021; 84(5): 1666–70.

47. Daoubi M, Marquez N, Mazoir N, Benharref A, Hernández-Galán R, Munoz E, et al. Isolation of new phenylacetylingol derivatives that reactivate HIV-1 latency and a novel spirotriterpenoid from Euphorbia officinarum latex. *Bioorganic Medicinal Chemistry.* 2007; 15(13): 4577–84.

48. Parhira S, Yang Z-F, Zhu G-Y, Chen Q-L, Zhou B-X, Wang Y-T, et al. In vitro anti-influenza virus activities of a new lignan glycoside from the latex of Calotropis gigantea. *PloS One.* 2014; 9(8): e104544.

49. Harraka GT, Magnuson JT, Du B, Wong CS, Maruya K, Schlenk D. Evaluating the estrogenicity of an effluent-dominated river in California, USA: comparisons of in vitro and in vivo bioassays. *Science of The Total Environment.* 2021; 758: 143965.

50. Shanle EK, Xu W. Selectively targeting estrogen receptors for cancer treatment. *Advanced Drug Delivery Reviews.* 2010; 62(13): 1265–76.

51. Michelini E, Cevenini L, Mezzanotte L, Coppa A, Roda A. Cell-based assays: fuelling drug discovery. *Analytical and Bioanalytical Chemistry.* 2010; 398(1): 227–38.

52. Wang Z, Kanda S, Shimono T, Enkh-Undraa D, Nishiyama T. The in vitro estrogenic activity of the crude drugs found in Japanese herbal medicines prescribed for menopausal syndrome was enhanced by combining them. *BMC Complementary and Alternative Medicine.* 2018; 18(1): 1–12.

53. Basu P, Meza E, Bergel M, Maier C. Estrogenic, antiestrogenic and antiproliferative activities of Euphorbia bicolor (Euphorbiaceae) latex extracts and its phytochemicals. *Nutrients.* 2020; 12(1): 59.

54. Yoon KS, Kwack SJJJoT, Environmental Health PA. In vitro and in vivo estrogenic activity of triclosan. *Journal of Toxicology Environmental Health, Part A,* 2021; 84(19): 800–9.

55. Desjeux P. Leishmaniasis: current situation and new perspectives. *Comparative Immunology, Microbiology and Infectious Diseases.* 2004; 27(5): 305–18.

56. Chappuis F, Sundar S, Hailu A, Ghalib H, Rijal S, Peeling RW, et al. Visceral leishmaniasis: what are the needs for diagnosis, treatment and control? *Nature Reviews Microbiology.* 2007; 5(11): 873–82.

57. Ganguly S, Das NK, Barbhuiya JN, Chatterjee M. Post-kala-azar dermal leishmaniasis–an overview. *International Journal of Dermatology.* 2010; 49(8): 921–31.

58. Koutsoni OS, Karampetsou K, Dotsika E. In vitro screening of anti-leishmanial activity of natural product compounds: determination of IC50, CC50 and SI values. *Bio Protocol.* 2019; 9(21): e3410.

59. Hommel M, Jaffe C, Travi B, Milon G. Experimental models for leishmaniasis and for testing anti-leishmanial vaccines. *Annals of Tropical Medicine & Parasitology.* 1995; 89(sup1): 55–73.

60. Tewabe Y, Kefarge B, Belay H, Bisrat D. Antileishmanial evaluation of the leaf latex of Aloe macrocarpa, aloin A/B, and its semisynthetic derivatives against two leishmania species. *Evidence-Based Complementary and Alternative Medicine.* 2019; 2019: 4736181.

61. Abeje F, Bisrat D, Hailu A, Asres K. Phytochemistry and antileishmanial activity of the leaf latex of Aloe calidophila Reynolds. *Phytotherapy Research.* 2014; 28(12): 1801–5.

62. Soares DC, Andrade AL, Delorenzi JC, Silva JR, Freire-de-Lima L, Falcão CA, et al. Leishmanicidal activity of Himatanthus sucuuba latex against Leishmania amazonensis. *Parasitology International.* 2010; 59(2): 173–7.

63. Alam M, Bano N, Ahmad T, Sharangi AB, Upadhyay TK, Alraey Y, et al. Synergistic role of plant extracts and essential oils against multidrug resistance and gram-negative bacterial strains producing extended-spectrum β-lactamases. *Antibiotics.* 2022; 11(7): 855.

64. Kurhekar J, Tupas GD, Otero MCB. In vitro assays for antimicrobial Assessment. In Kumar S, Egbuna C, eds. *Phytochemistry: An in silico and in vitro Update.* Springer, Singapore. 2019; 279–98 pp.

65. Choma IM, Jesionek W. TLC-direct bioautography as a high throughput method for detection of antimicrobials in plants. *Chromatography.* 2015; 2(2): 225–38.

66. Wiegand I, Hilpert K, Hancock RE. Agar and broth dilution methods to determine the minimal inhibitory concentration (MIC) of antimicrobial substances. *Nature Protocols.* 2008; 3(2): 163–75.

67. Asamenew G, Bisrat D, Mazumder A, Asres K. In vitro antimicrobial and antioxidant activities of anthrone and chromone from the latex of Aloe harlana Reynolds. *Phytotherapy Research : PTR.* 2011; 25(12): 1756–60.

68. Hiruy M, Bisrat D, Mazumder A, Asres K. Two chromones with antimicrobial activity from the leaf latex of Aloe monticola Reynolds. *Natural Product Research.* May 2019: 1–5.

69. Emiru YK, Siraj EA, Teklehaimanot TT. Antibacterial potential of Aloe weloensis (Aloeacea) leaf latex against gram-positive and gram-negative bacteria strains. *International Journal of Microbiology.* 2019; 2019: 5328238.

70. Megeressa M, Bisrat D, Mazumder A, Asres K. Structural elucidation of some antimicrobial constituents from the leaf latex of Aloe trigonantha L.C. Leach. *BMC Complementary and Alternative Medicine.* 2015; 15: 270.

71. Oumer A, Bisrat D, Mazumder A, Asres K. A new antimicrobial anthrone from the leaf latex of Aloe trichosantha. *Natural Product Communications.* 2014; 9(7): 949–52.

72. Lazreg-Aref H, Mars M, Fekih A, Aouni M, Said K. Chemical composition and antibacterial activity of a hexane extract of Tunisian caprifig latex from the unripe fruit of Ficus carica. *Pharmaceutical Biology.* 2012; 50(4): 407–12.

73. Taira T, Ohdomari A, Nakama N, Shimoji M, Ishihara M. Characterization and antifungal activity of gazyumaru (Ficus microcarpa) latex chitinases: both the chitin-binding and the antifungal activities of class I chitinase are reinforced with increasing ionic strength. *Bioscience, Biotechnology, and Biochemistry.* 2005; 69(4): 811–8.

74. Kanokwiroon K, Teanpaisan R, Wititsuwannakul D, Hooper AB, Wititsuwannakul R. Antimicrobial activity of a protein purified from the latex of Hevea brasiliensis on oral microorganisms. *Mycoses.* 2008; 51(4): 301–7.

75. Daruliza K, Yang K, Lam K, Priscilla J, Sunderasan E, Ong M. Anti-Candida albicans activity and brine shrimp lethality test of Hevea brasiliensis latex B-serum. *European Review for Medical and Pharmacological Sciences.* 2011; 15(10): 1163–71.

76. Giordani R, Regli P, Buc J. Antifungal effect of Hevea brasiliensis latex with various fungi. Its synergistic action with amphotericin B against Candida albicans. *Mycoses.* 2002; 45(11–12): 476–81.

77. Siritapetawee J, Limphirat W, Wongviriya W, Maneesan J, Samosornsuk W. Isolation and characterization of a galactose-specific lectin (EantH) with antimicrobial activity from Euphorbia antiquorum L. latex. *International Journal of Biological Macro-molecules.* 2018; 120(Pt B): 1846–54.

78. Giordani R, Trebaux J, Masi M, Regli P. Enhanced antifungal activity of ketoconazole by Euphorbia characias latex against Candida albicans. *Journal of Ethnopharmacology.* 2001; 78(1): 1–5.

79. Ibrahim EH, Kilany M, Mostafa OMS, Shaker KH, Alshehri M, Alsyaad KM, et al. TH1/TH2 chemokines/cytokines profile in rats treated with tetanus toxoid and Euphorbia tirucalli. *Saudi Journal of Biological Sciences.* 2019; 26(7): 1716–23.

80. Aliyu RM, Abubakar MB, Kasarawa AB, Dabai YU, Lawal N, Bello MB, et al. Efficacy and phytochemical analysis of latex of Calotropis procera against selected dermatophytes. *Journal of Intercultural Ethnopharmacology.* 2015; 4(4): 314–7.

81. Ramos MV, de Oliveira RS, Pereira HM, Moreno FB, Lobo MD, Rebelo LM, et al. Crystal structure of an antifungal osmotin-like protein from Calotropis procera and its effects on Fusarium solani spores, as revealed by atomic force microscopy: insights into the mechanism of action. *Phytochemistry.* 2015; 119: 5–18.

82. Ishnava KB, Chauhan JB, Garg AA, Thakkar AM. Antibacterial and phytochemical studies on Calotropis gigantia (L.) R. Br. latex against selected cariogenic bacteria. *Saudi Journal of Biological Sciences.* 2012; 19(1): 87–91.

83. Sequeira BJ, Vital MJS, Pohlit AM, Pararols IC, Caúper GSBJMdIOC. Antibacterial and antifungal activity of extracts and exudates of the Amazonian medicinal tree Himatanthus articulatus (Vahl) Woodson (common name: sucuba). *Memórias do Instituto Oswaldo Cruz.* 2009; 104: 659–61.

84. Moulin-Traffort J, Giordani R, Régli P. Antifungal action of latex saps from Lactuca sativa L. and Asclepias curassavica L. *Mycoses.* 1990; 33(7–8): 383–92.

85. Pham-Huy LA, He H, Pham-Huy C. Free radicals, antioxidants in disease and health. *International Journal of Biomedical Science.* 2008; 4(2): 89.

86. Alok S, Jain SK, Verma A, Kumar M, Mahor A, Sabharwal M. Herbal antioxidant in clinical practice: a review. *Asian Pacific Journal of Tropical Biomedicine.* 2014; 4(1): 78–84.

87. Gracz-Bernaciak J, Mazur O, Nawrot R. Functional studies of plant latex as a rich source of bioactive compounds: focus on proteins and alkaloids. *International Journal of Molecular Sciences.* 2021; 22(22): 12427.

88. Pinchuk I, Shoval H, Dotan Y, Lichtenberg D. Evaluation of antioxidants: scope, limitations and relevance of assays. *Chemistry and Physics of Lipids.* 2012; 165(6): 638–47.

89. Schaich K, Tian X, Xie J. Hurdles and pitfalls in measuring antioxidant efficacy: a critical evaluation of ABTS, DPPH, and ORAC assays. *Journal of Functional Foods*. 2015; 14: 111–25.

90. Akar Z, Küçük M, Doðan H. A new colorimetric DPPH$^{\bullet}$ scavenging activity method with no need for a spectrophotometer applied on synthetic and natural antioxidants and medicinal herbs. *Journal of Enzyme Inhibition and Medicinal Chemistry*. 2017; 32(1): 640–7.

91. Shahidi F, Zhong Y. Measurement of antioxidant activity. *Journal of Functional Foods*. 2015; 18: 757–81.

92. Al-Amiery AA, Al-Majedy YK, Kadhum AAH, Mohamad AB. Hydrogen peroxide scavenging activity of novel coumarins synthesized using different approaches. *PloS One*. 2015; 10(7): e0132175.

93. Guzik T, Korbut R, Adamek-Guzik T. Nitric oxide and superoxide in inflammation. *Journal of Physiology and Pharmacology*. 2003; 54(4): 469–87.

94. Mahmud ZA, Bachar SC, Hasan CM, Emran TB, Qais N, Uddin MMN. Phytochemical investigations and antioxidant potential of roots of Leea macrophylla (Roxb.). *BMC Research Notes*. 2017; 10(1): 1–9.

95. Ivanova A, Gerasimova E, Gazizullina E. Study of antioxidant properties of agents from the perspective of their action mechanisms. *Molecules*. 2020; 25(18): 4251.

96. Tekulu GH, Araya EM, Mengesha HG. In vitro α-amylase inhibitory effect of TLC isolates of Aloe megalacantha baker and Aloe monticola Reynolds. *BMC Complementary and Alternative Medicine*. 2019; 19(1): 206.

97. Amare GG, Meharie BG, Belayneh YM. Evaluation of antidiabetic activity of the leaf latex of Aloe pulcherrima gilbert and sebsebe (Aloaceae). *Evidence-Based Complementary and Alternative Medicine*. 2020; 2020: 8899743.

98. Yadav YC, Srivastava DN. Nephroprotective and curative effects of Ficus religiosa latex extract against cisplatin-induced acute renal failure. *Pharmaceutical Biology*. 2013; 51(11): 1480–5.

99. Basu P, Hornung RS, Averitt DL. Euphorbia bicolor (Euphorbiaceae) latex extract reduces inflammatory cytokines and oxidative stress in a Rat model of orofacial pain. *Oxidative Medicine and Cellular Longevity*. 2019; 2019: 8594375.

100. Li P-H, Lin Y-W, Lu W-C, Hu J-M, Huang D-W. In vitro hypoglycemic activity of the phenolic compounds in longan fruit (Dimocarpus

Longan var. Fen ke) shell against α-glucosidase and β-galactosidase. *International Journal of Food Properties*. 2016; 19(8): 1786–97.

101. Senhaji S, Lamchouri F, Boulfia M, Lachkar N, Bouabid K, Toufik H. Mineral composition, in vitro inhibitory effects of α-amylase, α-glucosidase, β-galactosidase enzymes and antibacterial activity of Ajuga Iva Subsp. Pseudoiva (DC.) Bric. *Biointerface Research in Applied Chemistry*. 2021; 12(2): 2373–91.

102. Singh MP, Pathak K. Animal models for biological screening of anti-diabetic drugs: an overview. *European Journal of Experimental Biology*. 2015; 5(5): 37–48.

103. Gilbert ER, Liu D, Anti-diabetic functions of soy isoflavone genistein: mechanisms underlying its effects on pancreatic β-cell function. *Food Function*. 2013; 4(2): 200–12.

104. Hammeso WW, Emiru YK. Antidiabetic and antihyperlipidemic activities of the leaf latex extract of Aloe megalacantha baker (Aloaceae) in streptozotocin-induced diabetic model. *Evidence-Based Complementary and Alternative Medicine*. 2019; 2019: 8263786.

105. Amare GG, Meharie BG, Belayneh YM. Evaluation of antidiabetic activity of the leaf latex of Aloe pulcherrima gilbert and sebsebe (Aloaceae). *Evidence-based Complementary and Alternative Medicine*. 2020; 1–9.

106. Morais FS, Canuto KM, Ribeiro PR, Silva AB, Pessoa OD, Freitas CD, et al. Chemical profiling of secondary metabolites from Himatanthus drasticus (Mart.) Plumel latex with inhibitory action against the enzymes α-amylase and α-glucosidase: in vitro and in silico assays. *Journal of Ethnopharmacology*. 2020; 253: 112644.

107. Azari-Hamidian S, Norouzi B, Harbach RE. A detailed review of the mosquitoes (Diptera: culicidae) of Iran and their medical and veterinary importance. *Acta Tropica*. 2019; 194: 106–22.

108. Huang H-T, Lin C-C, Kuo T-C, Chen S-J, Huang R-N. Phytochemical composition and larvicidal activity of essential oils from herbal plants. *Planta*. 2019; 250: 59–68.

109. Hailesillassie T, Bisrat D, Asres K. Larvicidal effect of the leaf latex of Aloe yavellana Reynolds and its major compounds against Amblyomma variegatum (Ixodidae). *Veterinary Parasitology*. 2018; 263: 23–6.

110. Ramos MV, Bandeira GdP, Freitas CDTd, Nogueira NAP, Alencar NMN, Sousa PASd, et al. Latex constituents from Calotropis procera (R. Br.) display toxicity upon egg hatching and larvae of Aedes aegypti (Linn.). *Memórias do Instituto Oswaldo Cruz*. 2006; 101(5): 503–10.

111. Hannoodee S, Nasuruddin DN. Acute inflammatory response. In *StatPearls [Internet]. Treasure Island (FL): StatPearls Publishing.* 2022.

112. Coura CO, Souza RB, Rodrigues JAG, Vanderlei EdSO, de Araújo IWF, Ribeiro NA, et al. Mechanisms involved in the anti-inflammatory action of a polysulfated fraction from Gracilaria cornea in rats. *PloS One.* 2015; 10(3): e0119319.

113. Verma S, Ojha S, Raish M. Anti-inflammatory activity of Aconitum heterophyllum on cotton pellet-induced granuloma in rats. *Journal of Medicinal Plants Research.* 2010; 4(15): 1566–69.

114. Chinnasamy V, Subramaniyan V, Chandiran S, Kayarohanam S, Kanniyan DC, Velaga VSSR, et al. Antiarthritic activity of achyranthes aspera on formaldehyde-induced arthritis in rats. *Open Access Macedonian Journal of Medical Sciences.* 2019; 7(17): 2709–14.

115. Basu P, Tongkhuya SA, Harris TL, Riley AR, Maier C, Granger J, et al. Euphorbia bicolor (Euphorbiaceae) latex phytochemicals induce long-lasting non-opioid peripheral analgesia in a rat model of inflammatory pain. *Frontiers in Pharmacology.* 2019; 10: 958.

116. Fernandez-Arche A, Saenz M, Arroyo M, De la Puerta R, Garcia M. Topical anti-inflammatory effect of tirucallol, a triterpene isolated from Euphorbia lactea latex. *Phytomedicine.* 2010; 17(2): 146–8.

117. Bani S, Kaul A, Jaggi BS, Suri KA, Suri OP, Sharma OP. Anti-inflammatory activity of the hydrosoluble fraction of Euphorbia royleana latex. *Fitoterapia.* 2000; 71(6): 655–62.

118. Kumar VL, Pandey A, Verma S, Das P. Protection afforded by methanol extract of Calotropis procera latex in experimental model of colitis is mediated through inhibition of oxidative stress and pro-inflammatory signaling. *Biomedicine & Pharmacotherapy.* 2019; 109: 1602–9.

119. Lucetti DL, Lucetti EC, Bandeira MAM, Veras HN, Silva AH, Leal LKA, et al. Anti-inflammatory effects and possible mechanism of action of lupeol acetate isolated from Himatanthus drasticus (Mart.) Plumel. *Journal of Inflammation.* 2010; 7(1): 1–11.

120. Santos GJL, Oliveira ES, Pinheiro ADN, da Costa PM, de Freitas JCC, de Araújo Santos FG, et al. Himatanthus drasticus (Apocynaceae) latex reduces oxidative stress and modulates CD^{4+}, CD^{8+}, $FoxP^{3+}$ and HSP-60+ expressions in Sarcoma 180-bearing mice. *Journal of Ethnopharmacology.* 2018; 220: 159–68.

121. de Moraes Oliveira NV, da Silva Souza B, Moita LA, Oliveira LES, Brito FC, Magalhães DA, et al. Proteins from Plumeria pudica latex exhibit

protective effect in acetic acid induced colitis in mice by inhibition of pro-inflammatory mechanisms and oxidative stress. *Life Sciences*. 2019; 231: 116535.

122. Alencar NMNd, Pinheiro RSP, Figueiredo ISTd, Luz PB, Freitas LBN, Souza TdFGd, et al. The Preventive effect on ethanol-induced gastric lesions of the medicinal plant plumeria rubra: involvement of the latex proteins in the NO/cGMP/KATP signaling pathway. *Evidence-Based Complementary and Alternative Medicine*. 2015; 2015: 706782.
123. Kumar VL, Roy S. Calotropis procera latex extract affords protection against inflammation and oxidative stress in Freund's complete adjuvant-induced monoarthritis in rats. *Mediators of Inflammation*. 2007; 2007: 47523.
124. Viana CA, Ramos MV, Filho JDBM, Lotufo LVC, Figueiredo IST, de Oliveira JS, et al. Cytotoxicity against tumor cell lines and anti-inflammatory properties of chitinases from Calotropis procera latex. *Naunyn-Schmiedeberg's Archives of Pharmacology*. 2019; 390(10): 1005–13.
125. Soares PM, Lima SR, Matos SG, Andrade MM, Patrocínio MC, de Freitas CD, et al. Antinociceptive activity of Calotropis procera latex in mice. *Journal of Ethnopharmacology*. 2005; 99(1): 125–29.
126. Saratha V, Subramanian SP. Lupeol, a triterpenoid isolated from Calotropis gigantea latex ameliorates the primary and secondary complications of FCA induced adjuvant disease in experimental rats. *Inflammopharmacology*. 2012; 20(1): 27–37.
127. de Souza LS, Puziol LC, Tosta CL, Bittencourt ML, Santa Ardisson J, Kitagawa RR, et al. Analytical methods to access the chemical composition of an Euphorbia tirucalli anticancer latex from traditional Brazilian medicine. *Journal of Ethnopharmacology*. 2019; 237: 255–65.
128. Petrunov B, Nenkov P, Shekerdjiisky R. The role of immuno-stimulants in immunotherapy and immunoprophylaxis. *Biotechnology & Biotechnological Equipment*. 2007; 21(4): 454–62.
129. Ralph MT, Silva AFB, da Silva DL, do Nascimento DCO, da Silva DMF, Gomes-Filho MA, et al. Peptidases from latex of Carica candamarcensis upregulate COX-2 and IL-1 mRNA transcripts against Salmonella enterica ser. Typhimurium-mediated inflammation. *Mediators of Inflammation*. 2014; 2014: 819731.
130. Albuquerque TM, Alencar NM, Figueiredo JG, Figueiredo IS, Teixeira CM, Bitencourt FS, et al. Vascular permeability, neutrophil migration

and edematogenic effects induced by the latex of Cryptostegia grandiflora. *Toxicon*. 2009; 53(1): 15–23.

131. Ibrahim EH, Shaker KH, Kilany M, Ghramh HA, Engineering. Cytokines/chemokines profile in rats treated with Euphorbia tirucalli extract. *Arabian Journal for Science*. 2018; 43: 3443–51.

132. Santana SS, Gennari-Cardoso ML, Carvalho FC, Roque-Barreira MC, Santiago AdS, Alvim FC, et al. Eutirucallin, a RIP-2 type lectin from the latex of Euphorbia tirucalli L. presents proinflammatory properties. *PLOS One*. 2014; 9(2): e88422.

133. Seddek AlS, Mahmoud ME, Shiina T, Hirayama H, Iwami M, Miyazawa S, et al. Extract from Calotropis procera latex activates murine macrophages. *Journal of Natural Medicines*. 2009; 63: 297–303.

134. Shivkar YM, Kumar VL. Histamine mediates the pro-inflammatory effect of latex of Calotropis procera in rats. *Mediators of Inflammation*. 2003; 12(5): 299–302.

135. de Oliveira Nascimento DC, Ralph MT, Batista JEC, Silva DMF, Gomes-Filho MA, Alencar NM, et al. Latex protein extracts from Calotropis procera with immunomodulatory properties protect against experimental infections with Listeria monocytogenes. *Phytomedicine*. 2016; 23(7): 745–53.

136. Oliveira RS, Figueiredo IS, Freitas LB, Pinheiro RS, Brito GAC, Alencar NM, et al. Inflammation induced by phytomodulatory proteins from the latex of Calotropis procera (Asclepiadaceae) protects against Salmonella infection in a murine model of typhoid fever. *Inflammation Research*. 2012; 61: 689–98.

137. Raza A, Muhammad F, Bashir S, Aslam B, Anwar M, Naseer M. In vitro and in-vivo anthelmintic potential of different medicinal plants against Ascaridia galli infection in poultry birds. *World's Poultry Science Journal*. 2016; 72(1): 115–24.

138. Demeler J, Kleinschmidt N, Küttler U, Koopmann R, von Samson-Himmelstjerna G. Evaluation of the egg hatch assay and the larval migration inhibition assay to detect anthelmintic resistance in cattle parasitic nematodes on farms. *Parasitology International*. 2012; 61(4): 614–18.

139. Kotze A, Coleman G, Mai A, McCarthy J. Field evaluation of anthelmintic drug sensitivity using in vitro egg hatch and larval motility assays with Necator americanus recovered from human clinical isolates. *International Journal for Parasitology*. 2005; 35(4): 445–53.

140. Evans CC, Moorhead AR, Storey BE, Blagburn BL, Wolstenholme AJ, Kaplan RM. Evaluation of the larval migration inhibition assay for detecting macrocyclic lactone resistance in Dirofilaria immitis. *Veterinary Parasitology*. 2017; 246: 76–81.

141. Ancheta P, Dumilon R, Venturina V, Cerbito W, Dobson R, LeJambre L, et al. Efficacy of benzimidazole anthelmintics in goats and sheep in the Philippines using a larval development assay. *Veterinary Parasitology*. 2004; 120(1–2): 107–21.

142. Štrbac F, Bosco A, Maurelli MP, Ratajac R, Stojanović D, Simin N, et al. Anthelmintic properties of essential oils to control gastrointestinal nematodes in sheep—In vitro and in vivo studies. *Veterinary Sciences*. 2022; 9(2): 93.

143. Stepek G, Lowe A, Buttle DJ, Duce I, Behnke JM. In vitro and in vivo anthelmintic efficacy of plant cysteine proteinases against the rodent gastrointestinal nematode, Trichuris muris. *Parasitology*. 2006; 132(5): 681–89.

144. Wanderley LF, Soares AMdS, Silva CR, Figueiredo IMd, Ferreira ATdS, Perales J, et al. A cysteine protease from the latex of Ficus benjamina has in vitro anthelmintic activity against Haemonchus contortus. *Revista Brasileira de Parasitologia Veterinária*. 2018; 27: 473–80.

145. Moraes D, Levenhagen MA, Costa-Cruz JM, Costa APd, Rodrigues RM. In vitro efficacy of latex and purified papain from Carica papaya against Strongyloides venezuelensis eggs and larvae. *Revista do Instituto de Medicina Tropical de Sao Paulo*. 2017; 59: e7.

146. Cavalcante GS, de Morais SM, Andre WP, Ribeiro WL, Rodrigues AL, De Lira FC, et al. Chemical composition and in vitro activity of Calotropis procera (Ait.) latex on Haemonchus contortus. *Veterinary Parasitology*. 2016; 226: 22–5.

147. Merrouni IA, Elachouri M. Anticancer medicinal plants used by Moroccan people: ethnobotanical, preclinical, phytochemical and clinical evidence. *Journal of Ethnopharmacology*. 2021; 266: 113435.

148. Supraja N, Prasad T, Gandhi AD, Anbumani D, Kavitha P, Babujanarthanam R. Synthesis, characterization and evaluation of antimicrobial efficacy and brine shrimp lethality assay of Alstonia scholaris stem bark extract mediated ZnONPs. *Biochemistry and Biophysics Reports*. 2018; 14: 69–77.

149. Ghasemi M, Turnbull T, Sebastian S, Kempson I. The MTT assay: utility, limitations, pitfalls, and interpretation in bulk and single-cell

analysis. *International Journal of Molecular Sciences.* 2021; 22(23): 12827.

150. Ngamwongsatit P, Banada PP, Panbangred W, Bhunia AK. WST-1-based cell cytotoxicity assay as a substitute for MTT-based assay for rapid detection of toxigenic Bacillus species using CHO cell line. *Journal of Microbiological Methods.* 2008; 73(3): 211–5.

151. Kamiloglu S, Sari G, Ozdal T, Capanoglu E. Guidelines for cell viability assays. *Food Frontiers.* 2020; 1(3): 332–49.

152. Hatakeyama S, Yamamoto H, Ohyama C. Tumor formation assays. In *Methods in Enzymology*, Vol. 479, Elsevier; 2010. pp. 397–411.

153. Al-Qahtani MA, Farah MA, Abou-Tarboush FM, Al-Anazi KM, Al-Harbi NO, Ali MA, et al. Anticancer effects of Calotropis procera latex extract in MCF-7 breast cancer cells. *Pharmacognosy Magazine.* 2020; 16(71): 550–56.

154. de Oliveira JS, Bezerra DP, de Freitas CDT, Marinho Filho JDB, de Moraes MO, Pessoa C, et al. In vitro cytotoxicity against different human cancer cell lines of laticifer proteins of Calotropis procera (Ait.) R. Br. *Toxicology in Vitro.* 2007; 21(8): 1563–73.

155. Oliveira JS, Costa-Lotufo LV, Bezerra DP, Alencar NM, Marinho-Filho JD, Figueiredo IS, et al. In vivo growth inhibition of sarcoma 180 by latex proteins from Calotropis procera. *Naunyn-Schmiedeberg's Archives of Pharmacology.* 2010; 382(2): 139–49.

156. Choedon T, Mathan G, Arya S, Kumar VL, Kumar V. Anticancer and cytotoxic properties of the latex of Calotropis procera in a transgenic mouse model of hepatocellular carcinoma. *World Journal of Gastroenterology.* 2006; 12(16): 2517–22.

157. Parhira S, Zhu GY, Chen M, Bai LP, Jiang ZH. Cardenolides from Calotropis gigantea as potent inhibitors of hypoxia-inducible factor-1 transcriptional activity. *Journal of Ethnopharmacology.* 2016; 194: 930–36.

158. Ghandehari F, Fatemi M. The effect of Ficus carica latex on 7, 12-dimethylbenz (a) anthracene-induced breast cancer in rats. *Avicenna Journal of Phytomedicine.* 2018; 8(4): 286–95.

159. Tezcan G, Tunca B, Bekar A, Yalcin M, Sahin S, Budak F, et al. Ficus carica latex prevents invasion through induction of let-7d expression in GBM cell lines. *Cellular and Molecular Neurobiology.* 2015; 35(2): 175–87.

160. Hashemi SA, Abediankenari S, Ghasemi M, Azadbakht M, Yousefzadeh Y, Dehpour AA. The effect of fig tree latex (Ficus carica)

on stomach cancer line. *Iranian Red Crescent Medical Journal*. 2011; 13(4): 272–5.

161. Khodarahmi GA, Ghasemi N, Hassanzadeh F, Safaie M. Cytotoxic effects of different extracts and latex of Ficus carica L. on HeLa cell line. *Iranian Journal of Pharmaceutical Research*. 2011; 10(2): 273–7.

162. Menichini G, Alfano C, Provenzano E, Marrelli M, A Statti G, Somma F, et al. Fig latex (Ficus carica L. cultivar Dottato) in combination with UV irradiation decreases the viability of A375 melanoma cells in vitro. *Anti-Cancer Agents in Medicinal Chemistry (Formerly Current Medicinal Chemistry-Anti-Cancer Agents)*. 2012; 12(8): 959–65.

163. Rubnov S, Kashman Y, Rabinowitz R, Schlesinger M, Mechoulam R. Suppressors of cancer cell proliferation from fig (Ficus carica) resin: isolation and structure elucidation. *Journal of Natural Products*. 2001; 64(7): 993–6.

164. Wang J, Wang X, Jiang S, Lin P, Zhang J, Lu Y, et al. Cytotoxicity of fig fruit latex against human cancer cells. *Chemical Toxicology*. 2008; 46(3): 1025–33.

165. Lam K, Yang K, Sunderasan E, Ong M. Latex C-serum from H evea brasiliensis induces non-apoptotic cell death in hepatocellular carcinoma cell line (HepG2). *Cell Proliferation*. 2012; 45(6): 577–85.

166. Basu P, Meza E, Bergel M, Maier C. Estrogenic, antiestrogenic and antiproliferative activities of Euphorbia bicolor (Euphorbiaceae) latex extracts and its phytochemicals. *Nutrients*. 2019; 12(1): 59.

167. Jadranin M, Pešić M, Aljančić IS, Milosavljević SM, Todorović NM, Podolski-Renić A, et al. Jatrophane diterpenoids from the latex of Euphorbia dendroides and their anti-P-glycoprotein activity in human multi-drug resistant cancer cell lines. *Phytochemistry*. 2013; 86: 208–17.

168. Cruz LS, de Oliveira TL, Kanunfre CC, Paludo KS, Minozzo BR, Prestes AP, et al. Pharmacokinetics and cytotoxic study of euphol from Euphorbia umbellata (Bruyns) Pax latex. *Phytomedicine*. 2018; 47: 105–12.

169. Martins CG, Appel MH, Coutinho DS, Soares IP, Fischer S, de Oliveira BC, et al. Consumption of latex from Euphorbia tirucalli L. promotes a reduction of tumor growth and cachexia, and immunomodulation in Walker 256 tumor-bearing rats. *Journal of Ethnopharmacology*. 2020; 255: 112722.

170. Brunetti RL, Da Paz DPA, Mackowiak Da Fonseca II, Nagamine MK, Mori CMC, Del Grande MP, et al. Inhibitory effects of Euphorbia tirucalli latex on murine B16/F10 melanoma cells and lung metastasis. *Molecular Clinical Oncology*. 2019; 11(5): 511–16.

171. Fu Z-Y, Han X-D, Wang A-H, Liu X-B. Apoptosis of human gastric carcinoma cells induced by Euphorbia esula latex. *World Journal of Gastroenterology*. 2016; 22(13): 3564.

172. Ravikanth V, Reddy VLN, Reddy AV, Ravinder K, Rao TP, Ram TS, et al. Three new ingol diterpenes from Euphorbia nivulia: evaluation of cytotoxic activity. *Chemical Pharmaceutical Bulletin* 2003; 51(4): 431–34.

173. Mousinho KC, Oliveira CdC, Ferreira JRdO, Carvalho AA, Magalhães HIF, Bezerra DP, et al. Antitumor effect of laticifer proteins of Himatanthus drasticus (Mart.) Plumel–Apocynaceae. *Journal of Ethnopharmacology*. 2011; 137(1): 421–26.

174. Chaniad P, Phuwajaroanpong A, Techarang T, Horata N, Chukaew A, Punsawad C. Evaluation of the antimalarial activity and toxicity of Mahanil-Tang-Thong formulation and its plant ingredients. *Complementary Medicine Therapies*. 2022; 22(1): 51.

175. Hintsa G, Sibhat GG, Karim AJE-BC, Medicine A. Evaluation of antimalarial activity of the leaf latex and TLC isolates from Aloe megalacantha Baker in Plasmodium berghei infected mice. *Evidence-Based Complementary and Alternative Medicine*. 2019; 2019: https://doi.org/10.1155/2019/6459498.

176. Teka T, Bisrat D, Yeshak MY, Asres KJ. Antimalarial activity of the chemical constituents of the leaf latex of Aloe pulcherrima Gilbert and Sebsebe. *Molecules*. 2016; 21(11): 1415.

177. Girma B, Bisrat D, Asres KJ. Antimalarial evaluation of the leaf latex of Aloe citrina and its major constituent. *Ancient Science of life*. 2015; 34(3): 142.

178. Mu K, Liu Y, Liu G, Ran F, Zhou L, Wu Y, et al. A review of hemostatic chemical components and their mechanisms in traditional Chinese medicine and ethnic medicine. *Journal of Ethnopharmacology*. 10 May 2023: 116200.

179. Edemeka D, Ogwu AJJoh, spices. Blood coagulation activities of the leaf extracts of Ocimum gratissimum plant in man. *Journal of Herbs, Spices & Medicinal Plants*. 2000; 7(4): 9–14.

180. Sirisidthi K, Kosai P, Jiraungkoorskul K, Jiraungkoorskul W. Antithrombotic activity of turmeric (Curcuma longa): a review. *Indian Journal of Agricultural Research*. 2016; 50(2): 101–6.

181. Bilheiro RP, Braga AD, Limborço Filho M, Carvalho-Tavares J, Agero U, das Graças Carvalho M, et al. The thrombolytic action of a proteolytic fraction (P1G10) from Carica candamarcensis. *Thrombosis Research*. 2013; 131(4): 175–82.

182. Richter G, Schwarz HP, Dorner F, Turecek PLJBJoH. Activation and inactivation of human factor X by proteases derived from Ficus carica. *British Journal of Haematology*. 2002; 119(4): 1042–51.

183. Patel GK, Kawale AA, Sharma AK. Purification and physicochemical characterization of a serine protease with fibrinolytic activity from latex of a medicinal herb Euphorbia hirta. *Plant Physiology and Biochemistry* 2012; 52: 104–11.

184. Siritapetawee J, Teamtisong K, Limphirat W, Charoenwattanasatien R, Attarataya J, Mothong N. Identification and characterization of a protease (EuRP-61) from Euphorbia resinifera latex. *International Journal of Biological Macromolecules*. 2020; 145: 998–1007.

185. Rajesh R, Gowda CR, Nataraju A, Dhananjaya B, Kemparaju K, Vishwanath B. Procoagulant activity of Calotropis gigantea latex associated with fibrin (ogen) olytic activity. *Toxicon*. 2005; 46(1): 84–92.

186. Gregory J, Vengalasetti YV, Bredesen DE, Rao RV. Neuroprotective herbs for the management of Alzheimer's disease. *Biomolecules*. 2021; 11(4): 543.

187. Kumar A, Mehta V, Raj U, Varadwaj PK, Udayabanu M, Yennamalli RM, et al. Computational and in vitro validation of natural molecules as potential acetylcholinesterase inhibitors and neuroprotective agents. *Current Alzheimer Research*. 2019; 16(2): 116–27.

188. Malabade R, Taranalli AD. Calotropis procera: a potential cognition enhancer in scopolamine and electroconvulsive shock-induced amnesia in rats. *Indian Journal of Pharmacology*. 2015; 47(4): 419.

189. Akkol EK, Ilhan M, Karpuz B, Genç Y, Sobarzo-Sánchez E. Sedative and anxiolytic activities of Opuntia ficus indica (L.) Mill.: an experimental assessment in mice. *Molecules*. 2020; 25(8): 1844.

190. Alnamer R, Alaoui K, Bouidida EH, Benjouad A, Cherrah YJ Aips. Sedative and hypnotic activities of the methanolic and aqueous extracts of Lavandula officinalis from Morocco. *Advances in Pharmacological Sciences*. 2011; 2012: 270824.

191. Liao Y-J, Zhai H-F, Zhang B, Duan T-X, Huang J-M. Anxiolytic and sedative effects of dehydroeffusol from Juncus effusus in mice. *Planta Medica*. 2011; 77(05): 416–20.

192. Rabbani M, Sajjadi S, Vaseghi G, Jafarian A. Anxiolytic effects of Echium amoenum on the elevated plus-maze model of anxiety in mice. *Fitoterapia*. 2004; 75(5): 457–64.

193. Gould TD, Dao DT, Kovacsics CE. The open field test. *Mood and Anxiety Related Phenotypes in Mice*. 1 January 2009: 1–20.

194. Khan AW, Khan A-U, Ahmed T. Anticonvulsant, anxiolytic, and sedative activities of Verbena officinalis. *Frontiers in Pharmacology*. 2016; 7: 499.

195. Ilgün S, Küpeli Akkol E, Ilhan M, Çiçek Polat D, Baldemir Kýlýç A, Coşkun M, et al. Sedative effects of latexes obtained from some Lactuca L. species growing in Turkey. *Molecules*. 2020; 25(7): 1587.

Chapter 6

Drug Discovery, Commercial Applications, Regulations, Safety and Toxicity of Latexes

Arya Lakshmi Marisetti,[a] Ajay Sharma,[a] Abhishek Tiwari,[b] Varsha Tiwari,[b] Ritu Karwasra,[c] Vineet Mittal,[d] Saurabh Bhatia,[e,f] Ahmed Al-Harrasi,[e] and Laleh Khodaie[g,h]

[a]*Department of Pharmacognosy & Phytochemistry,*
School of Pharmaceutical Sciences,
Delhi Pharmaceutical Sciences and Research University,
Mehrauli-Badarpur Road, New Delhi, India
[b]*Faculty of Pharmacy, Pharmacy Academy, IFTM University,*
Lodhipur Rajput, Moradabad, Uttar Pradesh, India.
[c]*Central Council for Research in Unani Medicine (CCRUM),*
Ministry of AYUSH, Governmentt of India, New Delhi, India
[d]*Department of Pharmaceutical Sciences, Maharshi Dayanand University,*
Rohtak, India
[e]*Natural & Medical Sciences Research Center, University of Nizwa,*
Birkat Al Mauz, Nizwa, Oman
[f]*School of Health Sciences, University of Petroleum and Energy Studies,*
Dehradun, Uttarakhand, India
[g]*Department of Traditional Pharmacy, Faculty of Traditional Medicine,*
Tabriz University of Medical Sciences, Tabriz, Iran
[h]*Department of Pharmacognosy, Faculty of Pharmacy,*
Tabriz University of Medical Sciences, Tabriz, Iran

asharmadpsru@gmail.com

Introduction to Plant-Based Latex: Phytochemistry, Ethnopharmacology, Biological, and Pharmacological Activities
Edited by Laleh Khodaie, Ajay Sharma, Saurabh Bhatia, and Ahmed Al-Harrasi

ISBN 978-981-5129-12-0 (Hardcover), 978-1-003-57071-4 (eBook)
www.jennystanford.com

6.1 General Introduction

Drug discovery process is currently facing considerable obstacles, including a reduction in the number of newly authorized medications and rising costs. Despite advances in combinatorial chemistry, which appeared to hold promise for raising the success rates of novel chemical entities (NCEs), the overall success rate in the drug development has remained constant. In light of this, a new strategy known as integrated drug development is being investigated, which entails the synergistic combination of Ayurveda wisdom and the research of medicinal characteristics in plant based sources. The process of developing new drugs is intricate, time consuming and expensive. It typically takes around 12 years from the initial discovery of a new drug to its approval for clinical use, involving substantial investments of over 1 billion US dollars in today's context. The key aspect of new drug discovery revolves around identifying NCEs that possess the necessary characteristics for drug development and medicinal chemistry. These NCEs can be obtained through either chemical synthesis or isolation from natural products. There are several instances of novel medications being developed from plant sources. About 200 years ago, opium (a latex containing plant) was made from chopped seed pods of *Papaver somniferum* and was found to contain morphine. It is time for major pharmaceutical companies to disclose their research and development plans. Alternative strategies, such as the development of herbal latex extracts need to be seriously considered in light of the rising cost of developing new drugs. The development costs for herbal extracts will be undoubtedly much lower. This approach is thought to reduce the risk of post-marketing withdrawals as well as enhance the acceptability of these medicines. Such a complementary situation would significantly advance the interests of both the pharmaceutical sector and the general public [1].

Nowadays, the research for a different lead chemical is drawing attention to the development of latex-bearing plants and to promote drug discovery. Around the globe, latex-producing plants have a long history of being used medicinally and to improve human health [2]. Plant latex is a whitish, viscous secretion secreted by the laticifers and is released in response to plant injury [3]. Approximately thirty five thousand of the plant species and forty plant families produce latex. These plant includes angiosperms and conifers [4]. Latex has been reported to contain a variety of phytoconstituents [2]. In numerous generations, more than 40 families generate latex. The spurge/castor family (Euphorbiaceae), milk weed family (Asclepiadaceae), mulberry family (Moraceae), dogbane family (Apocynaceae) and chicory tribe (Lactuceae) of the sunflower family (Asteraceae) are among the plant families that produce an abundance of latex. Numerous latex-producing plants, including *Euphorbia antiquorum, Euphorbia antisyphilitica, Euphorbia caducifolia, Euphorbia neerifolia, Euphorbia royleana, Calotropis procera, Calotropis gigantea, Cryptostegia grandiflora, Plumeria alba, Nerium indicum*, and *Mimusops elengi* were reported as a source of potential phytoconstituents [5].

Earlier human beings were in search of medicinal plants that could treat illnesses. This process has involved a lot of trial and error. Quinine (*Cinchona spp.*) and artemisinin (*Artemisia annua* are antimalarial while vinblastine (*Catharanthus roseus*) and taxol (*Taxus brevifolia*) are anticancer drug that are developed from natural products and are helpful in treating diseases. Most medications utilize concentrated plant extracts rather than isolated phytoconstituents. Moreover, there are numerous issues with their standardization procedure as well as their accuracy in characterizing lead compounds. Natural products research and development (R&D) may be crucial in the creation of novel drugs to address issues of global public health. The prophylaxis and treatment of diseases necessitate the use of products of higher safety, quality and efficacy. The application of molecular modelling and drug designing techniques gives an idea about the structure and binding affinity of novel lead molecules. Also, the use of sophisticated analytical techniques gives information about the quantitative and qualitative analysis of lead compounds. However, natural products based drug discovery is a tedious, lengthy and costly process. Future drug research will have fewer false positive leads because of the use of quantum computing, computer software, and databases to model chemical interactions and anticipate characteristics and parameters like pharmacokinetics and pharmacodynamics [6]. Commercially latex has been used for a variety of purposes in the dairy industry and in the textile industry for the manufacturing of antivibration mounts, adhesives, dry coupling, springs, toys and rubber bands. In the health care industry, a wide range of applications for latex bearing plants are reported as anticancer, antiallergic, antiviral, anti-inflammatory and wound healing. Plants that yield latex have long been used as medicine and to prevent diseases in several nations, cultures, and civilizations. It has a variety of bioactive substances that are advantageous to both plants and people's health such as morphine or codeine which are obtained from poppy latex [2–7]. In addition to this, *Cannabis sativa* L. suffered a divergence roughly 10,000 years ago, indicating that it was the oldest plant grown by our ancestors. *Papaver somniferum* L. (Opium Poppy) was being grown and their cultivation is done as per Government regulations [7].

6.2 Perspectives of Natural Product Drug Discovery

In nature, a plant system constantly serves as an example of symbiosis phenomena. Medicinal plants not only used in the management of diseases but also provide potential molecules used for a variety of purposes. Plants, plant products and plant parts have been used therapeutically as human civilization has progressed. Indigenous herbs have been utilized for generations to treat a number of diseases that have been clearly linked to their pharmacological properties. Recent growth in the usage of natural products has led to a huge demand of natural products for agricultural, medical and industrial uses. Many different lineages of 20,000 species from 40 families, including mono-cotyledons and dicotyledons, generate latex [4]. Around 10% of the flowering plants produce latex as a part of their defense mechanism [8].

Since the dawn of time and in folklore, natural remedies have been used to heal a range of diseases and disorders [9]. Natural compounds (secondary metabolites) have been shown to be the most promising sources of possible medicinal leads [10, 11]. However, a drop in interest from early initiatives using them in drug discovery and development has been seen [11]. Herbal products, on the other hand, continue to offer remarkable structural variability as compared to conventional combinatorial chemistry, which has the potential to uncover mostly novel low molecular weight lead molecules. Plants can be found in any livable setting, with the majority of them growing on land. Plants have developed a number of compounds to defend themselves from predators and the environment stress since they are sedentary and vulnerable to a variety of hazards [12]. Plants' ability to produce scents, colors, and even toxins is derived from it. There is historical proof that people have used plants as medicine [13]. Natural product research and development (R&D) may be crucial for the creation of novel drugs to address issues with impact on global public health. A multitude of natural and synthetic molecular parameters, including safety, pharmacokinetics and efficacy, must be taken into account when choosing potential therapeutic candidates because the creation of therapeutics is a complex problem [6]. The natural latex sector is made up of businesses with extensive capacities and sizes. This includes both huge estates with modern, well-equipped facilities and small holdings that could use outdated processing techniques or share equipment.

The creation of plant medicines is based on a multidisciplinary approach that combines biology, phytochemistry, ethnobotany and botany. This method identifies new chemical entities or lead compounds from plants that have potential therapeutic effects. These substances have demonstrated promise in the creation of drugs to treat diseases like metabolic disorders, neurological disorders, AIDS, cancer, inflammation and diseases caused by protozoa. Some of the plant species such as *Apocynum canabinum, Euphorbia antisyphilitica, Hevea brasiliensis* and *Papaver somniferum* have been investigated for such effects. Some of these drugs and their equivalents are now being tested in clinical trials as well as being utilized in clinical settings. Despite obstacles, plant-based drug development programs continue to produce valuable pharmacological leads. It is necessary to scale up the manufacturing of bioactive lead compounds, acquire real plant materials and use high-throughput screening bioassays among other things. The possibility for developing plant-based drugs in the future will be increased by overcoming these challenges [14]. India is rich with huge biodiversity, cultural heritage and ancient knowledge in the form of Indian system of medicine. This results in the exploration and value addition of medicinal plants and their products as potential therapeutic agents.

Plants have served as the basis of complimentary and alternative systems of medicine and are useful in the management of diseases. They have been prescribed either as crude medications or novel therapeutics [15]. Drug discovery and development is a multi- and inter-disciplinary process and it includes traditional sciences like taxonomy and ethnobotany in addition to the main disciplines associated with pharmaceutical research. Over the course of history, the traditional knowledge and information about the usage of medicinal plants inherited globally. This information has greatly influenced the growth of diverse traditional systems of medicine. During the initial stages of 19th century, the isolation of bioactive compounds from the plant based drugs has been started. This leads to the isolation of alkaloids from opium, cinchona and digitalis [16].

Today, pharmaceutically active plant molecules are still being isolated and characterized. More recently, methods for drug discovery have been used to standardize natural medicines and identify their analytical markers. Few of the plant latex is used as a pesticide and has immunological allergenic qualities. Numerous bio-active compounds like alkaloids, terpenoids, glycosides, steroids, saponins and phenolics with potential biological activities abundantly found in latex.

According to data, tropical plant species and (14%) of temperate plant species generate latex. One of the largest and most diverse angiosperm family, Euphorbiacae contains 317 genera arranged into five subfamilies, totaling 7500 species producing latex [8]. Over 40 of the plant families produce latex. Among the plant families that produce latex are spurge/castor family (Euphorbiaceae), milk weed family (Asclepiadaceae), mulberry family (Moraceae), dogbane family (Apocynaceae) and chicory tribe (Lactuceae) of the sunflower family (Asteraceae) are the major producer of latex. Numerous latex producing plants such as *Cryptostegia grandiflora, Plumeria alba, Nerium indicum* and *Mimusops elengi* have been considered as potential renewable energy crops. Drug discovery and development from the plant sources is diverse in nature and involves expertise of botany, ethnobotany, ethnopharmacology, ecology and chemistry. A taxonomist or botanist typically collect and identify the particular plant material. This is first and prominent step of plant based drug discovery. Taxa may be randomly selected either for a large screening program or species with established biological activity. Intellectual property rights of a particular country is the major concern in the present time [17]. Pharmacognosists or phytochemists deal with the extraction, isolation and characterization of phytoconstituents as well as their biological screening using in vitro and in vivo models.

6.2.1 Important Latex Bearing Plants in Drug Discovery

Recently, the pharmaceutical companies showing an interest in natural product based drug discovery that leads to the generation of new drugs or new chemical entities. The process of drug discovery involves the application of high throughput screening, computer aided drug design as well as combinatorial chemistry [18, 19]. Approximately 61% of the new chemical entities (NCEs) between 1981 and 2002 were derived from natural sources [20]. Structure of phytoconstituents from latex bearing medicinal plants are given in Figure 6.1.

Some metabolites of morphine such as morphine-6-glucuronide considered as analgesic with fewer side effects [1, 21].

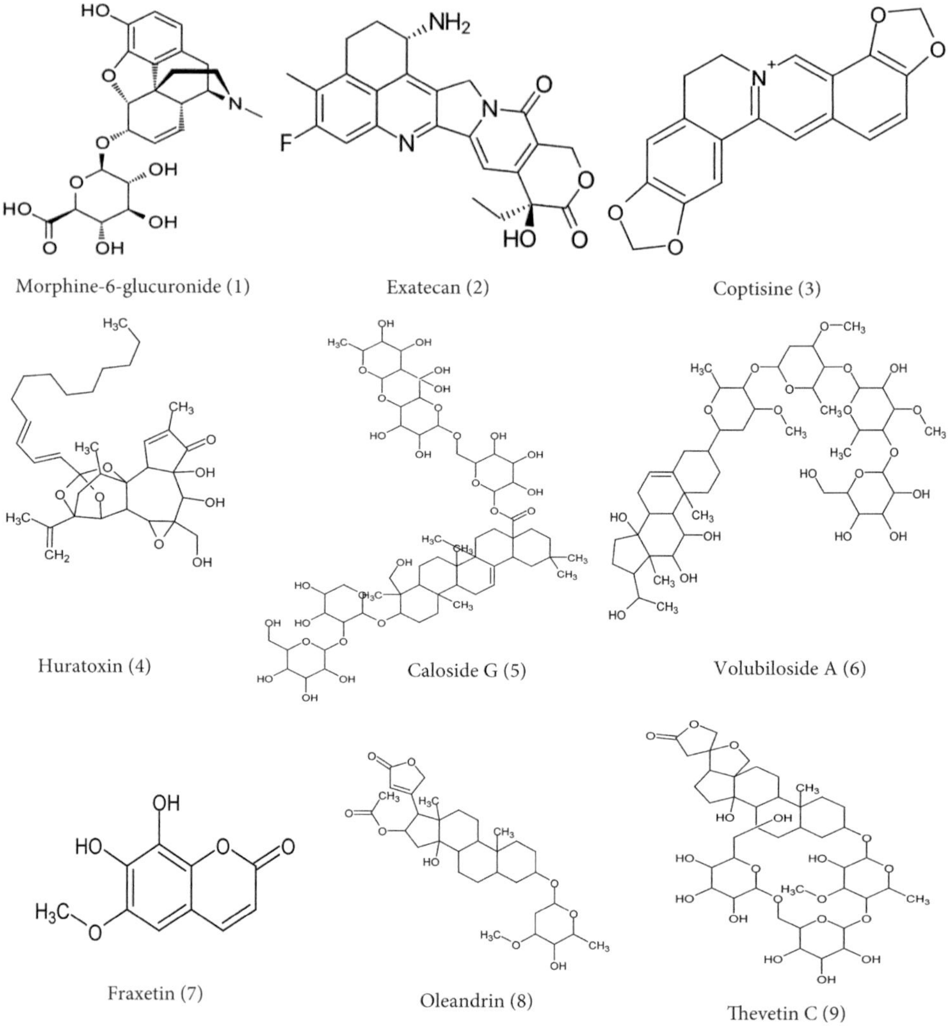

Figure 6.1 Chemical structures of plant-derived drugs from latex bearing plants.

Exatecan is a camptothecin analogue from *Camptotheca acuminata* Decne. (Nyssaceae) reported for their anticancer effect [22]. The derivatives of coptisine, an isoquinoline alkaloid from *Chelidonium majus L.* (Papaveraceae), are comparable to berberine for their anticancer effects [23]. Huracrepitans (Euphorbiaceae) contain Huratoxin and its derivatives are known for their termite control, termite-killing, insecticidal and bactericidal properties [24]. Caloside G, a cardenolide obtained from the plant *Calotropis gigantea* (L.) R. Br. (Apocynaceae), is considered as a potential anticancer agent [25]. Volubiloside A, a saponin derivative, derived from *Dregea volubilis (L.f.)* Benth. Ex Hook (Apocynaceae) with potent α-glucosidase inhibitory action has been reported [26]. Fracetin (coumarin derivative) from *Jatropha integerrima Jacq* (Euphorbiacae) showed effectiveness in neurological disorders. Oleandrin and its derivatives from *Nerium indicum Mill. Gard. Dict* (Apocynaceae) known for their anticancer properties [28]. Thevetin C from *Thevetia peruviana (Pers.)* K. Schum. (Apocynaceae) showed anticancer, anticoagulant, antioxidant and antimicrobial activities [29].

6.2.2 Challenges in Drug Discovery from Medicinal Plants

It has been noticed that natural product based drug discovery faced numerous obstacles with respect to their quantity and quality. As per the literature available, this process is tedious, lengthy and costly and may take an average of more than ten years with the expense of approximately several hundred millions of American dollars [30].

The identification of hits, their development and clinical trials are the important steps in drug discovery. It is assumed that single compound out of several thousands can be predicted and approved for its use (Fig. 6.2 and Fig. 6.3).

Natural product based drug discovery includes selection of plant material on the basis of their ethnopharmacological claims, preparation of standardized extracts, bioactivity guided fractionation, phytochemical screening of the plant constituents and their biological screening.

The ultimate aim of drug discovery is to isolate novel bioactive compounds and elucidates its structure and mechanism of action [31]. The most common issue encountered here is lack of availability of bioactive phytoconstituents. The availability problem can be solved utilizing synthetic, semi-synthetic or biotechnological approaches.

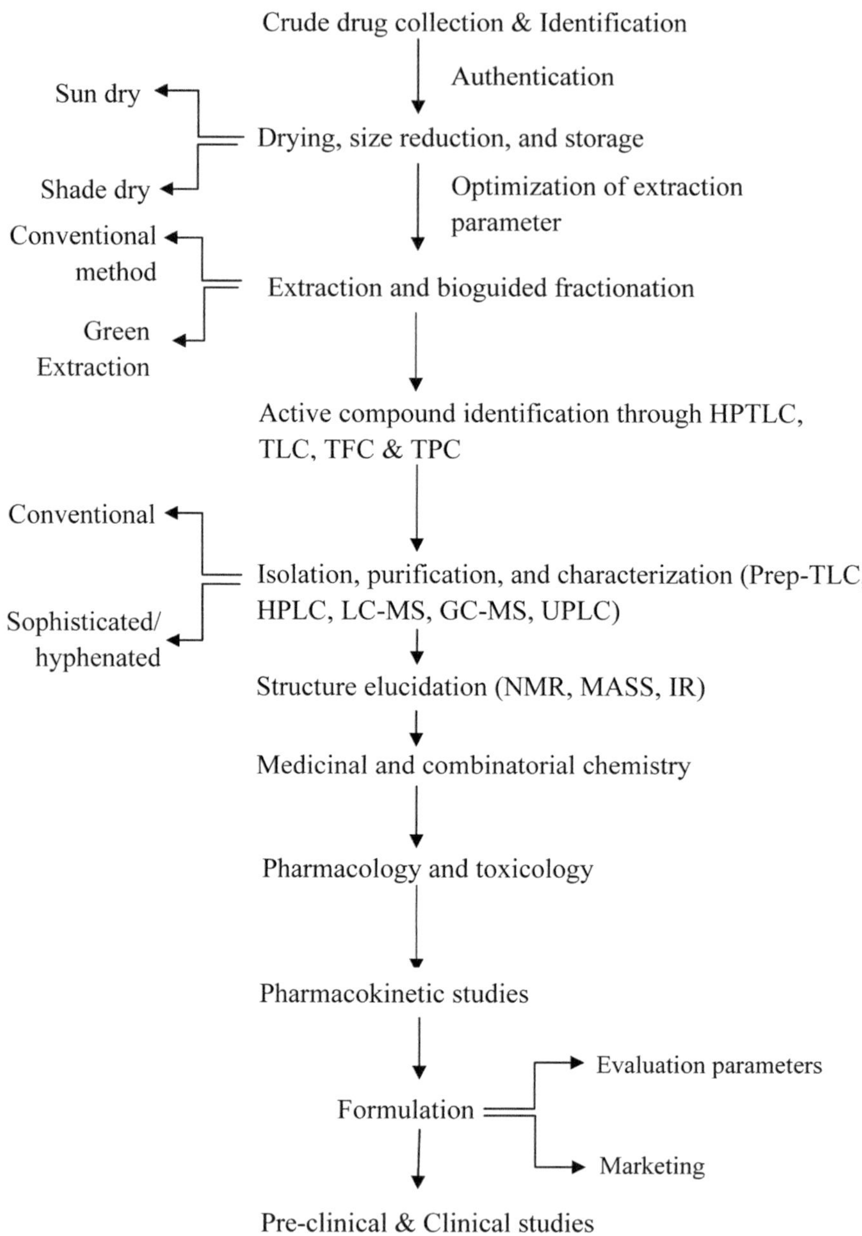

Figure 6.2 Drug discovery and development of latex bearing plant.

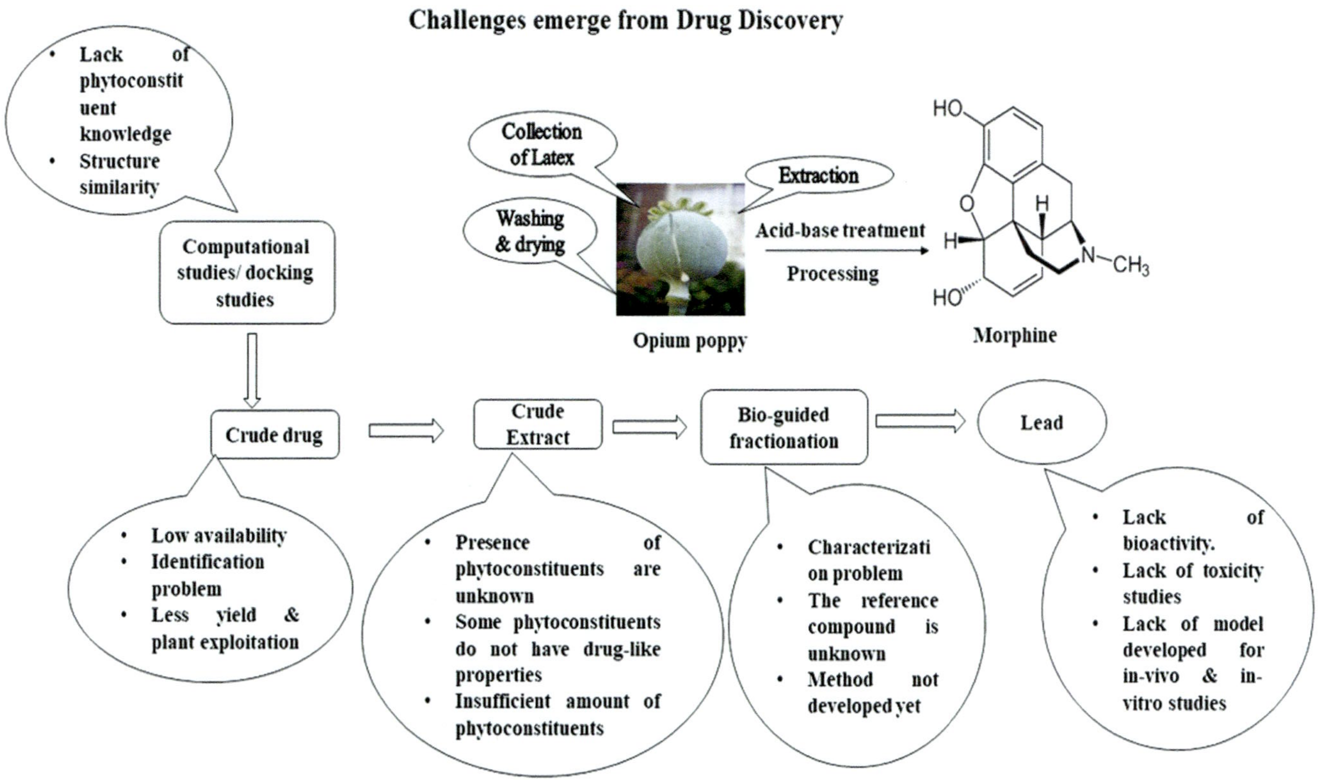

Figure 6.3 Various challenges of drug discovery and development of latex bearing plant.

Innovative techniques of plant collection considering legal and political problems are required in natural product based drug discovery as it is a tedious and lengthy process [32, 33]. Design, selection and implementation of acceptable, clinically relevant, high-throughput screening needs to be implemented [34, 35], although designing of high-throughput screening assays is difficult [36]. The biological activity of both the compounds and the plant extracts can be determined once the screening assay is in place. The most prevalent challenge encountered during extract screening is solubility and screening of extract libraries [31]. Bioassay screening challenges will continue to be an issue in plant based drug discovery.

Using hyphenated techniques such as GC-MS, LC-MS/MS, FT-NMR and LC-MS, the process of drug discovery becomes easier and it saves the time as well. One more challenge in natural product based drug discovery is the lower yield of isolated compounds which is not sufficient to conduct their optimization and clinical trials. Hence, it is needed to utilize synthetic, semi-synthetic and biotechnological tools [37, 38]. Further, the process of drug discovery can be improved by developing natural product library which integrates with their properties and combinatorial chemistry. A team of researchers including pharmacognosist, pharmacologist, medicinal chemist and biotechnologist etc can significantly contributes in this drug discovery process.

Identifying and fixing the issues during the processing of latex concentrate is one of the challenges for the latex industry. Natural latex is particularly important in the rubber business for the production of goods and other articles. Natural latex is considerably more delicate liquid and difficult to handle. This makes latex unsuitable for extraction. Other associated procedures are also taken into account during the manufacturing of concentrated latex. Problems encountered in concentrated latex processing can be classified into several categories including stability, handling, storage and environmental as well. The latex processing sector is essential for producing the raw materials needed to make different rubber goods. Natural rubber is used to make a variety of rubber products including condoms, hand gloves and other thin walled materials because of high tensile strength and resilience to strain from wear like chipping, cutting or tearing. The production of concentrate from natural latex is a straightforward procedure, but if certain operational parameters are not properly managed, complication can readily occur. The most crucial factor is that in a plant where processing concentrate latex is the primary concern. It is crucial to take necessary precautions at every level of the operation to ensure that nothing happens to throw the latex out of order and prevent the production of the intended output. Otherwise, there could be a significant loss of raw materials, energy and money. The most frequent problems encountered in the latex industry are related to cultivation, adulteration, handling and storage [39].

6.2.3 Opportunities in Natural Product Drug Discovery

Natural product drug discovery requires knowledge of human pursuits, including international laws and legal comprehension, social sciences, politics, anthropology and a team of scientific personnel from all the multidisciplinary subjects. Rich genetic resources and associated ethnomedical knowledge from Ayurveda and other traditional medical systems are essential for sustainable bioprospecting and value addition in the entire world. An industry partner is required for drug-targeted bioprospecting as they will be important in turning the discovery into a marketable product. The creation and execution of a contract or memorandum of understanding is crucial for any bioprospecting process. It should address issues related to access to biodiversity, intellectual property rights, benefit sharing during the process (short term), benefit sharing in the event that a product is discovered and commercialized (long term) and the preservation of biological resources for future generations. These issues should be taken into consideration while using ethnobotanical or ethnopharmacological approach. We need a vast collection of extracts in order to screen thousands of plant species simultaneously for a wide range of bioassays. The creation of natural products extract library is necessary. High-throughput screening offers biological screening of numerous extracts within a short period of time. Another benefit is the reduction in cost and time for repetitive plant collection. Even though several institutions in India have tiny plant extract libraries that are not accessible to the general population. Such libraries could be an effective resource and a source of extracts for high-throughput tests that screen for biological activity. The best combinatorial chemist is nature and nature holds the key to cure every illness known to man. Natural compounds derived from medicinal plants produced a large number of clinically relevant medications. Natural compounds isolated from plants will continue to be a crucial component in the hunt for novel medicines despite the numerous difficulties encountered in the discovery of medical plant-based drugs. Among all the plant species available very few are being explored for their pharmacological potential. Bioprospecting has a solid foundation to the rich traditional knowledge connected with the different genetic resources that are available globally. By utilizing current drug discovery methods and the combined efforts of numerous disciplines, proper use of these resources and tools in bioprospecting would undoubtedly aid in the discovery of novel lead compounds from plants [40, 41].

6.3 Contribution of Traditional Usages to the Modern Therapy

The use of natural products and conventional therapies is essential for the benefit of the mankind. In several regions of the world, Traditional Korean Medicine, Traditional Chinese Medicine (TCM), Ayurveda, Kampo and Unani have developed into well-organized, regulated medical systems [42]. TCM is supported by a wealth of data from "clinical tests" that testify to its efficacy and is based on traditional knowledge and clinical practice. Techniques have been created for things like the right dosage, how to produce and process materials and the best times to extract the various components of medicinal plants [43]. It is interesting to notice the convergence of TCM and modern medicine. The toxicological and therapeutic mechanisms of Chinese medicinal herbs may now be identified with the use of modern, sophisticated technology and TCM can now be comprehended in terms of contemporary medicine [44, 45].

The introduction of "modern" pharmaceuticals was heralded at the turn of the 19th century. The first pharmacologically valuable component of the opium poppy was successfully isolated in 1805 by a young German chemist by the name of Friedrich Serturner [46, 47]. Many active compounds have since been identified from natural sources. Some of them continue to use them in the conventional way, while others don't. Later, the value of natural products was substantially diminished by the development of synthetic processes and the use of plant based natural products would be outrightly prohibited because of their complex nature and lack of pharmacokinetic and pharmacodynamic data. However, the use of natural medicines dates back a long time and is crucial for the development of new medications [20, 48].

Since 1980s, the pharmaceutical industry and combinatorial chemistry based drug discovery have used high-throughput screening (HTS) more frequently. Even though significant effort has been made in this direction, the anticipated drug manufacturing has not materialized. It is difficult for many large pharmaceutical companies to develop new medications. Drug discovery and development from medicinal plants have been tremendously increased since a decade. High throughput screening technology considered as an important approach in the discovery of novel phytopharmaceuticals [49, 50]. The plant based natural products are very crucial in the creation of novel medicines that effectively cure serious illnesses [51].

Traditional medicine (TM), also known as complementary and alternative medicine or ethnic medicine, is the oldest form of healthcare in existence and is still of utmost significance in many contemporary countries. It is used to identify, treat and ward off both physical and mental illnesses. Utilizing a variety of efficient therapeutic techniques, various cultures have historically battled a number of health related and life-threatening ailments [52–54].

The majority of drugs used in TM are created with natural substances. "Clinical experiments" have been conducted in TM from the dawn of time. Clinical testing data from TM helped to develop modern pharmaceuticals. In the areas including identification of drug candidates and lead compounds, exploring drug like action and investigating pharmacokinetic and pharmacodynamic issues using natural products, TMs offer advantages over standard drug therapy [55]. It is estimated that around 150 millions peoples are using TCM throughout the globe. This shows the acceptability of TCM as an alternative to modern medicine in primary health care [56].

Kampo is Japanese system of medicine and was originated from Traditional Chinese medicine. This system of medicine has been practiced by Japanese practitioner and was routinely prescribed for cancer patients along with the modern medications [42, 57–58]. With a 2,500 year history, Unani is a complete medicinal technique that originated in ancient Greece [59]. Since the WHO began to concentrate more on Unani in the middle of the 1970s, it has attracted a lot of attention globally, particularly in India where it is a component of the country's national health care system [60].

According to WHO, numerous individuals still depend on conventional drugs [61]. Traditional medicines products also have important economic consequences. According to predictions, the market for these products will be valued at $5 trillion and grow at 7% annual rate by 2050 [62]. According to research, 80% of Africans use TMs either exclusively or in conjunction with contemporary medical care [63]. Traditional medicines in Australia is not well accepted by the community because of the popularity of modern medicine [64]. Modern medicine is taking over and TM is waning in Israel, a nation with a multiethnic population [65]. When producing new medications and treatments, it is feasible to generate amazing synergy and gain considerable advantages by combining efficient contemporary scientific methodologies and TM's tenable ethnobotanical and ethnomedical experiences and techniques [66].

6.3.1 Drugs Made from Traditional Medicines

A single plant or recipe may contain several phytochemical components in traditional medicines. Usually these substances act synergistically or individually to get the desired pharmacological effects [67]. It has been proven that the use of plants in TMs contributed to the development of various useful drugs. The traditional medicines have the advantage of acceptability, viability and accessibility.

Antimalarial drugs such as artemisinin are contemporary drugs based on TMs. Physician Hong Ge (AD 284–384) wrote about the effectiveness of *Artemisia annua* L. in treating malaria during the beginning of China's Jin Dynasty [68–70]. Significant progress has been made in the study of artemisinin (qinghaosu) and its analogue. The multiple modes of action of artemisinin make it effective against a variety of strains for the treatment of malaria including cerebral and falciparum resistant to chloroquine [70].

6.3.2 Ethnomedicinal Properties of Plant Latex

The term "ethnomedicine" refers to healthcare systems that incorporate cultural practices and beliefs about illnesses and health. It is used by the indigenous communities for their cure and spiritual purposes. *Apocynum cannabinum* milky sap is used as a folk remedy to treat venereal warts [71]. The juice of *Euphorbia hirta* is beneficial against colic, dysentery, diarrhea and particularly amoebiasis. It is also used as a tonic and antiasthmatic. To treat arthritis, *C. procera* latex is combined with turmeric powder and sesame oil and then applied to the affected area [72]. The antibacterial characteristics of *Cryptostegia grandiflora* latex has been examined. The leaf extract has also been employed for their anticancer, antiulcer and antiscabies effects and as a potential source of industrial raw materials and a substitute for conventional oil [73–75].

Table 6.1 List of latex bearing plants along with their folklore and pharma-cological uses.

No.	Botanical name	Family	Common name	Uses	Refs
1.	*Apocynum cannabinum*	*Apocynaceae*	Indian hemp	Served as food and tea, eliminates dampness through diuresis, clearing heat, cooling blood, stimulating diuresis, easing dryness, reducing inflammation, hepatoprotective	[78]
2.	*Croton bonplandianum*	Euphorbiaceae	Bonpland's croton	Treatment of bacterial, viral and protozoal infections like malaria	[79–81]
3.	*Calotropis gigantea*	*Apocynaceae*	Crown flower	Leprosy, worm infestation, piles, cough, abdominal and nervine disorders, pruritis, snake-bite, convulsions, joint swelling, dyspnea, paralysis, asthma, dropsy	[82–84]
4.	*Calotropis procera*	*Apocynaceae*	King's crown	Amoebic dysentery, antifertility, antipoison, ascites, cholera, extracting guinea worms, eye diseases, gonorrhea, hysteria	[85, 86]
5.	*Cryptostegia grandiflora*	*Apocynaceae*	Rubber vine	Anti-inflammatory	[87]
6.	Euphorbia antisyphilitica	Euphorbiaceae	Wax plant	Whitening of skin and effects on blood clotting	[88]
7.	*Euphorbia antiquorum*	Euphorbiaceae	Malayan spurge	Emetic, purgative, diuretic	[89, 90]
8.	*Euphorbia caducifolia*	Euphorbiaceae	Leafless milk hedge	Treatment of bleeding wound, skin diseases	[91–96]
9.	*Euphorbia hirta*	Euphorbiaceae	Asthma weed	Dengue, viral infection	[97]

No.	Botanical name	Family	Common name	Uses	Refs
10.	*Euphorbia nerifolia*	Euphorbiaceae	Indian spurge tree	Arthritis, local anesthetic, antibacterial, antiviral, respiratory stimulant, bronchitis, loss of consciousness, leukoderma, piles, paronychia inflammation, delirium, spleen enlargement, ulcers, anemia, fever	[98–101]
11.	*Euphorbia nivulia*	Euphorbiaceae	Leafy milk hedge	Bronchitis, asthma, leprosy, jaundice, tumors, purgative	[102]
12.	*Euphorbia pulcherima*	Euphorbiaceae	Poensettia	Analgesic, antipyretic, anti-inflammatory, sedation potential, anti-depressant, cure of skin diseases	[103]
13.	*Euphorbia royleana*	Euphorbiaceae	Royel's spurge	Piscicidal, muricidal	[104, 105]
14.	*Euphorbia tirucali*	Euphorbiaceae	Milk bush	Hemorrhoids, sexual impotence, warts, epilepsy, toothache, snake bite, cough	[106, 107]
15.	*Euphorbia thymifolia*	Euphorbiaceae	Chicken weed	Dysentery, diarrhea	[108–110]
16.	Ficus benghalensis	Moraceae	Banyan tree	Anti-inflammatory, anthelmintic, antioxidant, antimicrobial, antitumor, antipyretic, antidiabetic, immunomodulatory, wound healing	[111, 112]
17.	*Ficus benjamina*	Moraceae	Weeping fig	Antidiabetic, antimicrobial, antioxidant, cytotoxic, anti-inflammatory, antiulcer, anticonvulsant	[113]
18.	*Ficus carica*	Moraceae	Common fig	Gastrointestinal and respiratory problems, cardiovascular disorders, laxative, antispasmodic, anti-inflammatory	[114–116]

(*Continued*)

Table 6.1 (*Continued*)

No.	Botanical name	Family	Common name	Uses	Refs
19.	*Ficus elastica*	Moraceae	Indian rubber tree	Anthelmintic	[117, 118]
20.	*Ficus hispida*	Moraceae	Opposite leaf fig	Ulcer, anemia, diabetes, piles, jaundice, hemorrhage, convulsion, dysentery, hepatitis, biliousness, diarrhea, cardioprotective	[119–123]
21.	*Ficus racemosa*	Moraceae	Indian fig tree	Aphrodisiac, in boils, diarrhea, dysentery, hemorrhoids, cholera and mumps	[124–126]
22.	*Ficus religiosa*	Moraceae	Pipal tree	GIT and respiratory disorders, antiarthritic, skin diseases, allergies, anti-inflammatory, bone fracture, antidiabetic	[127–130]
23.	*Hevea brasiliensis*	Euphorbiaceae	Para rubber tree	Anti-fungal	[131, 132]
24.	*Mimusops elengi*	Sapotaceae	Spanish cherry	Wound healing, oral care	[133]
25.	*Nerium indicum*	*Apocynaceae*	Oleander	Stomach pain, ear pain, eyes problems	[134, 135]
26.	Papaver somniferum	Papaveraceae	Opium poppy	Flower infusion to enhance sleep, hemorrhoids, cosmetics	[136–139]
27.	*Parthenium argentatum*	Asteraceae	Guayule	Anti-inflammatory	[140]
28.	*Pergularia daemia*	Asclepiadaceae	Trellis vine	Dermatosis, diarrhea, agalactia, abscess, chronic alcoholism, anemia, asthenia, vaginal candidiasis, tuberculosis, malaria, cough	[141]

6.3.3 Natural Latex and Modern Medications

Plant latex has the potential to be source of therapeutically useful constituents in diseases treatment and industrially relevant commodities. The biological function of latex is to provide plant protection against invaders [76]. The chemical composition of latex is complex and diverse and it is affected by a variety of factors like species, developmental stage, external and internal stress [77]. Table 6.1 lists some of the latex-containing plants used in TMs and for pharmaceutical purposes.

6.4 Commercialization of Herbal Latexes

Researchers, innovators, and entrepreneurs in many emerging economies encounter numerous obstacles to successfully commercializing inventions. Because of limited access to the commercial market, underdeveloped technology and a prevailing emphasis on academic publishing over intellectual property (IP), it is frequently simpler and less dangerous for businesses to buy technology from overseas than to develop it domestically. Since many of these economies also depend on agricultural commodities. It is challenging for agricultural industry participants to take a chance on a locally developed technology that is still in its infancy.

6.4.1 Closing the Gap between the Laboratory and the Market and Research and Development

The agriculture industry is vital to the economy of the country because it accounts for more than 10% of the GDP of many countries. Due to the importance of latex to the national economy, various countries invest a significant amount of money in research and development (R&D) in all aspects of latex production from the growth of latex-containing trees through to final processing. Natural latex has steadily grown in popularity and price by the year 2020 it was twice as expensive as it was in 2010. Several toxic and flammable materials like ammonia, acids and formaldehyde are used in the processing of natural rubber. Such chemicals affect the health of workers and are toxic to environment as well. Hence, the latex industry tends to focus on increasing latex yield and minimizing the use of such toxic chemicals during processing.

Natural rubber is a flexible hydrocarbon polymer derived from the latex of para rubber tree. By carefully cutting an angle into the tree bark the latex is released from the incision. This is then collected in a vessel linked to the tree. Raw latex spontaneously coagulates when it comes into contact with air, thus it needs to be handled or processed in a way that minimizes coagulation and maximizes the desirable elastic qualities. Before being given to rubber producers, raw latex is treated into one of three semi-finished forms: (1) Concentrated latex kept in a liquid condition; (2) Smoked solid latex sheets; (3) Solid latex blocks. Greater usage of concentrated latex is necessary for the creation of numerous rubber and chemical products. When rubber latex is harvested from the para tree, it contains between 25% and 40% solid rubber. The concentration is raised to the industry standard of 35% by blending with latex from several different sources. This mixture is then centrifuged to remove water, bringing the concentration to 60%. The various compounds indicated above are then added to control coagulation and microbial contamination. Manufacturers of rubber goods purchase the final concentrated rubber latex to create dipped rubber goods [142].

There are untapped leftover rubber solids as waste products. Three categories can be used to describe the rubber solids in the garbage. The first is "skim" rubber, which is rubber after water has been skimmed off during centrifugation. The rubber solids in this water still range between 3% and 8%. The second is "sludge," which collects at the bottom of latex reservoirs and cakes within centrifuges and other equipment. During processing, sludge is eliminated at regular intervals. The third category is "washing water" which is used to clean centrifuges. The washing water has a tiny amount of rubber granules dissolved in it. Skim rubber has the most potential for commercial viability and possesses the most rubber solids of the three categories. In order to do this, National Metal and Materials Technology Center

(MTEC) started an R&D program to create a novel method of extracting rubber solids from skim rubber. Initial study indicated that it would be challenging to commercialize such a technique. In light of this, the focus is shifted towards the recovery of rubber solids from sludge. This process would be much simpler but the yield would be lesser.

6.4.2 Invention

Successful R&D cannot guarantee commercial success in advanced technology fields. Firms must also build or acquire capabilities for designing, manufacturing and marketing innovative goods, processes and services. They must create complementing technologies to make their idea more helpful as well as secure funding to sustain their work. Numerous capable organizations' attempts to introduce new inventions to the market, and many firms fail in their attempts. Improvements to the initial innovation resulted in increased economic value. Using wastewater in place of fresh water and old chemical versus new enhances output without increasing costs in a latex factory. According to the findings, around 15% of the sludge can be recovered as marketable latex solids by utilizing current water and chemical waste. A high quality fertilizer can also be created and supplied to offset any costs incurred by a rubber processor due to the new technique. Furthermore, contaminants in wastewater are minimized, resulting in cost savings for wastewater treatment. Employees have a safer working environment since they must manage fewer volatile chemicals [143].

6.4.3 IP and Commercialization

Intellectual property (IP) is a valuable asset that needs to be handled just like the other crucial assets that provide a company with a competitive edge (such as investment capital, cash flow, people, technology and research). An entrepreneur with limited resources can start patent filings and perhaps even prosecution, but without the support of a wealthy business partner, banker, or investor, they risk spending a lot of money before even the first patent is approved. Based on spending capital and organizational philosophy there is a need to make a plan. One of the reasons a researcher works with their research institute is to lower the cost of filing and prosecuting patents because the institute has the financial resources to support an inexpensive patent prosecution plan, delivered to the department in charge of commercialization. The National Science and Technology Development Agency (NSTDA) works to strategically employ the IP system to secure all of its produced technology. The technology license office, IP Management Group (IPM) and IP Policy Group (IPP) is responsible for protecting the IPRs of research. Additional IPRs and other forms of protection such as trade secrets and non-disclosure agreements (NDAs) are also the responsibility of IPM. IPP provides advice regarding the creation, protection and exploitation of intellectual property.

6.4.4 Patents and Trade Secrets

The procedure to recover residual rubber solids from sludge waste was developed by NSTDA as a patent application due to significant technology recognized in the latex and rubber industry. Although patents are the main method by which NSTDA safeguards its innovations. It may occasionally be essential to combine additional forms of protection, such as trade secrets. The sludge rubber recovery technique, which presented various IP issues, serves as an example of this. First off, the procedure is simple to follow and simple to duplicate. Second, enforcement is challenging since there is no way to tell if the recovered rubber came from this procedure or from any other source. Last but not least, since a key advantage of the technology is the use of currently available chemicals and manufacturing process, potential rivals and customers already have a wealth of pertinent information.

IPM came to the conclusion that it was crucial to preserve the process know-how through a trade secret after taking this into account. Several efficient acidic compounds were described in the patent application. However, the precise quantity of each substance required for peak performance was omitted. The NSTDA has designated this material as a trade secret. Moreover, the formulation must be modified to meet the particular needs of each industry because one key benefit of the technology is cost savings through the use of current waste streams. This is important for the researchers to know how the technology can be commercialized successfully and why it must be kept a trade secret.

6.4.5 Technology Transfer

Aiming to manage ties between commercial and scientific R&D and expediting technology development in multiple nations, NSTDA established the Technology Management Centre (TMC). Given the importance of benefit sharing to NSTDA, researchers at NSTDA organizations interested in commercialized inventions earn 70% of all income upto a cap of one million (about US $30,000). Researchers are given 30% of any additional money after this point. NSTDA receives 21% whereas the researcher's organization receives 64% when research is financed by NSTDA but R&D is carried out at a different organization. No matter where the R&D is conducted, the Technology Licensing Office (TLO) is compensated with 15% of the overall revenue, which includes both upfront and royalty payments, in order to cover operating expenses. This opens up possibilities beyond the TLO's efforts to acquire and market NSTDA research. In the long term, it enables the TLO to charge outside businesses or organizations for commercial services. The TLO anticipates being able to construct technological portfolios and commercialization strategies for R&D carried out by numerous research centers in various countries as its capabilities grow [144].

6.4.6 Licensing and Commercialization

The TLO concluded that licensing was the most effective method of commercializing the technology after considering th technology transfer and IP management regulations. The IP Business Group (IPB) is in charge of managing licensing agreements.

The business received a reduced initial licensing charge due to its early involvement. The corporation was also exempt from paying royalties because it was challenging to track the use and sales of products derived from this technology.

Rubber is grown from Andes to the Brazilian border in the north west of Bolivia. To avoid damaging the tree, latex is removed by making shallow cuts into the surface layers of the bark rather than the cambial layer. The harvesting is done overnight and in the early morning. Along forest trails (estradas), when it is possible to access certain rubber trees, rubber is gathered [145, 146]. Liquid latex, locally produced artisan goods, and semi-industrially produced goods are the three distinct products that are marketed. Liquid latex made from the rubber from Tomachi is sold in 200-liter drums. This is made feasible by the accessibility of reliable transportation: Tomachi is close to a sizable river in addition to everyday vehicular access to the community and back [145, 146]. *Hevea brasiliensis*, a Brazilian rubber tree, is used commercially to produce natural rubber latex. The proteins and lipids that wrap the rubber hydrocarbon (cis-1,4-polyisoprene) combine to produce the rubber particles. Proteins, lipids, inorganic components and dietary carbohydrates are examples of non-rubber compounds. Based on the rubber plants clonal origin, climatic variables, soil types, and fertilizers, the protein content ranges from 1% to 1.8% [145].

Following the introduction of synthetic latex in the 1940s, trade in natural rubber significantly decreased. However, there is a small, rural market for natural rubber, such as rubber boots. Natural rubber is still in relatively high demand in the country and is imported into Bolivia. Consistency in the wholesale pricing and the quality of the goods are the key barriers to competing with this imported natural rubber. Neither group has access to formal financing, nor is natural rubber supported by the government. Although there are no laws or regulations regarding manufacturing of rubber latex [146].

Natural rubber (cis-1,4-polyisoprene) is highly valuable because massive amounts of key raw materials utilized by industry, transportation, healthcare and the military are obtained from natural rubber. The United States consume natural rubber as the second largest user in the world. Natural rubber cannot be found domestically in the US, unlike fossil fuel and there is no longer a stockpile. Natural rubber's strong performance attributes cannot be matched by synthetic polymers and as a result synthetic rubber cannot replace natural rubber. Additionally, to regulate output and prices, the world's top three rubber producing nations (Thailand, Malaysia and Indonesia, who together account for 70% of the supply) created a cartel in 2002. Vietnam and India, two other nations that produce rubber recently applied for membership.

The majority of natural rubber currently produced comes from *Hevea brasiliensis* clonal plantations in tropical Southeast Asia and the clones have a similar common heritage, putting the natural rubber industry at risk. Also, lack of genetic diversity and disease resistance puts the industry at risk regarding supply of rubber raw material [147]. Natural rubber is a crucial raw material for many applications because it has key physical qualities that manufacturers of synthetic rubber are unable to replicate. In 2013, from roughly 10 million hectares of land, the globe generated 12 million tons of natural rubber with Asia accounting for 91.5% of that production [148].

The demand of herbal latex and its products has been increased throughout the world [149]. The Indian herbal industry has annual revenue of more than INR 80 billion. AYUSH (Ayurveda, Unani, Siddha, and homoeopathy) items that are herbal latex make about 3% of all Indian pharmaceutical exports. An estimated INR 10 billion worth of raw materials make up 70% of the exports from the herbal business each year. Herbal extracts are among the completed goods that make up 30% of the export [9]. However, India still accounts for less than 1% of the worldwide herbal export market [150]. Production and cultivation of herbal plant latex and their evaluation procedures are mentioned in Fig. 6.4.

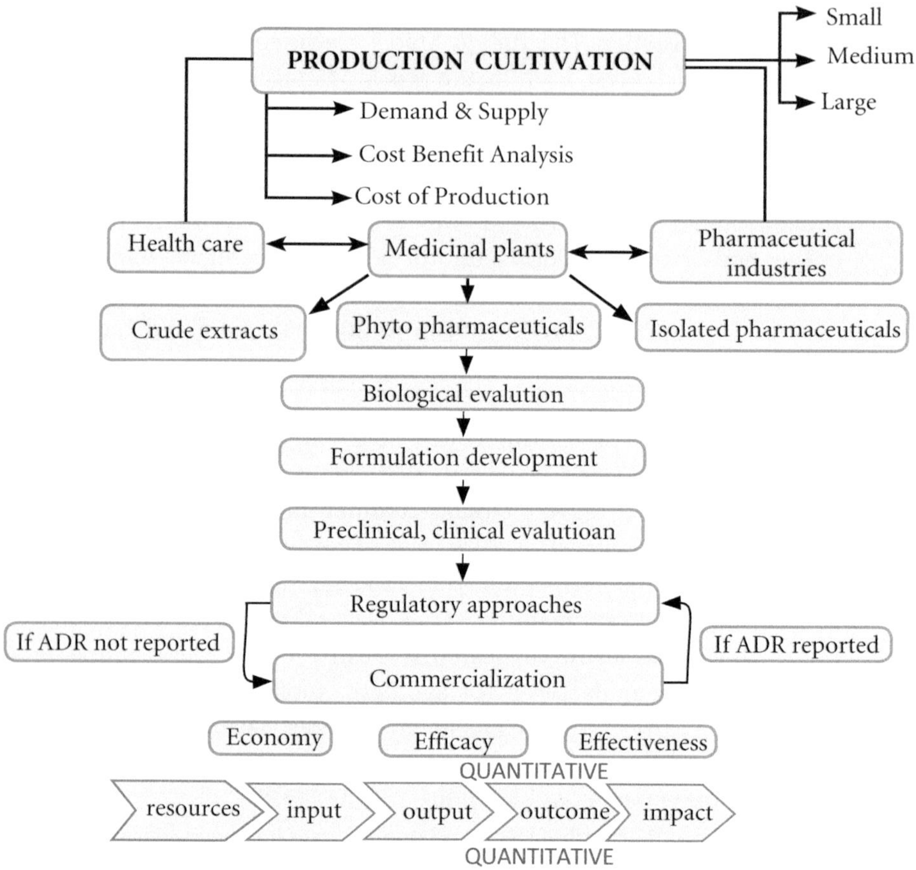

Figure 6.4 Evaluation processes of latex products from cultivation to commercialization.

6.4.7 Various Legal Standards

The results showed that delays in the application submission and review procedure as a result of different regulatory requirements were a top worry for the manufacturers of herbal latex. To comprehend the variations in approval procedures and submission specifications, a comparative analysis of the herbal latex registration standards for nations, including India, the United States and European nations, was carried out.

6.4.7.1 Latex products available

According to forecasts of the International Rubber Study Group (IRSG), in 2019 the global demand for natural rubber will increase to 14.2 million tons. They are valuable in the manufacture of tires for aircraft, space crafts, agricultural machinery and cars, thus, natural rubber cannot be replaced with synthetic rubber. In Russia, state corporation for assistance to development, production and export of advanced technology industrial products buys natural rubber for Russian tire manufacturers and the defense industry abroad. In 2014, importations of natural rubber and derived products from ASEAN countries amounted to $127.5 million. Considering both economic and political aspects, there is a strong necessity for Russia to find a domestic source of natural rubber able to substitute importations thereof.

Due to the growing demand for rubber, it is important to find and develop alternative crops in order to fill the gap between supply and demand. Some of the rubber preparations such as hats, swimsuits, chewing gum, beds, catheters, rubber bands, balloons, tennis shoes and sports goods are given in Table 6.2.

6.4.7.2 Business opportunities

The latex industry can be broken down into segments based on product (latex gloves, latex Foley catheters, latex probe covers, urine bags, and other products), end user (hospitals, diagnostic centers, ambulatory surgical centers, and other end users), end user (Asian, African and American). and geography. The report provides the value for the aforementioned sectors in millions of US dollars.

6.4.8 Herbal Latex Business Ideas

Here are some suggestions for starting one's own herbal business if someone has the entrepreneurial spirit rather than working for an already established herbal company:

- Make herbal extracts.

- Develop herbal skincare products.
- Open a clinical practice.
- Become an aroma therapist.
- Teach others about herbs.
- Grow or wild craft herbs.
- Build a monetized blog.
- Dive into social media.
- Write herbal content.
- Weave herbs into another profession.
- Sell herbs in bulk.

The ethnobotanical or folklore uses of plant latex and their phytochemical constituents are exploited to develop and commercialize modern medications. Here a few examples are given in Table 6.3.

6.4.9 Herbal Business Ideas: The Options Are Endless

Hopefully, this collection of herbal business concepts has inspired you to daydream and reflect. Simply keep in mind that this is simply a starting point as you decide where you want to go from here. There are too many alternatives to discuss in a single essay, and you may come across other company concepts that are not mentioned here. If you decide to pursue your passion for herbs whether by establishing a podcast or an apothecary then make sure you have a solid foundation and a ton of experience with them first. By 2035, the size of India's cosmetics and grooming market is projected to more than double to $35 billion with the use of cosmetics growing quickly. Table 6.4 shows natural rubber market report. The increased awareness among consumers of the harm done to their bodies after using chemical-based cosmetics products, their changing lifestyles, and their increasing desire to appear beautiful are driving factors that demand for herbal cosmetic goods [166].

Table 6.2 List of latex products and turnover per annum

No.	Rubber product	Turnover per annum
1	Rubberized cork sheet	INR 10–25 crores
2	Rubber compound toys	$36 million
3	Rubber sheet	INR 12000 crores
4	Rubber based adhesive	$1.10 billion
5	Rubber reclamation	$6999.00
6	Rubber gaskets	INR 12000 crores
7	Rubber auto parts	INR 100–500 crores
8	Rubber powder from waste tires	INR 100 crores
9	Condoms	$200 million
10	Ethylene propylene rubber	$4.45 billion
11	Rubber compounding for automobile industry	$12700 million
12	Latex rubber threads	$139.8 billion
13	Rubber floor mats	INR 252.71 million
14	Surgical gloves	$116.9 million
15	Bicycle tubes and motorcycle tubes	$14051.66 million
16	Waste tire pyrolysis	INR 3.5 - 4.0 crores
17	Butyl reclaim rubber	$1457.9 million
18	Radial tires for cars & trucks	INR 30000 crores
19	Rubber roller for printing machine	$12700 million
20	Auto tubes	INR 5986.79 crores
21	Butyl rubber-polyisobutylene rubber	$ 2.88 billion
22	Automobile hoses (air conditioned hose, fuel hose, hydraulic hose, petrol pump hose) and production of tires	$11.36 billion
23	Dental dams	INR 1.0 crore
24	Stethoscopes and blood pressure cuffs	INR 159 billion
25	Spandex	$8.3 billion
26	Pacifiers and baby bottle nipples	INR 25–50 crores
27	Mouse pads	$4200.00
28	Goggles	$7.64 billion

*Annual turnover of rubber product costing along with type of rubber product, 1 bn = 1000 million, 1 bn = 100 crore.

INR, Indian rupees; Cr, Crore, $, United States dollar.

Table 6.3 Commercial, pharmaceutical and traditional uses of latex

No.	Class	Latex bearing plants	Latex constituents	Uses	Res.
1.	Protein	*Calotropis procera Cryptostegia grandiflora*	Cysteine peptidases	**Pharmaceutical uses:** • Hydrolyze proteins in cow milk • Milk clotting activities like cheese production, curd formation • Meat tenderization, food processing, dehairing leather	[151, 152]
		Carica papaya	Chymopapain (cysteine peptidases)		
		Calotropis gigantea, Croton lechleri, Aloe megalacantha	Proteases	**Traditional uses:** • Induce clot formation, facilitate wound repair • Anti-inflammatory	[153]
		Plumeria alba, Calotropis procera		**Pharmaceutical Importance:** • Active against ectoparasite and endoparasite such as • Haemonchus contortus, nematode • Effective in scabies	[154]
2.	Tissue engineering, novel drug delivery systems	*Calotropis gigantean*	PAMAM (polydopamine dendrimers) Biocompatible material cellulose	**Pharmaceutical Aid:** • Berberine dendrimers exhibited better anticancer effects • Latex mediated green synthesis of silver nanoparticles anticancer drug, doxorubicin (DOX) • Nano careers mimic the anticancer action of doxorubicin • Silver nanoparticles as antimicrobial agents	[155]

No.	Class	Latex bearing plants	Latex constituents	Uses	Res.
		Euphorbia acruensis *Calotropis procera*	Latex bioactive compound/ bio-reductant	• AgNPs showed broad spectrum antimicrobial effect • Antiviral activity	[156, 157]
3.	Alkaloids	*Catharanthus roseus*	Vindoline, vinblastine	**Pharmaceutical importance:**	
		Papaver somniferum	Morphine, dopamine	• Morphine used as analgesic • Pharmaceutical industry relies on opium poppy to produce painkillers at commercial level	[158, 159]
		Papaver somniferum	Noscapine, codeine	**Commercial Importance:**	[160]
		Chelidonium majus	Berberine	• Anticancer • Analgesic • Neurodegenerative diseases	[161, 162]
4.	Cannabinoids	*Cannabis sativa*	Tetrahydrocannabinol (THC), cannabidiol (CBD)	**Commercial Importance:** • In 2018 FDA approved Epidiolex® against Dravet syndromes • Cannabinoids in the treatment of neurological ailments, cancer and metabolic disorders.	[163–165]

*Pharmaceutical products and their source along with their pharmacological effects latex plant.

Table 6.4 Natural rubber market report

Report Coverage	Details
Market size value in	US$ 21,410 million in 2020
Market size value by	US$ 30,100 million in 2027
Growth rate	CAGR of 1.3% from 2020 to 2027

6.4.10 Latex Medical Gloves for Varied Applications Are Expected to Capture Significant Revenue Share Over the Forecast Period

Due to their low cost, cleanliness and safety, disposable latex gloves have proven to be quite helpful in the medical industry. For their own protection, doctors who examine patients with infectious disorders must practice good cleanliness. The disposable gloves they wear serve as a barrier to keep them safe from spreadable illnesses.

6.4.11 Commercial Products from Latex Bearing Plants

Due to their widespread use as foods and medications, herbal goods are seen as low risk due to their reputation as being natural and secure. The quality of these products is ineffectively regulated and managed. Deep anxiety is being generated by the mounting evidence that these claims are untrue, but it is unknown how widespread this problem is on a global, continental or even a national level [166]. The use of plant based products has grown significantly in recent years and there is a huge market for latex containing items used for agricultural, medical and therapeutic applications. Pharmaceuticals, insecticides and immunological allergens can all be found in large amounts in plant latex [167]. Commercially marketed formulations from the latex bearing plants are listed in Tables 6.5 and 6.6.

Table 6.5 Commercially marketed formulation of latex bearing plant [168]

No.	Latex bearing plant	Chemical constituent	Marketed product	Generic name	Uses
1.	*Papaver somniferum*	1-deoxynojirimycin	Devoglustat HCl	Devoglustat	α-Glucosidase inhibitor, antiviral
2.	*Madagascar periwinkle*	Vincristine	Vincristine sulphate	Vincristine	In acute leukemia, malignant lymphoma
3.	*Madagascar periwinkle*	Vinblastine	Vinblastine sulphate	Vinblastine	Anticancer, Hodgkin's diseases
4.	*Cannabis sativa*	Cannabidiol	Epidiolex	Cannabidiol	In schizophrenia, epilepsy, Alzheimer's disease
5.	*Cannabis sativa*	δ-9-tetrahydrocannabinol	Dronabinol	δ-9-tetrahydrocannabinol	Antiviral, in anorexia
6.	*Carica papaya*	Papain	Axcel Papain	Papain	Anti-inflammatory
7.	*Croton lecheri*	Crofelemer	Mytesi	Crofelemer	Antidiarrheal
8.	*Anabasis aphylla*	Nicotine	Nicorette Nicotine	Nicotine	Analgesic
9.	*Papaver sominferum*	Opium alkaloids	Opium tincture	Opium	Analgesic, anticancer
10.	*Carica papaya*	Pancrelipase protease	Cotazym	Pancrelipase protease	Anti-coagulant
11.	*Anisophyllum peplis*	Ingenol mebutate	Picato	Ingenol mebutate	Anticancer
12.	*Thevetia peruviana*	Rutin	Rutoside trihydrate	Rutin	Anticancer, anticoagulant, antioxidant, antimicrobial
13.	*Cryptostegia grandiflora*	Lauric acid	Jayjun	Lauric acid	Milk clotting
14.	*Ficus benjam*	α-pinene	Rawachol	α-pinene	Antimycobacterial, antifungal

*Commercially marketed formulation of latex bearing plant and their uses.

Table 6.6 Commercially marketed synthetic derivatives of opium latex

No.	Synthetic drugs derive from latex	Mode of delivery	Uses	Refs
1	Hydromorphone	Oral, normal release and controlled release	Cancer related pain	[169]
2	Fentanyl	Transdermal	Severe or chronic cancer pain	[170]
3	Methadone	Oral, intramuscular (IM)	Severe cancer pain	[171]
4	Oxycodone	Oral, normal release and controlled release, IM and intravenous (IV)	Cancer related pain	[172]
5	Buprenorphine	Sub-lingual, sub-dermal, transdermal, rectal, epidural, IM, and IV	Severe cancer pain	[173]
6	Codeine	Oral, normal release and controlled release, and IM	Analgesic	[174]
7	Tapentadol	Oral	Analgesic	[175]
8	Morphine	Oral, instant release and modified release, rectal, epidural	Analgesic	[176]
9	Tramadol	Oral	Analgesic	[177]
10	Epidiolex	Oral	Pediatric epilepsy	[178]
11	Noscapine	Oral	Antitussive, anticancer, anti-inflammatory	[178]
12	Oxymorphone	Parenteral	Analgesic	[168]
13	Tapentadol	Oral	Analgesic	[168]

*Commercially marketed formulation of opium plant and their uses.

6.5 Guidelines of Latex Utilization

It has been reported that some individuals showed allergic reactions to natural rubber latex proteins. Hence, some regulations are required to safeguard the general public health with respect to latex allergy.

(a) All the devices comprising of natural rubber or their packaging containing natural rubber is covered under this section. Natural rubber includes natural rubber latex or dried form of natural rubber or a formulation containing synthetic rubber and natural rubber [179]. Natural rubber is a concentrated colloidal solution of natural rubber latex [180]. Dried natural rubber uses dried or milled sheets of coagulated natural latex and can be converted into products through compression molding, extrusion, or immersion process [181].

(b) The phrase "contacts humans" denotes that the user or patient will likely come into contact with the natural rubber present in a gadget or in patient's environment.

(c) Devices containing natural rubber shall be labeled as in section 502(c) of the Federal Food, Drug and Cosmetic Act (21 U.S.C. 352(c)).

(d) Devices containing natural rubber latex shall be labelled, and the label in bold print indicates "Caution." The product contains natural rubber latex and may cause allergy.

(e) Devices containing dry natural rubber that contacts humans and are not described in paragraphs (b) and (d) of this section, shall bear the following statement in bold print on the device labeling: "This product contains dry natural rubber."

(f) Devices with natural rubber latex shall not contain the term "hypoallergenic" on their labeling.

(g) Any affected person may request an exemption or variance from the requirements of this section.

(h) Any device subjected to this section that is not labeled in accordance with paragraphs (d) through (h) of this section and that is initially introduced or initially delivered for introduction into interstate commerce after the effective date of this regulation is misbranded under sections 201(n) and 502(a), (c), and (f) of the act [21 U.S.C. 321(n) and 352(a), (c) and (f).

6.6 Concerns of the Safety and Toxicity of Latexes

Natural rubber latex products such as latex gloves can cause allergy affecting around 8% to 12% of healthcare professionals

[182]. People without symptoms showed a greater prevalence of allergy [183].

In response to the discovery and development of HIV and hepatitis B virus, the centers for disease control and prevention (CDC) suggested extensive measures regarding the use of biodegradable natural rubber gloves. As a result, there have been more reports of latex allergies. Sensitivity develops in medical staff who frequently use latex gloves or inhaling latex aerosols. In 2000 reports of more than 600 products involving latex allergies were published. Hence, protein and glove powder levels should be kept to a minimum, according to the FDA *Medical Glove Guideline Manual* [184].

The role of plant based medicines in disease management is increasing globally particularly in developing countries. These medicines are affordable, easily available and accessible locally. Consumers around the world have a widespread misconception that herbal medications are always safer than modern medications. Even though there is little evidence to support that mild to moderate harmful effects are linked with herbal medications. Recent data revealed that some of the herbs that have long been believed to be safe have actually been linked to health hazards. Herbal treatments have the potential to enhance the effects of various pharmacological therapies by acting as agonists or antagonists. Therefore, knowledge of conventional medications is a requirement for efficient herbal therapies. With the advancement of technology, it is now possible to identify potential hazardous compounds of herbal drugs even if they are present in trace amount. There are reports of major adverse effects from medicinal plant species that are widely utilized to treat specific ailments. Numerous medications have their roots in biologically active plant compounds contributing to their therapeutic benefits.

Synthetic medicines frequently just include one component, in contrast to medicinal plants, which might have a complex diversity of 400 chemicals or more. There is simply no way for scientists to map all the intricate interactions and synergies that may exist between every single molecule that may be identified in a plant, or between the unprocessed plant extract that includes every single component and is often utilized. The activity and side effects of a single drug may be identified quite easily. There are four or five less harmful compounds that might counteract the first chemical lethal effects. The plant extract may perhaps even possess some anti-cancerous qualities. The majority of medicinal plants contain a wide range of complex components, such as mucilage, tannins, polysaccharides, and other elements that might modify and change the effects of any "active principles" [185].

The registration of drug, drug abuse, pharmacovigilance of drug, counterfeit and substandard medications, drug resistance, storage and disposal of medications and any other safety concerns have been discussed [186].

6.6.1 Safety of Medicinal Plant Latex Used in Traditional Medicines (TM)

It is believed that plant latex used in traditional medicines is considered as safe and effective. The long history of usage and evidence accumulated support this claim. Despite the fact that hundreds of individuals pass away every year due to the use of over-the-counter medications. Even though the United States national poison control centers do not include a database for adverse responses to herbal drugs and their products. Herbal related hospitalizations or fatalities are extremely rare making them hard to locate [187]. This means that any substance toxicity including food and medicinal plants depends greatly on the quantity or dose utilized. At high doses, non-toxic substances can become toxic whereas at low doses highly toxic substances can become safe. Overdosing during treatment will undoubtedly cause safety issues [188]. Aside from an overdose, mistakes are made by customers and healthcare professionals due to overuse and prolonged usage of herbal remedies even at a safe level. Drugs (allopathic medications) and plant latex (herbal medicines) can interact with each other in a synergistic or antagonistic manner. Even though herbal drug-food interaction has also been reported. Herbs traditionally used to treat diabetes may theoretically cause hypoglycemia if taken at the same time with other traditional antihyperglycemic drugs [189]. The toxicity issues with medicinal plants in the official herbal sector may be caused by inadequate quality assurance and noncompliance with the guidelines of good manufacturing practice [190].

Additionally, there is a lack of access to the knowledge necessary for using herbal medicines effectively as well as incorrect usage strategies. Additionally, the issue might be made worse by the adulteration of herbal treatments, which involves the addition of synthetic pharmaceuticals and other potentially harmful substances like botanicals, microbes, poisons, agrochemicals and heavy metals [191].

Poor quality control measures result in poor product quality and may leads to side effects associated with the use of natural products. Poor regulatory controls and mainly unregulated distribution of drugs may contribute to such incidents. This leads to the production of low-quality goods [192, 193]. It is found that adverse reactions or side effects with respect to natural products are typically hard to spot unless the root cause of such events has been determined. Herbal remedies are generally secure when used properly. They have a long history and their safety and effectiveness have been proven by experience passed down from generation to generation [194]. It is important to remember that traditional practices erroneously interpret the toxicity data of numerous therapeutic herbs and draw incorrect conclusions. In addition to aqueous extracts utilized in traditional medicinal practice, several toxicity studies have been reported on medicinal plants and their phytoconstituents extracted in organic solvents.

6.6.2 Plant Latex Regulatory Control

Papaver somniferum latex produces therapeutically important alkaloids like morphine, codeine, papaverine, noscapine and thebaine. The major producers of poppy products are Australia, Turkey and India. The cultivation of opium in India is legalized and its use for medical conditions is permissible. Australia, Turkey and India are the top three nations producing items made from poppies. The Convention on Narcotic Drugs 1961 legalized growing opium in India and permitted its use for medicinal ailments. A few locations where opium is produced lawfully in India include

Barabanki, Bareilly, Lucknow and Faizabad in Uttar Pradesh; Mandsaur, Ratlam and Neemuch in Madhya Pradesh; and Pratapgarh in Rajasthan. India relies on opium imports to meet the demand for opium products for both medicinal and culinary uses, despite a large-scale poppy industry. Only imports of codeine can satisfy the demand (30%). Farmers are granted licenses to produce opium poppies by the Central Bureau of Narcotics (CBN), Gwalior (Madhya Pradesh), under the supervision of the Narcotics Commissioner. The total number of permits given for the crop year 2008–09 was 44,821 and the minimum qualifying yield (MQY) for Rajasthan, Madhya Pradesh, and Uttar Pradesh was set at 56 kg/ha, 56 kg/ha and 49 kg/ha, respectively. CBN officers measure each field and implement procedures to make sure that no extra cultivation occurs. According to Section 6.8 of the Narcotics Drugs and Psychotropic Substances (NDPS) Act, 1985, it is illegal to cultivate opium poppies in India unless a license has been granted by the CBN in Canada, any activity involves the use of opium and its derivatives are controlled under the Controlled Drugs and Substances Act (CDSA) and Narcotic Control Regulations (NCR). These laws allow Canada to meet its obligations under international drug control treaties. Except in cases of national crises or if the drug is a narcotic raw material, importation of any substance classified in Schedule I or II of the United States Controlled Substances Act is prohibited by United States laws (21 CFR 1312). The Drug Enforcement Agency (DEA) is prohibited from issuing import licenses for opiate raw materials from nations that are not specified under this law, commonly known as 80/20 rule and permission must be provided in order to preserve the required 80/20 ratio. Except for low dosage codeine products that are placed on Schedule V of the United States controlled substances act, this is true of all opiates [195, 196].

6.6.3 Concern of Latex Allergy

Among health care professionals, one of the major health concerns is allergy caused by latex obtained from natural rubber. As per United States Occupational Safety and Health Administration around 8–12% health care professionals are sensitive to allergy.

Powdered natural rubber latex source is considered a major allergy source. Natural rubber latex is obtained from *H. brasiliensis* and is treated with ammonium hydroxide, zinc oxide or formaldehyde. This treatment results in protein hydrolysis and prevention of microbial growth. Hence, latex along with the chemicals used for its treatment acts as an allergen for the human body. During manufacturing, the latex gloves can be dipped into powder slurry of corn starch. This powder acts as an adjuvant. Additionally, this powder may become aeroallergen as well. The complications are type I and type IV immediate and delayed type of hypersensitivity reactions like allergic contact dermatitis, bronchospasm, conjunctivitis, rhinitis and angioedema. These reactions are mediated by Immunoglobulin-E. Moreover, cross reactivity is also observed in latex allergic individuals. The laboratory tests can elucidate the mechanism of allergy and give an idea about the importance of the route of allergy caused by natural rubber latex. Various animal models like mouse, rabbits and guinea pig have been developed to elucidate latex allergy. Such understanding can lead to providing a strategy to overcome natural rubber latex allergy.

Allergens, physiological role and their molecular weight are depicted in Table 6.7, the risk groups and medical conditions associated with latex allergy are given in Table 6.8 and latex allergy among healthcare workers (HCWs) are listed in Table 6.9 [197–199].

6.6.4 Food/Drug Interactions

It has been noted that latex allergy can be brought on by either drugs that have been contaminated by vials, packing equipment medical devices or by food that has been contaminated by employees using latex gloves. According to the USFDA, natural rubber latex has been allowed for use in packaging, medical equipment and food additives. An updated list of packaging material for vials, vial stoppers, tip caps and syringes was recommended by the Centers for Disease Control and Prevention (USA). As a result, latex sensitivity and allergy pose a serious threat to health [212–214].

Table 6.7 Some characteristic of allergens from *Hevea brasiliensis* and their cross reactivity

Allergen	Biochemical name	MW (kDa)	Predicted physiological role	Cross reactivity	Allergic reactions
Hev b 1	Rubber elongation factor	14	Biosynthesis of rubber	Papain	Immediate hypersensitivity reaction
Hev b 2	β-1,3-Glucanase	35	In defense mechanism	Glucanase	IgE hypersensitivity
Hev b 3	Particle protein	24	Biosynthesis of rubber	NA	IgE hypersensitivity
Hev b 4	Lecithinase homologue	53–55	Structural component	NA	IgE hypersensitivity
Hev b 5	Acid latex	16	Structural component	Kiwi protein	IgE hypersensitivity
Hev b 6	Hevein	20	Biosynthesis of rubber	Food chitinase	IgE hypersensitivity
Hev b 7	Patatin-like protein	42	In defense mechanism	Solanaceous proteins	IgE hypersensitivity
Hev b 8	Profilin	15	Structural component	Panallergen	IgE hypersensitivity
Hev b 9	Enolase	51	Glycolytic enzyme	NA	IgE hypersensitivity
Hev b 10	Class I chitinase	30	Antioxidant mechanism	NA	IgE hypersensitivity
Hev b 11	Legumin	55	In defense mechanism	Panallergen	IgE hypersensitivity
Hev b 12	Type 1 lipid transfer protein	9	In defense mechanism	Panallergen	IgE hypersensitivity
Hev b 13	Esterase	42	In defense mechanism	NA	IgE hypersensitivity
Hev b 14	Hevamine	30	In defense mechanism	NA	IgE hypersensitivity
Hev b 15	Serine protease inhibitor	7.5	In defense mechanism	PR-6	IgE hypersensitivity

*Characteristic of allergens from *Hevea brasiliensis* along with biochemical name.

Table 6.8 List of risk groups and medical conditions associated for latex allergy

No	List of risk groups	Medical conditions for latex allergy
1	Barbers	Tracheoesophageal fistula
2	Construction workers	Paralysis of limbs
3	Contraceptive users	Cerebral palsy
4	Cremation home workers	Allergic reactions
5	Domestic workers	Anorectal malformations
6	Florists	Skin problems
7	Food handlers, restaurant workers	Urogenital abnormalities
8	Health care professionals	Neural tube defects
9	Portraitists	Premature baby
10	Rubber industry workers	Multiple congenital anomalies
11	Security personnels, police officers or firefighters	Cerebral shunt

Table 6.9 Latex allergy among healthcare workers (HCWs)

Latex allergy	Number of health care workers	Country of origin	Year reported	References
4%	619	Italy	2015	[200]
13.3%	4,529	Thailand	2014	[201]
4.2%	1,115	Turkey	2014	[202]
8.8%	8,485	China	2013	[203]
9.8%	178	Netherlands	2013	[25]
5.9%	337	South Africa	2013	[204]
8.3%	144	South Africa	2013	[205]
18%	899	Thailand	2013	[206]
22.4%	66	Brazil	2012	[207]
5%	804	USA	2012	[208]
5.0%	617	Bangkok	2011	[22]
5.9%	620	Spain	2011	[31]
16%	163	India	2010	[209]
16.3%	324	Sri Lanka	2010	[210]
17.9%	104	Iran	2009	[211]

In real life, the interactions between prescription drugs, other dietary supplements and even foods are what lead to the majority of adverse reactions to herbal remedies. Certain fruits and vegetables might trigger latex responses because they contain similarly structured proteins that the body incorrectly interprets as latex. People who are allergic to latex get allergic responses. These allergic reactions might happen after touching, smelling, or even just consuming these foods. Even though they don't contain latex, many foods have proteins that are similar to latex [215]. A latex allergy sufferer who consumes these proteins may develop an allergic reaction. We refer to these foods as latex shows cross-reactivity with foods. When it comes to cross-reactive foods, prevention is crucial. A person with a latex allergy should avoid exposure to cross-reactive food. Carrot, coconut, apricot, strawberry, loquat, spinach, pineapple, cherimoya, passion fruit, papaya, mango and celery are foods that rarely trigger reactions in people with latex allergies. Some botanicals have been identified as the most dangerous or life-threatening botanicals/pharmaceutical interactions [216].

The chance of allergic responses exists and the people with allergies should read the research for contraindications and speak with their doctors. Critical components of a clinical history include symptoms of contact sensitivity to NRL gloves, work related latex allergy complaints, anaphylactic episodes during surgery and the existence of atopy and food allergy [217, 218].

A higher risk of latex allergy may exist in patients with food allergies. There is a cross-reaction between banana and latex. In addition to banana, the most frequently cross-reacting foods include avocado, chestnut, kiwi, potato and tomato [219, 220]. Apple, apricot, celery, cherry, fig, grape, melon, papaya and peach are less frequently cross-reacting foods. Clinical manifestations of the condition can include localized perioral irritation and rarely food-induced anaphylactic shock [221]. Although the presence of symptomatic cross-reactions to foods are frequently linked to latex sensitization, it is still unknown if sensitization to food represents a key gateway to latex allergy or vice versa.

6.6.5 Food–Latex Cross-Allergy

Plant latex is used in the manufacturing of a variety of products, hence latex allergy is of prime concern. More than 35 proteins obtained from latex of *H. brasiliensis* may be related to IgE mediated type 1 delayed type of hypersensitivity reactions. A cross reactivity between latex and food has been reported. Several enzymes like chitinase from latex is involved in these reactions. This has been reported in 30–70% of latex allergic individuals [222]. Individuals who have allergies to chestnuts, avocados and bananas are more likely to have latex allergies due to the presence of proteins identical to latex. Other foods including figs, apple, celery, melons, potatoe, papaya and fruits with pits like cherries and peach have also been known to trigger oral and cutaneous itching type of allergic reactions in certain people. It will be highly beneficial to consult with a nutritionist or nutritional expert or dietician [220, 223, 224]. The intensity of cross reactivity between latex and food items is listed in Table 6.10 and the list of natural rubber latex based products is given in Table 6.11 [199, 212–214].

Table 6.10 Cross reactivity between latex and food items

No	Cross reactivity with latex	Examples of food items
1	Low	Apricot, Buckwheat, Castor bean, Coconut, Cucumber, Dill, Fig, Grape, Hazelnut, Indian jujube, Jackfruit, Lychee, Mango, Nectarine, Oregano, Peanut, Rye, Sage, Soybean, Tobacco, Walnut, Zucchini
2	Medium	Apple, Carrot, Melons, Papaya, Tomato
3	High	Avocado, Banana, Chestnut, Kiwi

6.6.6 Improper Dosage/Usage

Correct and exact dosages are essential for drug safety and effectiveness. Poor effectiveness and negative effects are caused by improper dosage. For instance, morphine may result in significant or life-threatening respiratory issues, drowsiness or coma. Tetrahydrocannabinol has been associated with severe adverse effects. Cannabinoid hyperemesis syndrome (CHS) may causes extreme nausea and vomiting. People with epilepsy who consume cannabis in high dosages may get seizures.

Table 6.11 Natural rubber latex based products

No.	FDA approved products
1	Adhesives
2	Antioxidants and stabilizers for polymers
3	Chewing gum base
4	Components of paper
5	Components of paperboard
6	Condom
7	Resinous and polymeric coatings
8	Rubber articles intended for repetitive use
9	Rubber dam and accessories
10	Sensitive adhesives
11	Surgical drape and drape accessories
12	Surgical gloves
13	Urology devices

6.6.7 Adulteration, Contamination & Quality Control

Herbal drugs are adulterated with inferior-quality materials and this in turn affects the quality, safety and efficacy of drugs. Such a failure in quality control measures is extremely important and considered to be serious. These types of practices are seen in herbal supplements imported from India, China or other countries as per the reports from the University of Minnesota published in 2007–2011. Hence, the quality control measures or the standardization of herbal drugs including latex bearing plant materials are very important to maintain its quality, safety and efficacy. The concentration of opium alkaloids like morphine, codeine and thebaine reduced after thermal and washing treatment while baking didn't significantly affect the presence of alkaloids in poppy seeds [225].

The *Papaver somniferum* seeds can be adulterated with *Amaranthus paniculata* seeds. which can be detected by analyzing the presence of fatty acids [226].

Papaver somniferum contaminated with lead is a major health problems among chronic users in Iran. Quinine and quinidine have been reported as adulterants of heroin. Intravenous injection of heroin adulterated with quinine/quinidine resulted in cardiac disorders [227].

Phytoconstituents from *Cannabis sativa* like δ-9-tetrazydrocannabinol can be adulterated with synthetic cannabinoids. The usage of such type of adulterated compounds containing potent and toxic synthetic cannabinoids can result in increased health risks (psychological and cardiological adverse effects) [228].

Moreover, the impact of cannabis contaminants on the human being has been observed. The common cannabis contaminants include microbes, heavy metals and pesticides. Their effects include infection, carcinogenicity, reproductive and developmental toxicity. This may also affect dose, route as well as bioavailability of cannabis formulations. With respect to the contamination of cannabis or its products, there are some areas where attention and further investigations are required. These are

- Quantitative analysis of heavy metals, pesticides and microbial toxins (particularly aflatoxin) in cannabis formulations.
- Comparative analysis of contaminants in cannabis plant, crude extracts and phytopharmaceuticals and isolated cannabinoids from cannabis plants.
- Volatility and pyrolysis of pesticides and heavy metals while inhaling cannabinoids.
- Contaminant bioavailability and metabolism.
- Bioavailability of both the cannabinoids and metals.
- Metabolism of cannabinoids.

Adulteration of cannabis with tobacco, calamus, and other cholinergic compounds increases cannabimimetic effects and decreases the adverse effect of cannabis [229–232].

6.6.8 Latex Safety: from Prejudice to Evidence

A long time ago, natural herbal latex was used to heal human ailments. During that time little scientific data was available regarding the safety and efficacy of plant latex. Herbal medicine and other conventional therapies have experienced a significant contraction. China is not an exception because it has a wealth of herbal resources. The nature of diseases has been changed recently covering acute, chronic and metabolic disorders as well. The outcomes of Western medical therapies are generally dismal and difficulties with adverse drug reactions are particularly prominent. The use of complementary and alternative medicine, notably herbal medicine and natural latex has expanded in popularity and interest.

6.6.9 Safety Issues of Plant Latex

The enormous increase in global use has been accompanied by a focus on the safety of herbal latex. Currently, there is misunderstanding and prejudice around the safety of plant latex. Therefore, impartial and fair interpretation, as well as public knowledge, is necessary. Latex allergy is a serious concern associated with plant latex.

6.6.9.1 Plant latex is drug, not food

The use of plant latex is increasing day by day because of its easy availability, therapeutic efficacy and less side effects. However, misusing latex will result in a number of serious adverse outcomes. It is important to understand that although latex contains potential phytoconstituents which are considered to be safe, but toxicity or side effects are still a possibility. Therefore, latex should be treated as a drug in order to assure its safe use.

6.6.9.2 The relative property of plant latex safety

As the Chinese proverb says, "all medicines have their own side effects." In other words, medication is a double-edged weapon that may both treat illness and preserve health while also having the potential to harm the human body. Latex is no different from other effective medications in that they might cause adverse drug reactions. Instead of inappropriate use, herbal medicine should be used with the proper caution. Over dosage and improper treatment regimens result in adverse drug reactions and drug resistance as well.

6.6.9.3 The complexity of safety of plant latex

The safety and toxicity aspect of latex bearing plants material can be categorized as

Intrinsic toxicity: It is the inherent toxicity of several herbs whether taken in overdose or at a regular therapeutic dosage.

Extrinsic toxicity: It is because of inappropriate processing of herbal drugs or its products, adulteration, misidentification or substitution of herbal constituents, or contamination of products with hazardous metals.

Adulteration, improper dosage, allergic reactions and food drug interaction are the factors that affect safety of latex.

6.6.9.4 Weak basic research in safety of plant latex

To ensure safe and effective use of herbal medicine including latex, it is crucial to identify its harmful and negative effects. Lack of objective experimental evidence and scientific standards lead to the classification of toxicity. Information on poisonous herbs, target organs, safe dose ranges, effective doses and minimal toxic doses is insufficient. The majority of the things came from conventional wisdom and clinical experience for their efficacy and toxicity.

It is found that detailed studies on plant latex has not been done. Hence, It is important to note that the safety and efficacy of plant latex can be further improved by generating the scientific data with respect to toxicity, side effects, drug-drug and drug-food interaction. A regulatory framework for plant latex and latex containing products can provide consumers with more peace of mind. Latex was treated like medications, functional foods, health goods, or dietary supplements, leading to inconsistent standards and a disorganized industry.

The WHO can take the initiatives for settling and optimizing the global herbal standards regarding the cultivation, collection, harvesting, processing, storage, breeding and manufacturing of herbal drugs and their dosage forms for ensuring its quality, safety and efficacy. A method to guarantee the safety of herbal medications should also be established. This system should play a stronger role in safeguarding public health by including responsible clinical practice and risk management [233].

Examples of a Questionnaire for Latex Sensitivity Screening can be found in Appendix A.

Health Care Workers.

The following actions should be taken by healthcare professionals to safeguard oneself from being allergic to latex at work.

For tasks that are unlikely to involve coming into contact with harmful objects, use non-latex gloves.

1. When using latex gloves, choose gloves with less protein or without powder for activities where it is likely that you will come in contact with blood or other body fluids.
2. Recognize that latex allergy risk is not decreased by using hypoallergenic gloves. However, they might lessen reactivity to the latex chemical additions.
3. Use proper work procedures to lessen the possibility of latex reactions: Use oily-base creams or lotions sparingly when wearing gloves made of latex unless they have been shown to minimize latex-related issues. Dry your hands gently after taking off the latex gloves, then wash them with a mild soap.
4. Take advantage of all the training and knowledge about latex allergies.
5. To avoid skin contact with latex use cotton gloves liners and external barrier creams if you are allergic to latex gloves.
6. If you start to experience latex allergy symptoms, keep latex gloves and other products that contain latex away from your skin. Consult a doctor who specializes in the treatment of latex allergy.

Inform the FDA MedWatch program of allergic reactions to latex medical equipment

If you are known to be allergic to latex:

1. Avoid wearing latex gloves and other latex made products.
2. Steer clear of places where you might breathe in powder from other workers' latex gloves.
3. Let your employer and medical professionals know that you are allergic to latex.
4. Put on a medical bracelet.
5. Carefully adhere to your doctor's recommendations for handling latex allergies. This might entail always having auto-injectable epinephrine on hand.

Patients

A. Check every patient for allergies. The risk of an allergic reaction should be reduced for those who have been found to be sensitive or allergic to latex. For those who are sensitive, keeping a latex-free atmosphere and avoiding latex-containing products are necessary.

B. Check patients for a latex allergy or a latex predisposition. Preoperative patients should, whenever possible, be pre-screened before admission. Relevant details that may prove useful include:

    1. The presence of atopy, which includes hay fever, food allergies (particularly to bananas, avocados, chestnuts, and kiwis), eczema in childhood or adulthood and asthma.
    2. Several operations.
    3. Urticaria during surgery, angioedema, respiratory discomfort or breathing problems.
    4. The type of latex equipment, exposure and the duration of the exposure throughout history.
    5. Manifestations of a potential latex allergy at work such as dermatitis, eczema, and urticaria, along with rhinorrhea, itchiness, congestion, sneezing, dyspnea and breathlessness.

6. Additional symptoms, including itchy hands, regional angioedema, and potential systemic anaphylactic reactions to condoms, balloons and diaphragms.

C. Patients with latex allergy must have access to latex-free supplies.

D. Eliminate all latex containing items from the area around the patient such as elastic bands latex gloves, tourniquets etc.

E. If any allergic reaction is experienced by the patient, then remove the suspected allergen and administer emergency care as necessary and as instructed by the doctor. This could involve resuscitation drugs, airway control and intravenous fluids.

F. Observe institutional guidelines for puncturing latex containing vial stoppers. There is no evidence that the latex content of drugs reduced by filter needles used for medication withdrawal. Three primary methods of vial closure have been identified in the literature.

   a. For vial closures or stoppers, it is possible to utilize synthetic rubber or a blend of synthetic and natural rubber latex. Opening the vial stoppers could result in microbiological contamination and the solution may already be contaminated with latex that was leached during manufacture and storage.
   b. Once only, medication is pulled into a non-latex syringe, vial stoppers are ruptured, patient is monitored after each medication intake. Although the closed-vial system is kept in place, the issue of leached latex is not addressed.
   c. Staff members at the pharmacy examine, identify and catalogue drugs contained in latex-containing vials. The catalogue must be created and maintained over time, which could result in information that is erroneous or lacking.

6.7 Evaluation of Latex Safety and Toxicity

In the modern system of medicine there is no Xenobiotics that can be considered as safe. The complimentary and alternative systems of medicine involves the use of herbal drugs for the prophylaxis and treatment of diseases. Generally, the plant based drugs are considered as safe and effective when it is being used in proper dose for a prescribed period of time. The safety and efficacy of latex plants and the phytoconstituents isolated from them can be determined by several parameters.

Plant latex is the source of many phytoconstituents with potential biological activities [8, 234–236]. However, apart from these activities side effects and toxicities have also been observed for laticiferous plants [237]. Hence, the evaluation of their side effects and toxicities is of great importance (Table 6.12).

6.7.1 Toxicity from Intrinsic Sources

6.7.1.1 Intrinsic toxicity evaluation

It is reported that the side effects and toxicity of herbal drugs and their finished formulations depends on the presence of phytoconstituents present in them [1]. In case of complex, heterogeneous plant mixtures containing numerous phytoconstituents the situation is even more difficult for determining their toxicity [53].

Numerous toxic endogenous compounds obtained from latex bearing medicinal plants exhibits hepatotoxicity, nephrotoxicity, neurotoxicity, cardiotoxicity, teratogenicity and cytotoxicity etc [54]. The regulatory agencies are ensuring the quality, safety and efficacy of phytoconstituents with reported hazardous effects. In this regard, work has been done on creating technology to get rid of toxic substances and effective ways to find and optimize their concentration (Fig. 6.5) [55–57].

6.7.2 Acute/Sub-Acute/Chronic Toxicity Evaluation

The toxicity evaluation of medicinal plant latex and their dosage forms have been determined as per OECD (Organization for economic co-operation and development) guidelines.

Table 6.12 Toxicity evaluation of latex bearing medicinal plants and their effects on biochemical and other parameters

No.	Plants	Parts used	Extract types	Doses (mg/kg) body weight	Toxicity assessed	Effects
1	*Aloe ferox*	Leaves	Aqueous	50–200	07 days in rats	Safe, regulates the level of cellular enzymes like transaminases [238, 239]
2	*Croton membranaceus*	Root	Ethanol	1500–3000	Acute in rats	No acute toxicity, hematological and renal parameters (urea and creatinine) are normal [240]
3	*Jatropha curcas*	Leaves	Methanol	50–200	21 days in rats	Non-toxic [241]
4	*Phyllanthus muellerianus*	Stem	Methanol	4000–20000	Acute in rats	Relatively toxic, hematological and renal parameters significantly affected [242]
5	*Tabernaemontana crassa*	Bark	Hydro ethanol	2000–8000 acute in rats; 100–1000 sub-acute in rats	Acute and subacute in rats	Fairly toxic, hematological and renal parameters like, AST, ALT, ALP, reduced glutathione (GSH) significantly affected [243]
6	*Calotropis gigantea*	Latex from aerial plant	Aqueous, alcoholic	1000	Sub-acute	No acute toxicity, hematological and renal parameters (urea and creatinine) are normal [244]
7	*Alstonia congensis*	Equal proportion of herbs	Mixture	1000–20000 for acute and 50–500 for subacute	Acute and sub-acute in rats and mice	Renal toxicity, significantly affects AST, ALT, ALP, creatinine, high density & low-density lipoprotein level [245]

(*Continued*)

Table 6.12 (*Continued*)

No.	Plants	Parts used	Extract types	Doses (mg/kg) body weight	Toxicity assessed	Effects
8	*Aloe buettneri*	Equal proportion of herbs	Mixture	2000–32000 for acute and 12.5–100 for subacute	Acute and sub-acute in rats and mice	Renal toxicity, harmful on long term usage & significant changes in AST, ALT, creatinine [246]
9	*Himatanthus drasticus*	Bark	Aqueous	2000	Acute in mice	Genotoxic, mutagenic [247]
10	*Plumeria zpudica*	Aerial part	Aqueous	40	Acute & sub-chronic in mice	No significant difference in hematological parameters, significant increase in AST and ALT level was observed in acute toxicity but no significant difference observed in sub-chronic toxicity. GSH content increased while MDA level did not differ significantly. [248]
11	*Carica papaya*	Unripe fruit	Alcohol	400	Acute toxicity in mice	Significant reduction in SGPT, SGOT, bilirubin and ALP while significant increase in protein level [249]
12	*Ficus benghalensis*	Unripe fruit	Alcohol	300	Acute toxicity in mice	Significant reduction in SGPT, SGOT, bilirubin and ALP while significant increase in protein level [249]

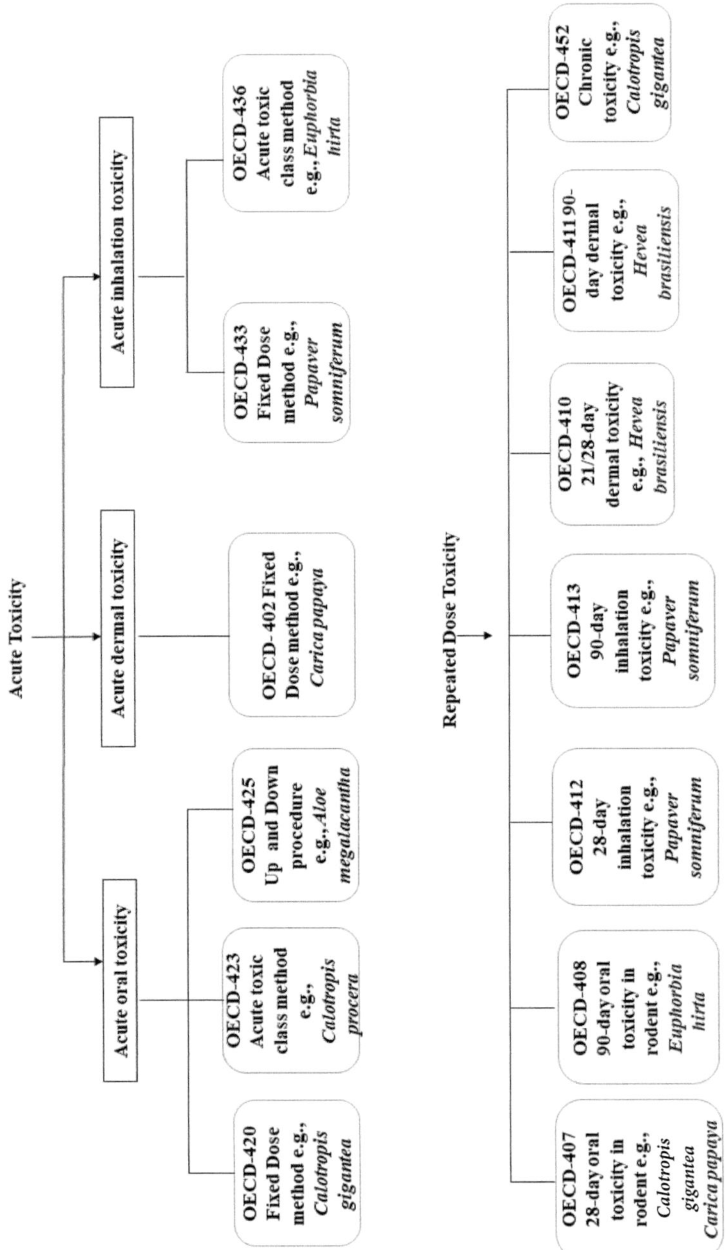

Figure 6.5 Evaluation of toxicity.

Various laboratory protocols including animal studies are available for toxicological evaluation. The effective dose and the lethal dose have been determined [58–60].

6.7.3 Oral Acute Toxicity Study

Acute toxicity study is conducted as per OECD guidelines 423. As per this guideline, a group of 3 animals are taken and are fasted overnight and weighed. The dose of test samples is determined based on body weight of animals. The animals are observed for behavioral changes and general signs of toxicity during the first 24 h and then for 14 days. The acute toxicity study of *Alstonia scholaris* was determined by this method.

6.7.4 Acute Dermal Toxicity

Acute dermal toxicity studies are conducted as per OECD 402 guidelines. In this study the highest concentration of the topical latex containing formulation (10% w/w) is applied topically to the shaved back area of skin of rats for a period of 24 h. The observations of sign and symptoms of toxicity is observed daily for the next 14 days. By this method, the toxicity studies of *Aloe megalacantha Baker* latex was evaluated.

6.7.5 For Subacute and Subchronic Toxicity Studies

The subacute and subchronic toxicity studies are conducted as per OECD 407 and OECD 408 guidelines. Different doses of latex plant extract are prepared and administered orally for the time period mentioned in protocol (14 days & 90 days respectively for subacute and subchronic toxicity). Toxicological manifestations are monitored and all the animals are monitored for the presence of morbidity or mortality. The toxicity study of latex containing plants like *Cannabis sativa, Calotropis procera (Aiton.), Himatanthus drasticus (Mart.), Euphorbia fusiformis buch.-Ham, Himatanthus drasticus (Mart.) Plumel* are evaluated by these guidelines [253–256].

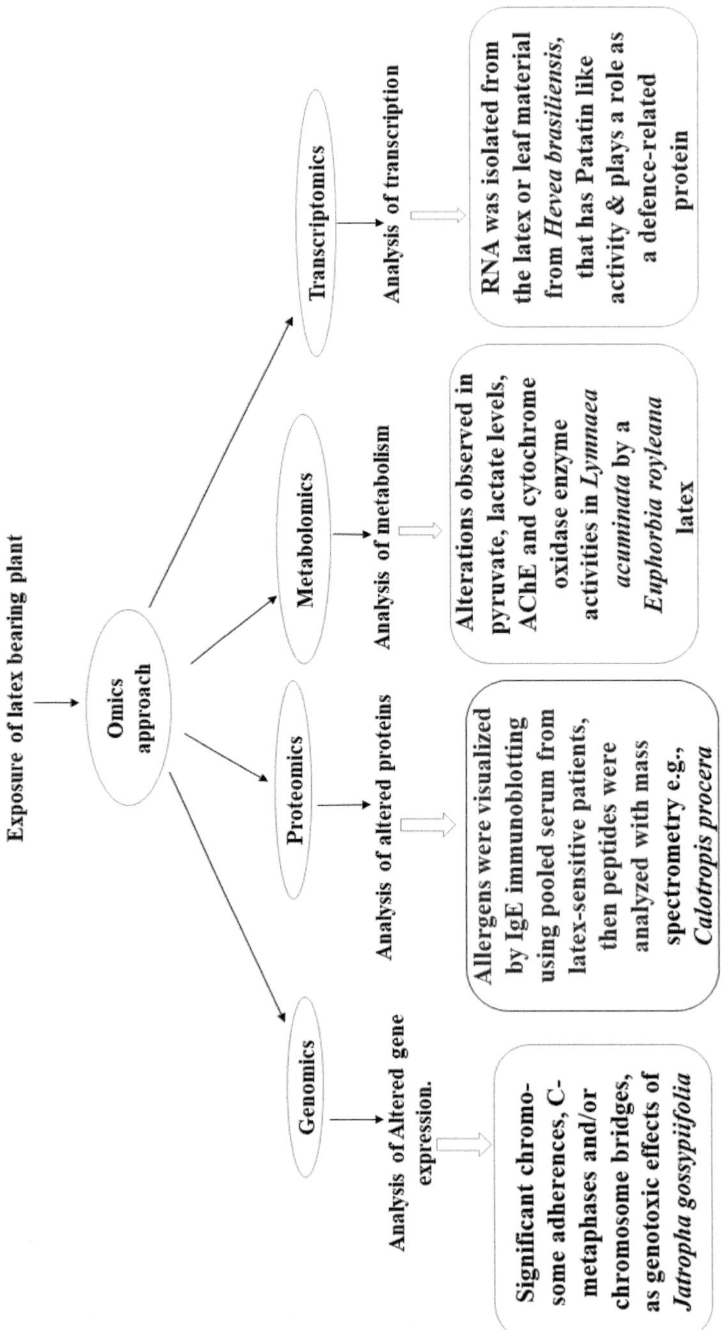

Figure 6.6 Omics toxicology approach for latex containing plants.

6.7.6 For Chronic Toxicity Studies

The chronic toxicity studies are performed according to WHO guidelines [193] and the OECD 452 guideline for testing of latex containing extract (OECD 1981). Different doses of latex plant extract are prepared and administered orally for the time period mentioned in protocol (270 days). Toxicological manifestations are monitored and all the animals are monitored for the presence of morbidity or mortality. For individual animals, both the physiological examinations, physical examinations and histopathological studies are performed [257].

It is possible to determine toxicity characteristics at different levels. The most common method of delivery for latex containing phytomolecule is oral but other options include cutaneous, intraperitoneal and inhalations [61–63]. Although rodents (mice or rats) are the favored species, rabbits can also be employed for cutaneous exposure [64–70]. Numerous ethical concerns surround in vivo research, and there is a strong trend to use the fewest number of animals possible in experiments. Preclinical toxicology was focused on different strategies in this situation and the use of in vitro models for simulating in vivo settings is a result of technological advancement.

6.7.7 Toxicity from Extrinsic Sources

6.7.7.1 Mycotoxins

Cannabinoids rich extract from the flower of *C. sativa* are used for the preparation of cannabis formulations intended for human use. The presence of microbial toxins like aflatoxins in cannabis preparations is of major concern as they have disease producing ability (carcinogenic, mutagenic, nephrotoxic and hepatotoxic). Higher amounts of aflatoxins and ochratoxin are reported from cannabis flowers. Hence, the cultivation, post-harvest treatment and storage of cannabis is performed under controlled conditions to avoid the formation of mycotoxins.

The analysis of mycotoxins in herbal preparations can be performed by highly sensitive sophisticated chromatographic techniques as recommended by International Pharmacopoeia [258–260].

6.7.8 Presence of Microorganisms

The presence of fungal contamination including *Aspergillus* in *C. sativa* samples leads to lung infection and the presence of Salmonella in cannabis may lead to respiratory problems [261, 262]. Meta genomic analysis detects the presence of microorganism like Aspergillus sp., and Penicillium sp., in cannabis samples. Cannabis smoking increases the chances of pulmonary infection by affecting alveolar macrophages [263], cilia and mucus-secreting cells of the respiratory system [264].

6.7.9 Heavy Metals

The presence of heavy metals is a serious concern among herbal drugs and their formulations. The heavy metals either added to the formulations because of their therapeutic values or accumulated in the plant system a result of cross contamination during formulation development. Heavy metals such as cadmium, chromium, lead and mercury are known to be carcinogenic. There are three pathways through which cannabis may be contaminated with heavy metal substances. The cannabis preparations can be detected with the presence of arsenic leads to cannabis arteritis. Adulteration of cannabis preparations with lead results in poisoning. With the advent of sophisticated analytical tools like atomic absorption spectroscopy and inductively coupled plasma mass spectroscopy the analysis of heavy metals in herbal drugs and their formulations becomes easier [265–270, 230–232].

6.7.10 Pesticides

A significant amount of pesticides has been reported in cannabis preparations. A wide range of chemical compounds including carcinogens endocrine disruptors, as well as developmental, reproductive and neurological toxins are well reported in cannabis preparation [230, 271].

6.7.11 Plant Growth Regulators

Daminozide and paclobutraxol are widely distributed in cannabis. Daminozide is considered as human carcinogen while paclobutraxol showed detrimental effects on aquatic species and mice [273–275].

6.7.12 Polycyclic Aromatic Hydrocarbons

Polycyclic aromatic hydrocarbons (PAH) are generated by the incomplete combustion of crude oil, coal and wood. Several PAH like benzoanthracene, chrysene, benzofluoranthene, benzopyrene are carcinogens found in cannabis. Higher levels of PAHs in cannabidiol cause immunosuppression as a result of DNA alteration [276, 277].

6.7.13 Other Foreign Matters

Foreign materials like dust particles, hairs, metal fragments, machine parts or insect parts can be found in cannabidiol. These foreign particles are not associated with serious health hazards but are important to affect the development of high quality standardized products [230].

6.8 Conclusions

This chapter discusses drug discovery, development, challenges and perspectives pertaining to latex bearing plants and their products. It explores the ethnomedicinal and therapeutic potential of latex as a resource for the modern medication. The chapter covers the various commercial applications of latex in various industries and health care. Also, it discusses various regulatory standards (WHO, USFDA) for commercial success in the latex advance technology fields. Concerns about the safety and toxicity of latex, latex allergy, and food and drug interactions of latex have also been covered.

References

1. Harvey AL. Natural products in drug discovery. *Drug Discov Today*. 2008; 13(19–20): 894–901.
2. Gracz-Bernaciak J, Mazur O, Nawrot R. Functional studies of plant latex as a rich source of bioactive compounds: focus on proteins and alkaloids. *Int J Mol Sci*. 2021; 22(22): 12427–51.

3. Konno K. Plant latex and other exudates as plant defense systems: roles of various defense chemicals and proteins contained therein. *Phytochemistry*. 2011; 72(13): 1510–30.
4. Lewinsohn T. The geographical distribution of plant latex. *Chemoecology*. 1991; 2: 64–8.
5. Mishra A, Parida S. Latex of plants: wonders of nature for its therapeutic potentials and a valuable resource towards new drug development. *Int J Botany Studies*. 2020; 5: 334–38.
6. Thomford NE, Senthebane DA, Rowe A, Munro D, Seele P, Maroyi A, Dzobo K. Natural products for drug discovery in the 21st Century: innovations for novel drug discovery. *Int J Mol Sci*. 2018; 19(6): 1578–1606.
7. Jesus A, Bonhomme V, Evin A, Ivorra S, Soteras R, Salavert A, Antolín F, Bouby L. A morphometric approach to track opium poppy domestication. *Sci Rep*. 2021; 11(1): 9778–88.
8. Agrawal AA, Konno K. Latex: a model for understanding mechanisms, ecology, and evolution of plant defense against herbivory. *Annu Rev Ecol Evol Syst*. 2009; 40: 311–31.
9. Dias DA, Urban S, Roessner U. A historical overview of natural products in drug discovery. *Metabolites*. 2012; 2(2): 303–36.
10. Mishra BB, Tiwari VK. Natural products: an evolving role in future drug discovery. *Eur J Med Chem*. 2011; 46(10): 4769–807.
11. Cragg GM, Newman DJ. Biodiversity: a continuing source of novel drug leads. *Pure Appl Chem*. 2005; 77(1): 7–24.
12. Weng JK, Philippe RN, Noel JP. The rise of chemodiversity in plants. *Science*. 2012; 336(6089): 1667–70.
13. Lietava J. Medicinal plants in a Middle Paleolithic grave Shanidar IV? *J Ethnopharmacol*. 1992; 35(3): 263–6.
14. Jachak S, Saklani A. Challenges and opportunities in drug discovery from plant. *Curr Sci*. 2007; 92: 1251–57.
15. Krueger RJ. Drugs of natural origin. In *A Textbook of Pharmacognosy*. 5th ed., Gunnar Samuelson. Swedish Pharmaceutical Press, Stockholm. 2005; 620–31.
16. Kinghorn AD. Pharmacognosy in the 21st century. *J Pharm Pharmacol*. 2001; 53(2): 135–48.
17. Baker JT, Borris RP, Carté B, Cordell GA, Soejarto DD, Cragg GM, Gupta MP, Iwu MM, Madulid DR, Tyler VE. Natural product drug discovery and development: new perspectives on international collaboration. *J Nat Prod*. 1995; 58(9): 1325–57.

18. Geysen HM, Schoenen F, Wagner D, Wagner R. Combinatorial compound libraries for drug discovery: an ongoing challenge. *Nat Rev Drug Discov.* 2003; 2(3): 222–30.
19. Lombardino JG, Lowe JA. The role of the medicinal chemist in drug discovery - Then and now. *Nat Rev Drug Discov.* 2004; 3(10): 853–62.
20. Newman DJ, Cragg GM, Snader KM. Natural products as sources of new drugs over the period 1981–2002. *J Nat Prod.* 2003; 66(7): 1022–37.
21. Lötsch J, Geisslinger G. Morphine-6-glucuronide: an analgesic of the future? *Clin Pharmacokinet.* 2001; 40(7): 485–99.
22. Cragg GM, Newman DJ. A tale of two tumor targets: topoisomerase I and tubulin. The Wall and Wani contribution to cancer chemotherapy. *J Nat Prod.* 2004; 67(2): 232–44.
23. Kong Y, Li L, Zhao LG, Yu P, Li DD. A patent review of berberine and its derivatives with various pharmacological activities (2016–2020). *Expert Opin Ther Pat.* 2022; 32(2): 211–23.
24. Liao SG, Chen HD, Yue JM. Plant orthoesters. *Chem Rev.* 2009; 109(3): 1092–140.
25. Nguyen MTT, Nguyen KDH, Dang PH, Nguyen HX, Awale S, Nguyen NT. Calosides A-F, cardenolides from Calotropis gigantea and their cytotoxic activity. *J Nat Prod.* 2020; 83(2): 385–91.
26. Thuy NT, Phuong PT, Hien NT, Trang DT, Huan NV, Anh PT, Tai BH, Nhiem NX, Hung NT, Kiem PV. Pregnane glycosides from the leaves of Dregea volubilis and their α-glucosidase and α-amylase inhibitory activities. *Nat Prod Res.* 2021; 35(21): 3931–8.
27. Ahmed Z, Tokhi A, Arif M, Rehman NU, Sheibani V, Rauf K, Sewell RD. Fraxetin attenuates disrupted behavioral and central neurochemical activity in a model of chronic unpredictable stress. *Front Pharmacol.* 2023; 14: 1135497–508.
28. Zhang Y, Chen X, Zhou Y, Hou J, Long H, Zhang Z, Lei M, Wu W. Synthesis of oleandrin derivatives and their cytotoxic activity. *Steroids.* 2020; 159: 108650.
29. Al-Rajhi AM, Yahya R, Abdelghany TM, Fareid MA, Mohamed AM, Amin BH, Masrahi AS. Anticancer, anticoagulant, antioxidant and antimicrobial activities of Thevetia peruviana latex with molecular docking of antimicrobial and anticancer activities. *Molecules.* 2022; 27(10): 3165–82.
30. Dickson M, Gagnon JP. Key factors in the rising cost of new drug discovery and development. *Nat Rev Drug Discov.* 2004; 3(5): 417–29.

31. Koehn FE, Carter GT. The evolving role of natural products in drug discovery. *Nat Rev Drug Discov.* 2005; 4(3): 206–20.

32. Rosenthal J. Curtain has fallen on hopes of legal bioprospecting. *Nature.* 2002; 416: 15.

33. Soejarto DD, Gyllenhaal C, Fong HH, Xuan LT, Hiep NT, Hung NV, Bich TQ, Southavong B, Sydara K, Pezzuto JM. The UIC ICBG (University of Illinois at Chicago international cooperative biodiversity group) memorandum of agreement: a model of benefit-sharing arrangement in natural products drug discovery and development. *J Nat Prod.* 2004; 67(2): 294–9.

34. Knowles J, Gromo G. A guide to drug discovery: target selection in drug discovery. *Nat Rev Drug Discov.* 2003; 2(1): 63–9.

35. Kramer R, Cohen D. Functional genomics to new drug targets. *Nat Rev Drug Discov.* 2004; 3(11): 965–72.

36. Walters WP, Namchuk M. Designing screens: how to make your hits a hit. *Nat Rev Drug Discov.* 2003; 2(4): 259–66.

37. Ley SV, Baxendale IR. New tools and concepts for modern organic synthesis. *Nat Rev Drug Discov.* 2002; 1(8): 573–86.

38. Federsel HJ, Federsel HJ. Logistics of process R&D: transforming laboratory methods to manufacturing scale. *Nat Rev Drug Discov.* 2003; 2: 654–64.

39. Afreen S, Haque K, Huda M. Troubleshooting for the observed problems in processing latex concentrate from natural resource. *IOP Publishing.* 2013; 16: 012007-11.

40. Patwardhan B. Ethnopharmacology and drug discovery. *J Ethnopharmacol.* 2005; 100(1–2): 50–2.

41. Soejarto DD, Fong HH, Tan GT, Zhang HJ, Ma CY, Franzblau SG, Gyllenhaal C, Riley MC, Kadushin MR, Pezzuto JM, Xuan LT. Ethnobotany/ethnopharmacology and mass bioprospecting: issues on intellectual property and benefit-sharing. *J Ethnopharmacol.* 2005; 100(1–2): 15–22.

42. Yuan H, Ma Q, Ye L, Piao G. The traditional medicine and modern medicine from natural products. *Molecules.* 2016; 21(5): 559–76.

43. Chan K, Shaw D, Simmonds MS, Leon CJ, Xu Q, Lu A, Sutherland I, Ignatova S, Zhu YP, Verpoorte R, Williamson EM. Good practice in reviewing and publishing studies on herbal medicine, with special emphasis on traditional Chinese medicine and Chinese materia medica. *J Ethnopharmacol.* 2012; 140(3): 469–75.

44. Zhang W, Huai Y, Miao Z, Qian A, Wang Y. Systems pharmacology for investigation of the mechanisms of action of traditional Chinese medicine in drug discovery. *Front Pharmacol.* 2019; 10: 1–22.

45. Zhang L, Yan J, Liu X, Ye Z, Yang X, Meyboom R, Chan K, Shaw D, Duez P. Pharmacovigilance practice and risk control of traditional Chinese medicine drugs in China: current status and future perspective. *J Ethnopharmacol.* 2012; 140(3): 519–25.

46. Hamilton GR, Baskett TF. In the arms of morpheus the development of morphine for postoperative pain relief. *Can J Anaesth.* 2000; 47(4): 367–74.

47. Joo YE. Natural product-derived drugs for the treatment of inflammatory bowel diseases. *Intest Res.* 2014; 12(2): 103–9.

48. Atanasov AG, Zotchev SB, Dirsch VM, Supuran CT. Natural products in drug discovery: advances and opportunities. *Nat Rev Drug Discov.* 2021; 20(3): 200–16.

49. Ngo LT, Okogun JI, Folk WR. 21st century natural product research and drug development and traditional medicines. *Nat Prod Rep.* 2013; 30(4): 584–92.

50. Zhu F, Ma XH, Qin C, Tao L, Liu X, Shi Z, Zhang CL, Tan CY, Chen YZ, Jiang YY. Drug discovery prospect from untapped species: indications from approved natural product drugs. *PLoS One.* 2012; 7(7): 1–9.

51. Newman DJ, Cragg GM. Natural products as sources of new drugs over the nearly four decades from 01/1981 to 09/2019. *J Nat Prod.* 2020; 83(3): 770–803.

52. Lemonnier N, Zhou GB, Prasher B, Mukerji M, Chen Z, Brahmachari SK, Noble D, Auffray C, Sagner M. Traditional knowledge-based medicine: a review of history, principles, and relevance in the present context of P4 systems medicine. *Prog Prev Med.* 2017; 2(7): e0011-25.

53. James PB, Wardle J, Steel A, Adams J. Traditional, complementary and alternative medicine use in Sub-Saharan Africa: a systematic review. *BMJ Glob Heal.* 2018; 3(5): e000895.

54. Abdullahi AA. Trends and challenges of traditional medicine in Africa. *African J Tradit Complement Altern Med.* 2011; 8(5 Suppl): 115–23.

55. Qi F, Wang Z, Cai P, Zhao L, Gao J, Kokudo N, Li A, Han J, Tang W. Traditional Chinese medicine and related active compounds: a review of their role on hepatitis B virus infection. *Drug Discov Ther.* 2013; 7(6): 212–24.

56. Kuchta K, Cameron S. Editorial: Kampo medicine in a modern context: ethnopharmacological perspectives. *Front Pharmacol.* 2022; 13: 1–3.
57. Watanabe S, Imanishi J, Satoh M, Ozasa K. Unique place of Kampo (Japanese traditional medicine) in complementary and alternative medicine: a survey of doctors belonging to the regional medical association in Japan. *Tohoku J Exp Med.* 2001; 194(1): 55–63.
58. Yakubo S, Ito M, Ueda Y, Okamoto H, Kimura Y, Amano Y, Togo T, Adachi H, Mitsuma T, Watanabe K. Pattern classification in kampo medicine. *Evid Based Complement Alternat Med.* 2014; 2014: 535146–50.
59. Shimizu M, Takayama S, Kikuchi A, Arita R, Ono R, Ishizawa K, Ishii T. Kampo medicine treatment for advanced pancreatic cancer: a case series. *Front Nutr.* 2021; 8: 702812–17.
60. Lone A, Ahmad T, Anwar M, Sofi G, Imam H, Habib S. Perception of health promotion in Unani herbal medicine. *J Herb Med.* 2012; 2: 1–5.
61. Jabin F. A guiding tool in Unani Tibb for maintenance and preservation of health: a review study. *African J Tradit Complement Altern Med.* 2011; 8(5): 140–3.
62. Parasuraman S, Thing GS, Dhanaraj SA. Polyherbal formulation: concept of ayurveda. *Pharmacogn Rev.* 2014; 8(16): 73–80.
63. Jadhav CA, Vikhe DN, Jadhav RS. Global and domestic market of herbal medicines: a review. *Res J Sci Technol.* 2020; 12(4): 327–30.
64. Boakye MK, Pietersen DW, Kotzé A, Dalton DL, Jansen R. Knowledge and uses of African pangolins as a source of traditional medicine in Ghana. *PLoS One.* 2015; 10(1): 1–14.
65. Oliver SJ. The role of traditional medicine practice in primary health care within Aboriginal Australia: a review of the literature. *J Ethnobiol Ethnomed.* 2013; 9: 1–8.
66. Lev E. Ethno-diversity within current ethno-pharmacology as part of Israeli traditional medicine – A review. *J Ethnobiol Ethnomed.* 2006; 2: 1–12.
67. Li Y. Qinghaosu (artemisinin): chemistry and pharmacology. *Acta Pharmacol Sin.* 2012; 33(9): 1141–6.
68. Li J, Zhou B. Biological actions of artemisinin: insights from medicinal chemistry studies. *Molecules.* 2010; 15(3): 1378–97.
69. Su XZ, Miller LH. The discovery of artemisinin and the Nobel prize in physiology or medicine. *Sci China Life Sci.* 2015; 58: 1175–9.

70. Wang J, Xu C, Wong YK, Li Y, Liao F, Jiang T, Tu Y. Artemisinin, the Magic drug discovered from traditional Chinese medicine. *Engineering*. 2019; 5(1): 32–9.

71. Felter HW. The eclectic materia medica, pharmacology and therapeutics. 1923; 12: 188–89.

72. Mishra A, Parida S. Wonders of nature for its therapeutic potentials and a valuable resource towards new drug development new drug development. *Int J Bot Studies.* 2020; 5: 334–8.

73. Doskotch RW, Malik MY, Hufford CD, Malik SN, Trent JE, Kubelka W. Antitumor agents. V. Cytotoxic cardenolides from Cryptostegia grandiflora (Roxb.) R. *Br J Pharm Sci*. 1972; 61(4): 570–3.

74. Augustus G, Jayabalan M, Seiler GJ. Cryptostegia grandiflora – A potential multi-use crop. *Ind Crop Prod*. 2000; 11: 59–62.

75. Mukherjee PK, Gunasekhran R, Subburaju T, Dhanbal SP, Duraiswamy B, Vijayan P, Suresh B. Studies on the antibacterial potential of Cryptostegia grandiflora R. Br. (Asclepiadaceae) extract. *Phytother Res*. 1999; 13(1): 70–2.

76. Ramos MV, Demarco D, da Costa Souza IC, de Freitas CDT. Laticifers, latex, and their role in plant defense. *Trends Plant Sci*. 2019; 24(6): 553–67.

77. Nawrot R, Wołuń-Cholewa M, Goździcka-Józefiak A. Nucleases isolated from Chelidonium majus L. milky sap can induce apoptosis in human cervical carcinoma HeLa cells but not in Chinese Hamster Ovary CHO cells. *Folia Histochem Cytobiol*. 2008; 46(1): 79–83.

78. Xie W, Zhang X, Wang T, Hu J. Botany, traditional uses, phytochemistry and pharmacology of Apocynum venetum L. (Luobuma): a review. *J Ethnopharmacol*. 2012; 141(1): 1–8.

79. Moremi MP, Makolo F, Viljoen AM, Kamatou GP. A review of biological activities and phytochemistry of six ethnomedicinally important South African Croton species. *J Ethnopharmacol*. 2021; 280: 114416.

80. Jones K. Review of sangre de drago (Croton lechleri)–a South American tree sap in the treatment of diarrhea, inflammation, insect bites, viral infections, and wounds: traditional uses to clinical research. *J Altern Complement Med*. 2003; 9(6): 877–96.

81. Saggoo MI, Walia S, Kaur R. Evaluation of genotoxic and anti-microbial potential of Croton bonplandianum Baill. *Arch Appl Sci Res*. 2010; 2(2): 211–6.

82. Mishra P, Yadav KS, Shrivastava P. An updated review on phytochemistry, pharmacological activity and medicinal uses of Calotropis gigantea R. Br. *Res J Pharmacog Phytochem.* 2017; 9(2): 135–8.

83. Kadiyala M, Ponnusankar S, Elango K. Calotropis gigantea (L.) R. Br (Apocynaceae): a phytochemical and pharmacological review. *J Ethnopharmacol.* 2013; 150(1): 32–50.

84. Alafnan A, Sridharagatta S, Saleem H, Khurshid U, Alamri A, Ansari SY, Zainal Abidin SA, Ansari SA, Alamri AS, Ahemad N, Anwar S. Evaluation of the phytochemical, antioxidant, enzyme inhibition, and wound healing potential of Calotropis gigantea (L.) Dryand: a source of a bioactive medicinal product. *Front Pharmacol.* 2021; 12: 701369.

85. Kaur A, Batish DR, Kaur S, Chauhan BS. An overview of the characteristics and potential of Calotropis procera from botanical, ecological, and economic perspectives. *Front Plant Sci.* 2021; 12: 690806.

86. Misra MK, Mohanty MK, Das PK. Studies on the method - ethnobotany of Calotropis gigantea and C. procera. *Anc Sci Life.* 1993; 13(1–2): 40–56.

87. Castro JP, Ocampo YC, Franco LA. In vivo and in vitro anti-inflammatory activity of Cryptostegia grandiflora Roxb. ex R. Br. leaves. *Biol Res.* 2014; 47(1): 1–8.

88. Ascacio-Valdés J, Burboa E, Aguilera-Carbo AF, Aparicio M, Pérez-Schmidt R, Rodríguez R, Aguilar CN. Antifungal ellagitannin isolated from Euphorbia antisyphilitica Zucc. *Asian Pac J Trop Biomed.* 2013; 3(1): 41–6.

89. Kumar A, Saikia D. Euphorbia antiquorum linn : a comprehensive review of ethnobotany, phytochemistry and pharmacology. *J Anal Pharm Res.* 2016; 2(4): 10–13.

90. Dong M, Chen XQ, Chen CH, Li RT. Terpenes from Euphorbia antiquorum and their in vitro anti-HIV activity. *Chem Biodivers.* 2018; 15(5): e1700560.

91. Kumar S, Nitesh G, Parihar R. Study of the ethno-medicinal – Importance of some plant species among flora of north – west Rajasthan materials and methods: 2015; 3(3): 723–6.

92. Patel YS, Joshi EP, Joshi PN. Ethnobotanical Study of Tapkeshwari Hill, Bhuj, Kachchh, India. *Life Sciences Leaflets.* 2010; 2: 22–31.

93. Goraya K, Iqbal Z, Sajid MS, Muhammad G, Ain Q, Saleem M. Diversity of flora used for the cure of equine diseases in selected peri-urban areas of Punjab, Pakistan. *J Ethnobiol Ethnomed.* 2013; 9: 1–9.

94. Nasim MJ, Asad MH, Sajjad A, Khan SA, Mumtaz A, Farzana K, Rashid Z, Murtaza G. Combating of scorpion bite with Pakistani medicinal plants having ethno-botanical evidences as antidote. *Acta Pol Pharm.* 2013; 70(3): 387–94.

95. Goyal M, Nagori BP, Sasmal D. Wound healing activity of latex of Euphorbia caducifolia. *J Ethnopharmacol.* 2012; 144(3): 786–90.

96. Qasim M, Gulzar S, Khan MA. Halophytes as medicinal plants. *Urbanization, Land Use, Land Degradation and Environment.* 2011; 330–43.

97. de Guzman GQ, Dacanay ATL, Andaya BA, Alejandro GJD. Ethnopharmacological studies on the uses of Euphorbia hirta in the treatment of dengue in selected indigenous communities in Pangasinan (Philippines). *J Intercult Ethnopharmacol.* 2016; 5(3): 239–43.

98. Kumar A, Mahanty B, Goswami RCD, Barooah PK, Choudhury B. In vitro antidiabetic, antioxidant activities and GC-MS analysis of Rhynchostylis Retusa and Euphorbia Neriifolia leaf extracts. *Biotech.* 2021; 11(7): 1–10.

99. Sultana A, Hossain MJ, Kuddus MR, Rashid MA, Zahan MS, Mitra S, Roy A, Alam S, Sarker MM, Naina Mohamed I. Ethnobotanical uses, phytochemistry, toxicology, and pharmacological properties of Euphorbia neriifolia Linn. against infectious diseases: a comprehensive review. *Molecules.* 2022; 27(14): 4374–404.

100. Mali PY, Panchal SS. Euphorbia neriifolia L.: Review on botany, ethnomedicinal uses, phytochemistry and biological activities. *Asian Pac J Trop Med.* 2017; 10(5): 430–8.

101. Bigoniya P, Rana AC. A comprehensive phyto-pharmacological review of Euphorbia neriifolia Linn. *Pharmacog Rev.* 2008; 2: 57–66.

102. De Britto AJ, Sujin RM, Mahesh R, Dharmar K. Ethnomedicinal wisdom of the Manavalakuruchi people in Kanyakumari district, Tamil Nadu. *Int J Biol Tech.* 2010;1: 25–30.

103. Singh KK, Rauniar GP, Sangraula H. Experimental study of neuropharmacological profile of Euphorbia pulcherrima in mice and rats. *J Neurosci Rural Pract.* 2012; 3(3): 311–9.

104. Bani S, Kaul A, Jaggi BS, Suri KA, Suri OP, Sharma OP. Anti-inflammatory activity of the hydrosoluble fraction of Euphorbia royleana latex. *Fitoterapia*. 2000; 71(6): 655–62.

105. Singh P, Singh A. Evaluation of latex extract of Euphorbia royleana for its piscicidal and muricidal activities. *World J Agric Sci*. 2012; 8(5): 520–4.

106. Mali PY, Panchal SS. Euphorbia tirucalli L.: review on morphology, medicinal uses, phytochemistry and pharmacological activities. *Asian Pac J Trop Biomed*. 2017; 7(7): 603–13.

107. Mwine J, Van Damme P, Hastilestari BR, Papenbrock J. Euphorbia tirucalli L. (euphorbiaceae) -The miracle tree: current status of knowledge. *ACS Symp Ser*. 2013; 1127(23): 3–17.

108. Gairola S, Sharma J, Gaur RD, Siddiqi TO, Painuli RM. Plants used for treatment of dysentery and diarrhoea by the Bhoxa community of district Dehradun, Uttarakhand, India. *J Ethnopharmacol*. 2013; 150(3): 989–1006.

109. Mali PY, Panchal SS. A review on phyto-pharmacological potentials of Euphorbia thymifolia L. *Ancient Sci Life*. 2013; 32(3): 165–72.

110. Kale MP. A Review Article on Euphorbia hirta uses and pharmacological activities. *Asian J Res Pharm Sci*. 2016; 6(3): 141–5.

111. Chaudhary S, Alok S, Jain SK, Chanchal D, Dongray A. Phytopharamacology and pharmacognostic properties of Ficus benghalensis-A review. *Int J Phytochem Res*. 2015; 2: 560–9.

112. Chakraborty A, Mahajan S, Bisht MS, Sharma VK. Genome sequencing and comparative analysis of Ficus benghalensis and Ficus religiosa species reveal evolutionary mechanisms of longevity. *iScience*. 2022; 25(10): 105100.

113. Salehi B, Prakash Mishra A, Nigam M, Karazhan N, Shukla I, Kiełtyka-Dadasiewicz A, Sawicka B, Głowacka A, Abu-Darwish MS, Hussein Tarawneh A, Gadetskaya AV. Ficus plants: state of the art from a phytochemical, pharmacological, and toxicological perspective. *Phytother Res*. 2021; 35(3): 1187–217.

114. Mawa S, Husain K, Jantan I. Ficus carica L. (Moraceae): phytochemistry, traditional uses and biological activities. *Evid Based Complement Alternat Med*. 2013; 2013: 974256–64.

115. Rahman IU, Ijaz F, Afzal A, Iqbal Z, Ali N, Khan SM. Contributions to the phytotherapies of digestive disorders: traditional knowledge

and cultural drivers of Manoor Valley, Northern Pakistan. *J Ethnopharmacol.* 2016; 192: 30–52.

116. Yabesh JEM, Prabhu S, Vijayakumar S. An ethnobotanical study of medicinal plants used by traditional healers in silent valley of Kerala, India. *J Ethnopharmacol.* 2014; 154(3): 774–89.

117. Sher Z, Khan Z, Hussain F. Ethnobotanical studies of some plants of Chagharzai valley, district Buner, Pakistan. *Pak J Bot.* 2011; 43(3): 1445–52.

118. de Amorin A, Borba HR, Carauta JP, Lopes D, Kaplan MA. Anthelmintic activity of the latex of Ficus species. *J Ethnopharmacol.* 1999; 64(3): 255–8.

119. Cheng JX, Zhang BD, Zhu WF, Zhang CF, Qin YM, Abe M, Akihisa T, Liu WY, Feng F, Zhang J. Traditional uses, phytochemistry, and pharmacology of Ficus hispida L.f.: a review. *J Ethnopharmacol.* 2020; 248: 112204.

120. Quesada-Granados JJ, Rufian-Henares JA, Chakradhari S, Sahu PK, Sahu YK, Patel KS. Comparative analysis of traditional oriental herbal fruits as potential sources of polyphenols and minerals for nutritional supplements. *Molecules.* 2023; 28(6): 2682–93.

121. Shahreen S, Banik J, Hafiz A, Rahman S, Zaman AT, Shoyeb A, Chowdhury MH, Rahmatullah M.. Antihyperglycemic activities of leaves of three edible fruit plants (Averrhoa carambola, Ficus hispida and Syzygium samarangense) of Bangladesh. *African J Tradit Complement Altern Med.* 2012; 9(2): 287–91.

122. Moraceae L. Distribution and development of the non-articulated branched laticifers of Morus nigra L. (Moraceae). *Acta Bot Neerl.* 1990; 39(3): 39: 285–96.

123. Arsyad AS, Nurrochmad A, Fakhrudin N. Phytochemistry, traditional uses, and pharmacological activities of Ficus elastica Roxb. ex Hornem: A review. *J Herbmed Pharmacol.* 2022; 12(1): 41–53.

124. Ahmed F, Urooj A. Traditional uses, medicinal properties, and phytopharmacology of Ficus racemosa: a review. *Pharm Biol.* 2010; 48(6): 672–81.

125. Seca AML, Grigore A, Pinto DCGA, Silva AMS. The genus Inula and their metabolites: from ethnopharmacological to medicinal uses. *J Ethnopharmacol.* 2014; 154(2): 286–310.

126. Ali M, Chaudhary N. Ficus hispida Linn.: A review of its pharmacognostic and ethnomedicinal properties. *Pharmacog Rev.* 2011; 5(9): 96–102.

127. Singh D, Singh B, Goel RK. Traditional uses, phytochemistry and pharmacology of Ficus religiosa: a review. *J Ethnopharmacol.* 2011; 134(3): 565–83.

128. Deepa P, Sowndhararajan K, Kim S, Park SJ. A role of Ficus species in the management of diabetes mellitus: a review. *J Ethnopharmacol.* 2018; 215: 210–32.

129. Mon AM, Shi Y, Yang X, Hein PP, Oo TN, Whitney CW, Yang Y. The uses of fig (Ficus) by five ethnic minority communities in Southern Shan State, Myanmar. *J Ethnobiol Ethnomed.* 2020; 16(1): 1–4.

130. Chetia D, Nath LK, Dutta SK. Pharmacological properties of a protease from Ficus hispiad linn. *Anc Sci Life.* 2001; 21(2): 99–110.

131. Singh AK, Raghubanshi AS, Singh JS. Medical ethnobotany of the tribals of Sonaghati of Sonbhadra district, Uttar Pradesh, India. *J Ethnopharmacol.* 2002; 81(1): 31–41.

132. Daruliza KMA, Lam KL, Yang KL, Priscilla JT, Sunderasan E, Ong MT. Anti-fungal effect of Hevea brasiliensis latex C-serum on Aspergillus niger. *Eur Rev Med Pharmacol Sci.* 2011; 15(9): 1027–33.

133. Gowda NGS, Raju NR, Silina E, Stupin V, Achar RR. Mimusops elengi (Bakula) gelatinolytic protease and its plasmin-like action on the blood clot. *Curr Protein Pept Sci.* 2022; 23(10): 706–12.

134. Bauer G, Gorb SN, Klein MC, Nellesen A, von Tapavicza M, Speck T. Comparative study on plant latex particles and latex coagulation in Ficus benjamina, Campanula glomerata and Three Euphorbia species. *PLoS One.* 2014; 9(11): 1–8.

135. Khan MF, Mashwani ZU, Mehmood A, Qureshi R, Sarwar R, Ahmad KS, Quave CL. An ethnopharmacological survey and comparative analysis of plants from the Sudhnoti District, Azad Jammu and Kashmir, Pakistan. *J Ethnobiol Ethnomed.* 2021; 17(1): 1–22.

136. Lucchetti L, Zitti S, Taffetani F. Ethnobotanical uses in the Ancona district (Marche region, Central Italy). *J Ethnobiol Ethnomed.* 2019; 15(1): 1–33.

137. Seema Rani, Khaleequr R. Molsari (Mimusops elengi Linn.): a boon drug of traditional medicine. *Int J Pharm Sci Res.* 2017; 8(1): 17–28.

138. Tetik F, Civelek S, Cakilcioglu U. Traditional uses of some medicinal plants in Malatya (Turkey). *J Ethnopharmacol.* 2013; 146(1): 331–46.

139. Kamali M, Kamali H, Doustmohammadi M, Sheikhbardsiri H, Moghadari M. Treatment of opium addiction in Persian medicine: a review study. *J Educ Health Promot.* 2021; 10: 157.

140. Romero JC, Martínez-Vázquez A, Herrera MP, Martinez-Mayorga K, Parra-Delgado H, Pérez-Flores FJ, Martínez-Vázquez M. Synthesis, anti-inflammatory activity and modeling studies of cycloartane-type terpenes derivatives isolated from Parthenium argentatum. *Bioorg Med Chem.* 2014; 22(24): 6893–8.

141. Hoekou YP, Tchacondo T, Karou SD, Koudouvo K, Atakpama W, Pissang P, Gbogbo AK, Woegan AY, Batawila K, Akpagana K, Gbeassor M. Ethnobotanical study of latex plants in the maritime region of Togo. *Pharmacog Res.* 2016; 8(2):128–34.

142. Darji D, Mohd Rasdi FR, Abdul Rahim MA, Danya MS. Epoxidised natural rubber (ENR) latex: an alternative raw material for latex dipped products. *J Rubber Res.* 2020; 23:375–85.

143. US Congress Office of Technology Assessment. Innovation and commercialization of emerging technologies. 1995; 31–49.

144. Smith M, Hansen F. Managing intellectual property: a strategic point of view. *J Intellect Cap.* 2002; 3: 366–74.

145. Nettis E, Colanardi MC, Ferrannini A, Tursi A. Latex hypersensitivity: personal data and review of the literature. *Immunopharmacol Immunotoxicol.* 2002; 24(2): 315–34.

146. Marshall E, Schreckenberg K, Newton AC. Commercialisation of non-timber forest products: factors influencing success. Lessons learned from Mexico and Bolivia and policy implications for decision-makers. *Int For Rev.* 2006; 8(3): 368–9.

147. Cornish K, Williams J, Martin JA. Commercialization of Yulex natural rubber latex™: the improving margin structure of guayule latex products. In 8th International latex conference, Charlotte, NC, 2005; 1–14.

148. Guyot J, Guen V Le. A review of a century of studies on South American leaf blight of the rubber tree. *Plant Dis.* 2018; 102(6): 1052–65.

149. Mukherjee PK, Wahile A. Integrated approaches towards drug development from Ayurveda and other Indian system of medicines. *J Ethnopharmacol.* 2006; 103(1): 25–35.

150. Singh H. Prospects and challenges for harnessing opportunities in medicinal plants sector in India. *Law, Environ Dev J.* 2006; 2: 196.

151. Freitas CD, Leite HB, Oliveira JP, Amaral JL, Egito AS, Vairo-Cavalli S, Lobo MD, Monteiro-Moreira AC, Ramos MV. Insights into milk-clotting activity of latex peptidases from Calotropis procera and Cryptostegia grandiflora. *Food Res Int.* 2016; 87: 50–9.

152. Domsalla A, Melzig MF. Occurrence and properties of proteases in plant latices. *Planta Med.* 2008; 74(7): 699–711.

153. Urs A, Manjuprasanna V, Rudresha G, Manjunath Y, Vishwanath BS. Plant latex proteases: natural wound healers. *Proteases in Physiology and Pathology*. 2017. pp. 297–323.

154. Nadu T. Investigating the possible mechanism for potency of latex of Plumeria alba L. in treating scabies. *Int J Rev Life Sci*. 2019; 9(1): 1–4.

155. Gupta L, Sharma AK, Gothwal A, Khan MS, Khinchi MP, Qayum A, Singh SK, Gupta U. Dendrimer encapsulated and conjugated delivery of berberine: a novel approach mitigating toxicity and improving in vivo pharmacokinetics. *Int J Pharm*. 2017; 528(1–2): 88–99.

156. Haggag EG, Elshamy AM, Rabeh MA, Gabr NM, Salem M, Youssif KA, Samir A, Bin Muhsinah A, Alsayari A, Abdelmohsen UR. Antiviral potential of green synthesized silver nanoparticles of Lampranthus coccineus and Malephora lutea. *Int J Nanomed*. 2019; 14: 6217–29.

157. Roy K, Srivastwa A, Ghosh C. Anticoagulant, thrombolytic and antibacterial activities of Euphorbia acruensis latex-mediated bioengineered silver nanoparticles. *Green Process Synth*. 2019; 8: 590–9.

158. Hagel JM, Facchini PJ. Benzylisoquinoline alkaloid metabolism: a century of discovery and a brave new world. *Plant Cell Physiol*. 2013; 54(5): 647–72.

159. Frick S, Kramell R, Kutchan TM. Metabolic engineering with a morphine biosynthetic P450 in opium poppy surpasses breeding. *Metab Eng*. 2007; 9(2): 169–76.

160. Ramos MV, Freitas CD, Morais FS, Prado E, Medina MC, Demarco D. Plant latex and latex-borne defense. *In Advances in Botanical Research*. 2020; 93: 1–25.

161. Hussain G, Rasul A, Anwar H, Aziz N, Razzaq A, Wei W, Ali M, Li J, Li X. Role of plant derived alkaloids and their mechanism in neurodegenerative disorders. *Int J Biol Sci*. 2018; 14(3): 341–57.

162. Barken I, Geller J, Rogosnitzky M. Noscapine inhibits human prostate cancer progression and metastasis in a mouse model. *Anticancer Res*. 2008; 28(6A): 3701–4.

163. Johnson R. Hemp as an agricultural commodity. *Washington, DC, USA: Congressional Research Service*. 2018; 2–48.

164. Huang P, Sun M, Zeng J. Low-molecular compounds of latex-bearing plants and their biosynthesis. *Adv Bot Res*. 2020; 93: 99–118.

165. Chen JW, Borgelt LM, Blackmer AB. Cannabidiol: a new hope for patients with Dravet or Lennox-Gastaut syndromes. *Ann Pharmacother*. 2019; 53(6): 603–11.

166. Ichim MC. The DNA-based authentication of commercial herbal products reveals their globally widespread adulteration. *Front Pharmacol.* 2019; 10: 1–9.

167. Upadhyay RK. Plant latex: a natural source of pharmaceuticals and pesticides. *Int J Green Pharm.* 2011; 5(3): 169–80.

168. Wishart DS, Feunang YD, Guo AC, Lo EJ, Marcu A, Grant JR, Sajed T, Johnson D, Li C, Sayeeda Z, Assempour N. Drug Bank 5.0: a major update to the Drug Bank database for 2018. *Nucleic Acids Res.* 2018; 46(D1): D1074–82.

169. Pigni A, Brunelli C, Caraceni A. The role of hydromorphone in cancer pain treatment: a systematic review. *Palliative Med.* 2011; 25(5): 471–7.

170. Gourlay GK. Treatment of cancer pain with transdermal fentanyl. *The Lancet Oncol.* 2001; 2(3): 165–72.

171. Leppert W. The role of methadone in cancer pain treatment–a review. *Int J Clin Prac.* 2009; 63(7): 1095–109.

172. Cairns R. The use of oxycodone in cancer-related pain: a literature review. *Int J Palliat Nurs.* 2001;7(11):522–7.

173. Schmidt-Hansen M, Taubert M, Bromham N, Hilgart JS, Arnold S. The effectiveness of buprenorphine for treating cancer pain: an abridged Cochrane review. *BMJ Support Palliat Care.* 2016;6(3): 292–306.

174. Straube C, Derry S, Jackson KC, Wiffen PJ, Bell RF, Strassels S, Straube S. Codeine, alone and with paracetamol (acetaminophen), for cancer pain. *Cochrane Database Syst Rev.* 2014; 2014(9): CD006601.

175. Bayonas CA, Fonseca JP, Echaburu VJ. Tapentadol for cancer pain management: a narrative review. *Pain Practice.* 2017; 17(8): 1075–88.

176. Walsh TD. Oral morphine in chronic cancer pain. *Pain.* 1984; 18(1): 1–1.

177. Abu-Sawwa R, Stehling C. Epidiolex (Cannabidiol) primer: frequently asked questions for patients and caregivers. *J Pediatr Pharmacol Ther.* 2020; 25(1): 75–7.

178. Rahmanian-Devin P, Rahimi BV, Jaafari MR, Golmohammadzadeh S, Sanei-Far Z, Askari VR. Noscapine, an emerging medication for different diseases: a mechanistic review. *Evidence-Based Complement Altern Med.* 2021; 2021: 2–17.

179. Czuppon AB, Chen Z, Rennert S, Engelke T, Meyer HE, Heber M, Baur X. The rubber elongation factor of rubber trees (Hevea brasiliensis) is the major allergen in latex. *J Allergy Clin Immunol.* 1993; 92(5): 690–7.

180. Yeang HY, Cheong KF, Sunderasan E, Hamzah S, Chewa NP, Hamidb S, Hamilton RG, Cardosa MJ. The 14.6 kd rubber elongation factor (Hev b 1) and 24 kd (Hev b 3) rubber particle proteins are recognized by IgE from patients with spina bifida and latex allergy. *J Allergy Clin Immunol.* 1996; 98(3): 628–39.

181. Elumalai S, Hamzah S, Hamid S, Ward M, Yeang H, Cardosa M. Latex B-Serum p-l,3-glucanase (Hev b II) and a component of the microhelix (Rev b IV) are major latex allergens. *J Rubber Res.* 1995; 10: 82–99.

182. Sussman GL, Beezhold DH. Allergy to latex rubber. *Ann Int Med.* 1995; 122(1): 43–6.

183. Brown RH, Schauble JF, Hamilton RG. Prevalence of latex allergy among anesthesiologists: identification of sensitized but asymptomatic individuals. *Anesthesiology.* 1998; 89(2): 292–9.

184. Kurtz S. Making your facility latex-safe. *Outpatient Surgery Mag.* 2000; 29–33.

185. George P. Concerns regarding the safety and toxicity of medicinal plants–An overview. *J Appl Pharm Sci.* 2011; 1(6): 40–4.

186. Al-Worafi YM. Herbal medicines safety issues. *In drug safety in developing countries: Achievements and challenges.* 2020; 163–178.

187. Nasri H, Shirzad H. Toxicity and safety of medicinal plants. *J Herb Med Pharmacol.* 2013; 2: 21–2.

188. Zahn-Waxler C, Klimes-Dougan B, Slattery MJ. Internalizing problems of childhood and adolescence: prospects, pitfalls, and progress in understanding the development of anxiety and depression. *Dev Psychopathol.* 2000; 12(3): 443–66.

189. Fugh-Berman A. Herb-drug interactions. *Lancet.* 2000; 355(9198): 134–8.

190. Palombo EA. Phytochemicals from traditional medicinal plants used in the treatment of diarrhoea: modes of action and effects on intestinal function. *Phytother Res.* 2006; 20(9): 717–24.

191. Gurib-Fakim A. Medicinal plants: traditions of yesterday and drugs of tomorrow. *Mol Aspects Med.* 2006; 27(1): 1–93.

192. World Health Organization (WHO). WHO guidelines on safety monitoring of herbal medicines in pharmacovigilance systems. *World Health Organization Geneva.* 2004; 82: 1–47.

193. World Health Organization (WHO). General guidelines for methodologies on research and evaluation of traditional medicine world health organization. 2000; 1–73.

194. Lv L, Huang W, Yu X, Ren H, Sun R. Comparative research of different Bupleurum chinense composition to influence of hepatotoxicity of rats and oxidative damage mechanism. Zhongguo Zhong yao za zhi = Zhongguo Zhongyao Zazhi = *China J Chinese Materia Medica*. 2009; 34(18): 2364–8.

195. Chaitanya MV, Ali HS, Usamo FB. Regulatory considerations of herbal biomolecules. *In Herbal Biomolecules in Healthcare Applications*. 2022; 669–76.

196. Government of Canada. Regulations amending the narcotic control regulations (Opium Poppy). 2016; 150(12): SOR/2016-123.

197. Toraason M, Sussman G, Biagini R, Meade J, Beezhold D, Germolec D. Latex allergy in the workplace. *Toxicol Sci*. 2000; 58(1): 5–14.

198. Parisi CA, Kelly KJ, Ansotegui IJ, Gonzalez-Díaz SN, Bilò MB, Cardona V, Park HS, Braschi MC, Macias-Weinmann A, Piga MA, Acuña-Ortega N. Update on latex allergy: new insights into an old problem. *World Allergy Organ J*. 2021; 14(8): 100569–83.

199. Wu M, McIntosh J, Liu J. Current prevalence rate of latex allergy: why it remains a problem? *J Occup Health*. 2016; 58(2): 138–44.

200. Lamberti M, Buonanno R, Ritonnaro C, Giovane G, Crispino V, Feola A, Medici N, Sannolo N, Di Carlo A, Di Domenico M. Molecular profile of sensitization in subjects with short occupational exposure to latex. *Int J Occup Med Environ Health*. 2015; 28(5): 841–8.

201. Boonchai W, Sirikudta W, Iamtharachai P, Kasemsarn P. Latex glove-related symptoms among health care workers: a self-report questionnaire-based survey. *Dermat Contact, Atopic, Occup Drug*. 2014; 25(3): 135–9.

202. Köse S, Mandiracioğlu A, Tatar B, Gül S, Erdem M. Prevalence of latex allergy among healthcare workers in Izmir (Turkey). *Cent Eur J Public Health*. 2014; 22(4): 262–5.

203. Liu QL, He XZ, Liang K, Xie R, Fang HP, Zhu KJ, Fan YM. Prevalence and risk factors for latex glove allergy among female clinical nurses: a multicenter questionnaire study in China. *Int J Occup Environ Health*. 2013; 19(1): 29–34.

204. Phaswana SM, Naidoo S. The prevalence of latex sensitisation and allergy and associated risk factors among healthcare workers using hypoallergenic latex gloves at King Edward VIII Hospital,

KwaZulu-Natal South Africa: a cross-sectional study. *BMJ Open.* 2013; 3(12): e002900.

205. Risenga SM, Shivambu GP, Rakgole MP, Makwela ML, Ntuli S, Malatji TA, Maligavhada NJ, Green RJ. Latex allergy and its clinical features among healthcare workers at Mankweng Hospital, Limpopo Province, South Africa. *S Afr Med J.* 2013; 103(6): 390–4.

206. Supapvanich C, Povey AC, de Vocht F. Respiratory and dermal symptoms in Thai nurses using latex products. *Occup Med (Lond).* 2013; 63(6): 425–8.

207. Gomes MJ, Barbosa RS, Dias FP, de Carvalho RB, de Oliveira ERA, Hebling E. [Sensitivity to latex and the dosage of specific antibodies in professionals in the area of health]. *Cien Saude Colet.* 2012; 17(2): 351–8.

208. Wang ML, Kelly KJ, Klancnik M, Petsonk EL. Self-reported hand symptoms: a role in monitoring health care workers for latex sensitization? *Ann Allergy, Asthma Immunol.* 2012; 109(5): 314–8.

209. Agrawal A, Bhatt N, Kk S, Singh K, Chaudhary H, Asawa K. Prevalence of allergy to latex gloves among dental professionals in Udaipur, Rajasthan, India. *Oral Health Prev Dent.* 2010; 8(4): 345–50.

210. Amarasekera M, Rathnamalala N, Samaraweera S, Jinadasa M. Prevalence of latex allergy among healthcare workers. *Int J Occup Med Environ Health.* 2010; 23(4): 391–6.

211. Nabavizadeh SH, Anushiravani A, Amin R. Natural rubber latex hypersensitivity with skin prick test in operating room personnel. *Iran J Allergy Asthma Immunol.* 2009; 8(4): 219–20.

212. Fuentes RMD, León MP, González MP, Monroy NS, Gómez RM, Lantigua PD. Drugs having latex and therapeutic alternatives in hospital formulary. *Farm Hosp.* 2015; 39(1): 44–58.

213. Condemi JJ. Allergic reactions to natural rubber latex at home, to rubber products, and to cross-reacting foods. *J Allergy Clin Immunol.* 2002; 110(2): S107–10.

214. Heitz JW, Bader SO. An evidence-based approach to medication preparation for the surgical patient at risk for latex allergy: is it time to stop being stopper poppers? *J Clin Anesth.* 2010; 22(6): 477–83.

215. Ferreira F, Hawranek T, Gruber P, Wopfner N, Mari A. Allergic cross-reactivity: from gene to the clinic. *Allergy.* 2004; 59(3): 243–67.

216. Brinker FJ. Herbal contraindications & drug interactions: Plus herbal adjuncts with medicines. *Eclectic Medical Publications.* 2010. p. 598.

217. Liss GM, Sussman GL. Latex sensitization: occupational versus general population prevalence rates. *Am J Ind Med.* 1999; 35(2): 196–200.

218. Brehler R, Theissen U, Mohr C, Luger T. "Latex-fruit syndrome": frequency of cross-reacting IgE antibodies. *Allergy.* 1997; 52(4): 404–10.

219. M'Raihi L, Charpin D, Pons A, Bongrand P, Vervloet D. Cross-reactivity between latex and banana. *J Allergy Clin Immunol.* 1991; 87(1 Pt 1): 129–30.

220. Beezhold DH, Sussman GL, Liss GM, Chang NS. Latex allergy can induce clinical reactions to specific foods. *Clin Exp allergy J Br Soc Allergy Clin Immunol.* 1996; 26(4): 416–22.

221. Fernandez de Corres L, Moneo I, Munoz D, Bernaola G, Fernandez E, Audicana M, Urrutia I. Sensitization from chestnuts and bananas in patients with urticaria and anaphylaxis from contact with latex. *Ann Allergy.* 1993; 70(1): 35–9.

222. Radauer C, Adhami F, Fürtler I, Wagner S, Allwardt D, Scala E, Ebner C, Hafner C, Hemmer W, Mari A, Breiteneder H. Latex-allergic patients sensitized to the major allergen hevein and hevein-like domains of class I chitinases show no increased frequency of latex-associated plant food allergy. *Mol Immunol.* 2011; 48(4): 600–9.

223. Lavaud F, Cossart C, Reiter V, Bernard J, Deltour G, Holmquist I. Latex allergy in patient with allergy to fruit. *Lancet.* 1992; 339: 492–3.

224. Blanco C, Carrillo T, Castillo R, Quiralte J, Cuevas M. Latex allergy: clinical features and cross-reactivity with fruits. *Ann Allergy.* 1994; 73(4): 309–14.

225. Shetge SA, Dzakovich MP, Cooperstone JL, Kleinmeier D, Redan BW. Concentrations of the opium alkaloids morphine, codeine, and thebaine in poppy seeds are reduced after thermal and washing treatments but are not affected when incorporated in a model baked product. *J Agric Food Chem.* 2020; 68(18): 5241–8.

226. Singhal R, Kulkarni P. Detection of adultration of the spice poppy seeds (Papaver somniferum) with Amaranthus paniculatas (Rajgeera) seeds. *J Food Qual.* 1990; 13(5): 375–81.

227. Pascale A. Drug adulterants and their effects on the health of users: a critical review. *OAS CICAD.* 2019; 1–163.

228. Monti MC, Zeugin J, Koch K, Milenkovic N, Scheurer E, Mercer-Chalmers-Bender K. Adulteration of low-delta-9-tetrahydrocannabinol products with synthetic cannabinoids: results from drug checking services. *Drug Test Anal.* 2022; 14(6): 1026–39.

229. McPartland JM. Adulteration of cannabis with tobacco, calamus, and other cholinergic compounds. *Cannabinoids.* 2008; 3(4): 16–20.

230. Dryburgh LM, Bolan NS, Grof CP, Galettis P, Schneider J, Lucas CJ, Martin JH. Cannabis contaminants: sources, distribution, human toxicity and pharmacologic effects. *Br J Clin Pharmacol.* 2018; 84(11): 2468–76.

231. Busse FP, Fiedler GM, Leichtle A, Hentschel H, Stumvoll M. Lead poisoning due to adulterated marijuana in leipzig. *Dtsch Arztebl Int.* 2008; 105(44): 757–62.

232. Combemale P, Consort T, Denis-Thelis L, Estival JL, Dupin M, Kanitakis J. Cannabis arteritis. *Br J Dermatol.* 2005; 152(1): 166–9.

233. Zhang J, Onakpoya IJ, Posadzki P, Eddouks M. The safety of herbal medicine: from prejudice to evidence. *Evid-Based Complement Altern Med.* 2015; 2015: 1–3.

234. Gupta M, Mazumder UK, Gomathi P, Selvan VT. Antiinflammatory evaluation of leaves of Plumeria acuminata. *BMC Complement Altern Med.* 2006; 6: 1–6.

235. Fernandes HB, Machado DL, Dias JM, Brito TV, Batista JA, Silva RO, Pereira AC, Ferreira GP, Ramos MV, Medeiros JV, Aragão KS. Laticifer proteins from Plumeria pudica inhibit the inflammatory and nociceptive responses by decreasing the action of inflammatory mediators and pro-inflammatory cytokines. *Rev Bras Farmacogn.* 2015; 25(3): 269–77.

236. Ali N, Ahmad D, Bakht J, Shah S, Ullah F, UrRahman M. Antimicrobial activity of leaves extracted samples from medicinally important Plumeria obtusa. *J Med Plants Res.* 2013; 7(17): 1146–53.

237. Ramos MV, Aguiar VC, Melo VM, Mesquita RO, Silvestre PP, Oliveira JS, Oliveira RS, Macedo NM, Alencar NM. Immunological and allergenic responses induced by latex fractions of Calotropis procera (Ait.) R.Br. *J Ethnopharmacol.* 2007; 111(1): 115–22.

238. Dzoyem JP, Kuete V, Eloff JN. Biochemical parameters in toxicological studies in Africa: significance, principle of methods, data interpretation, and use in plant screenings. *In Toxicological Survey of African Medicinal Plants.* 2014; 659–715.

239. Wintola OA, Sunmonu TO, Afolayan AJ. Toxicological evaluation of aqueous extract of Aloe ferox Mill. in loperamide-induced constipated rats. *Hum Exp Toxicol.* 2011; 30(5): 425–31.

240. Asare GA, Sittie A, Bugyei K, Gyan BA, Adjei S, Addo P, Wiredu EK, Nyarko AK, Otu-Nyarko LS, Adjei DN. Acute toxicity studies of

Croton membranaceus root extract. *J Ethnopharmacol.* 2011; 134(3): 938–43.

241. Igbinosa OO, Oviasogie EF, Igbinosa EO, Igene O, Igbinosa IH, Idemudia OG. Effects of biochemical alteration in animal model after short-term exposure of Jatropha curcas (Linn) leaf extract. *Sci World J.* 2013; 2013: 798096.
242. Assob JC, Kamga HL, Nsagha DS, Njunda AL, Nde PF, Asongalem EA, Njouendou AJ, Sandjon B, Penlap VB. Antimicrobial and toxicological activities of five medicinal plant species from Cameroon traditional medicine. *BMC Complement Altern Med.* 2011; 11: 1–11.
243. Kuete V, Manfouo RN, Beng VP. Toxicological evaluation of the hydroethanol extract of Tabernaemontana crassa (Apocynaceae) stem bark. *J Ethnopharmacol.* 2010; 130(3): 470–6.
244. Sanyal S, Maity P, Pradhan A, Bepari M, Dey SK, Roy T, Choudhurym SM. Sub-acute toxicity study of Calotropis gigantea latex extracts in male Swiss albino mice. *Toxicol Forensic Med Open J.* 2016; 1(2): 54–64.
245. Ogbonnia S, Adekunle AA, Bosa MK, Enwuru VN. Evaluation of acute and subacute toxicity of Alstonia congensis Engler (Apocynaceae) bark and Xylopia aethiopica (Dunal) A. Rich (Annonaceae) fruits mixtures used in the treatment of diabetes. *African J Biotechnol.* 2008; 7(6): 701–5.
246. Pone KB, Telefo PB, Tchouanguep FM. Acute and subacute toxicity of aqueous extract of leaves mixture of Aloe Buettneri (Liliaceae), Dicliptera Verticillata (Acanthaeae), Hibiscus Macranthus (Malvaceae) and Justicia Insularis (Acanthaceae) on Swiss mice and albinos Wistar female rats. *Indian J Forensic Med Toxicol.* 2013; 7(1): 138–43.
247. e Moura DF, Rocha TA, de Melo Barros D, da Silva MM, de Lira MA, dos Santos Souza TG, da Silva CJ, de Aguiar Júnior FC, Chagas CA, da Silva Santos NP, de Souza IA. Evaluation of the cytotoxicity, oral toxicity, genotoxicity, and mutagenicity of the latex extracted from Himatanthus drasticus (Mart.) Plumel (Apocynaceae). *J Ethnopharmacol.* 2020; 253: 112567.
248. da Silva Souza B, de Moraes Oliveira NV, Moita LA, Oliveira LE, de Brito FC, da Silva FR, de Carvalho França LF, Sousa FB, Pereira AC, Ferreira GP, Medeiros JV. Acute and subchronic toxicity assessment of proteins from Plumeria pudica latex in mice. *Res Soc Dev.* 2021; 10(6): e58610616174.

249. Yogendra kumar GG and PKM. Evaluation of hepatoprotective activity of Carica papaya and Ficus bengalensis latex on thioacetamide induced hepatotoxicity in rats. Yogendra. *Int J Adv Res.* 2018; 6(9): 300–6.

250. Jonsson M, Jestoi M, Nathanail AV, Kokkonen UM, Anttila M, Koivisto P, Karhunen P, Peltonen K. Application of OECD Guideline 423 in assessing the acute oral toxicity of moniliformin. *Food Chem Toxicol* 2013; 53: 27–32.

251. Sankhari JM, Jadeja RN, Thounaojam MC, Devkar RV, Ramachandran AV. Safety evaluation of Eugenia jambolana seed extract. *Asian Pac J Trop Med.* 2010; 3(12): 982–7.

252. Nana HM, Ngane RN, Kuiate JR, Mogtomo LK, Tamokou JD, Ndifor F, Mouokeu RS, Etame RE, Biyiti L, Zollo PA. Acute and sub-acute toxicity of the methanolic extract of Pteleopsis hylodendron stem bark. *J Ethnopharmacol.* 2011; 137(1): 70–6.

253. Marx TK, Reddeman R, Clewell AE, Endres JR, Béres E, Vértesi A, Glávits R, Hirka G, Szakonyiné IP. An assessment of the genotoxicity and subchronic toxicity of a supercritical fluid extract of the aerial parts of Hemp. *J Toxicol.* 2018; 2018: 8143582.

254. Kumar A, Kumar B, Kumar R, Kumar A, Singh M, Tiwari V, Trigunayat A, Paul P, Singh P. Acute and subacute toxicity study of ethanolic extract of Calotropis procera (Aiton) Dry and flower in Swiss albino mice. *Phytomedicine Plus.* 2022; 2(2): 100224.

255. Ozarde YS, Choudhari VP. Preliminary qualitative phytochemical analysis and acute oral toxicity study of latex of an ethnomedicinal plant Euphorbia fusiformis Buch. *Res J Pharm Technol.* 2022; 15: 1123–27.

256. Sousa EL, Grangeiro AR, Bastos IV, Rodrigues GC, Silva MJ, Anjos FB, Souza IA, Sousa CE. Antitumor activity of leaves of Himatanthus drasticus (Mart.) Plumel-Apocynaceae (Janaguba) in the treatment of Sarcoma 180 tumor. *Braz J Pharm Sci.* 2010; 46: 199–203.

257. Adeneye A. Subchronic and chronic toxicities of African medicinal plants. In: *Toxicological Survey of African Medicinal Plants.* 2014; 99–133.

258. Serafimovska T, Stefanovski S, Erler J, Keskovski Z, Stefkov G, Mitevska M, Darkovska Serafimovska M, Balkanov T, Tonic Ribarska J. Does Cannabis extract obtained from cannabis flowers with maximum allowed residual level of aflatoxins and ochratoxin a have an impact on human safety and health? *Front Med.* 2021; 8: 1–12.

259. WHO. World Health Organization, International Agency for Research on Cancer. Some naturally occurring substances: food items and constituents, heterocyclic aromatic amines and mycotoxins. IARC monographs on the evaluation of the carcinogenic risk of chemicals to humans. 1993; 56: 599.

260. World Health Organization. Evaluation of certain contaminants in food. World Health Organization technical report series. 2017; (1002): 1–66.

261. Tashkin DP. Smoked marijuana as a cause of lung injury. *Monaldi Arch Chest Dis.* 2005; 63(2): 93–100.

262. Taylor DN, Wachsmuth IK, Shangkuan YH, Schmidt EV, Barrett TJ, Schrader JS, Scherach CS, McGee HB, Feldman RA, Brenner DJ. Salmonellosis associated with marijuana: a multistate outbreak traced by plasmid fingerprinting. *N Engl J Med.* 1982; 306(21): 1249–53.

263. Baldwin GC, Tashkin DP, Buckley DM, Park AN, Dubinett SM, Roth MD. Marijuana and cocaine impair alveolar macrophage function and cytokine production. *Am J Resp Crit Care Med.* 1997; 156(5): 1606–13.

264. Fligiel SE, Roth MD, Kleerup EC, Barsky SH, Simmons MS, Tashkin DP. Tracheobronchial histopathology in habitual smokers of cocaine, marijuana, and/or tobacco. *Chest.* 1997; 112(2): 319–26.

265. Street RA. Heavy metals in medicinal plant products – An African perspective. *South African J Bot.* 2012; 82: 67–74.

266. Nkansah MA, Hayford ST, Borquaye LS, Ephraim JH. Heavy metal contents of some medicinal herbs from Kumasi, Ghana. *Cogent Environ Sci.* 2016; 2(1): 1234660.

267. Dghaim R, Al Khatib S, Rasool H, Ali Khan M. Determination of heavy metals concentration in traditional herbs commonly consumed in the United Arab Emirates. Factor-Litvak PR, editor. *J Environ Public Health.* 2015; 2015: 973878.

268. Abdulla NM, Adam B, Blair I, Oulhaj A. Heavy metal content of herbal health supplement products in Dubai – UAE: a cross-sectional study. *BMC Complement Altern Med.* 2019; 19(1): 276.

269. Annan K, Dickson RA, Amponsah IK, Nooni IK. The heavy metal contents of some selected medicinal plants sampled from different geographical locations. *Pharmacognosy Res.* 2013; 5(2): 103–8.

270. Bhagyashree B, Sanjib B. The importance of assessing heavy metals in medicinal herbs: a quantitative study. *Cellmed.* 2016; 6(1): 3.1–3.4.

271. Russo EB. Current therapeutic cannabis controversies and clinical trial design issues. *Front Pharmacol.* 2016; 7: 208672.

273. Neff RA, Goldman LR. Regulatory parallels to Daubert: Stakeholder influence, "sound science," and the delayed adoption of health-protective standards. *Am J Public Health.* 2005; 95(1): 81–91.

274. Li J, Sun L, Zuo Z, Chen M, Geng H, Wang C. Exposure to paclobutrazol disrupts spermatogenesis in male Sebastiscus marmoratus. *Aquat Toxicol.* 2012; 122: 120–24.

275. Xu M, Yang F. Integrated gender-related effects of profenofos and paclobutrazol on neurotransmitters in mouse. *Ecotoxicol Environ Saf.* 2020; 190(866): 110085.

276. Abdel-Shafy HI, Mansour MSM. A review on polycyclic aromatic hydrocarbons: source, environmental impact, effect on human health and remediation. *Egypt J Pet.* 2016; 25(1): 107–23.

277. Tesfatsion TT, Ramirez GA, Docampo-Palacios ML, Collins AC, Ray KP, Cruces W. Evaluation of preclinical in vitro cytotoxicity, genotoxicity, and cardiactoxicity screenings of hydrogenated cannabidiol. *Pharmacog Mag.* 2023; 19(4): 862–73.

Index

For Product Safety Concerns and Information please contact our
EU representative GPSR@taylorandfrancis.com Taylor & Francis
Verlag GmbH, Kaufingerstraße 24, 80331 München, Germany